Everyone's Guide to Planet Mars

Compiled by

Kiera Mcune

Scribbles

Year of Publication 2018

ISBN : 9789352979523

Book Published by

Scribbles

(An Imprint of Alpha Editions)

email - alphaedis@gmail.com

Produced by: PediaPress GmbH
Limburg an der Lahn
Germany
http://pediapress.com/

Contents

Introduction

Mars

<indicator name="pp-default"> 🔒 </indicator>

Mars

Mars in natural color in 2007[1]

Designations	
Pronunciation	UK: /ˈmɑːz/ US: /ˈmɑːrz/ (🔊 listen)
Adjectives	Martian
Orbital characteristics	
Epoch J2000	
Aphelion	249200000 km (154800000 mi; 1.666 AU)
Perihelion	206700000 km (128400000 mi; 1.382 AU)
Semi-major axis	227939200 km (141634900 mi; 1.523679 AU)

Eccentricity	0.0934
Orbital period	686.971 d (1.88082 yr; 668.5991 sols)
Synodic period	779.96 d (2.1354 yr)
Average orbital speed	24.007 km/s (86430 km/h; 53700 mph)
Mean anomaly	320.45776°
Inclination	• 1.850° to ecliptic; • 5.65° the Sun's equator; • 1.67° to invariable plane
Longitude of ascending node	49.558°
Argument of perihelion	286.502°
Satellites	2
Physical characteristics	
Mean radius	3389.5 ± 0.2 km (2106.1 ± 0.1 mi)
Equatorial radius	3396.2 ± 0.1 km (2110.3 ± 0.1 mi; 0.533 Earths)
Polar radius	3376.2 ± 0.1 km (2097.9 ± 0.1 mi; 0.531 Earths)
Flattening	0.00589±0.00015
Surface area	144798500 km^2 (55907000 sq mi; 0.284 Earths)
Volume	1.6318×10^{11} km^3 (0.151 Earths)
Mass	6.4171×10^{23} kg (0.107 Earths)
Mean density	3.9335 g/cm^3 (0.1421 lb/cu in)
Surface gravity	3.711 m/s^2 (12.18 ft/s^2; 0.376 g)
Moment of inertia factor	0.3662±0.0017
Escape velocity	5.027 km/s (18100 km/h; 11250 mph)
Sidereal rotation period	1.025957 d 24^h 37^m 22^s
Equatorial rotation velocity	241.17 m/s (868.22 km/h; 539.49 mph)
Axial tilt	25.19° to its orbital plane

North pole right ascension	317.68143° 21^h 10^m 44^s		
North pole declination	52.88650°		
Albedo	• 0.170 geometric • 0.25 Bond		

Surface temp.	min	mean	max
Kelvin	130 K	210 K	308 K
Celsius	–143 °C	–63 °C	35 °C
Fahrenheit	–226 °F	–82 °F	95 °F

Apparent magnitude	+1.6 to –3.0
Angular diameter	3.5–25.1"

Atmosphere	
Surface pressure	0.636 (0.4–0.87) kPa 0.00628 atm
Composition by volume	• 95.97% carbon dioxide • 1.93% argon • 1.89% nitrogen • 0.146% oxygen • 0.0557% carbon monoxide

Mars is the fourth planet from the Sun and the second-smallest planet in the Solar System after Mercury. In English, Mars carries a name of the Roman god of war, and is often referred to as the **"Red Planet"** because the reddish iron oxide prevalent on its surface gives it a reddish appearance that is distinctive among the astronomical bodies visible to the naked eye. Mars is a terrestrial planet with a thin atmosphere, having surface features reminiscent both of the impact craters of the Moon and the valleys, deserts, and polar ice caps of Earth.

The rotational period and seasonal cycles of Mars are likewise similar to those of Earth, as is the tilt that produces the seasons. Mars is the site of Olympus Mons, the largest volcano and second-highest known mountain in the Solar System, and of Valles Marineris, one of the largest canyons in the Solar System. The smooth Borealis basin in the northern hemisphere covers 40% of the planet and may be a giant impact feature. Mars has two moons, Phobos and Deimos, which are small and irregularly shaped. These may be captured asteroids, similar to 5261 Eureka, a Mars trojan.

There are ongoing investigations assessing the past habitability potential of Mars, as well as the possibility of extant life. Future astrobiology missions are planned, including the Mars 2020 and ExoMars rovers. Liquid water cannot exist on the surface of Mars due to low atmospheric pressure, which is less than 1% of the Earth's, except at the lowest elevations for short periods. The two polar ice caps appear to be made largely of water. The volume of water ice in the south polar ice cap, if melted, would be sufficient to cover the entire planetary surface to a depth of 11 meters (36 ft). In November 2016, NASA reported finding a large amount of underground ice in the Utopia Planitia region of Mars. The volume of water detected has been estimated to be equivalent to the volume of water in Lake Superior.

Mars can easily be seen from Earth with the naked eye, as can its reddish coloring. Its apparent magnitude reaches –2.91, which is surpassed only by Jupiter, Venus, the Moon, and the Sun. Optical ground-based telescopes are typically limited to resolving features about 300 kilometers (190 mi) across when Earth and Mars are closest because of Earth's atmosphere.

Physical characteristics

Mars is approximately half the diameter of Earth with a surface area only slightly less than the total area of Earth's dry land. Mars is less dense than Earth, having about 15% of Earth's volume and 11% of Earth's mass, resulting in about 38% of Earth's surface gravity. The red-orange appearance of the Martian surface is caused by iron(III) oxide, or rust. It can look like butterscotch; other common surface colors include golden, brown, tan, and greenish, depending on the minerals present.[2]

<templatestyles src="Multiple_image/styles.css" />

Comparison: Earth and Mars

Animation (00:40) showing major features of Mars

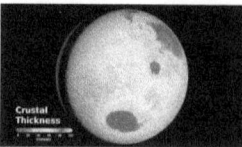

Video (01:28) showing how three NASA orbiters mapped the gravity field of Mars

Internal structure

Like Earth, Mars has differentiated into a dense metallic core overlaid by less dense materials. Current models of its interior imply a core with a radius of about 1,794 ± 65 kilometers (1,115 ± 40 mi), consisting primarily of iron and nickel with about 16–17% sulfur. This iron(II) sulfide core is thought to be twice as rich in lighter elements as Earth's. The core is surrounded by a silicate mantle that formed many of the tectonic and volcanic features on the planet, but it appears to be dormant. Besides silicon and oxygen, the most abundant elements in the Martian crust are iron, magnesium, aluminum, calcium, and potassium. The average thickness of the planet's crust is about 50 km (31 mi), with a maximum thickness of 125 km (78 mi). Earth's crust averages 40 km (25 mi).

Surface geology

Mars is a terrestrial planet that consists of minerals containing silicon and oxygen, metals, and other elements that typically make up rock. The surface of Mars is primarily composed of tholeiitic basalt, although parts are more silica-rich than typical basalt and may be similar to andesitic rocks on Earth or silica glass. Regions of low albedo suggest concentrations of plagioclase feldspar, with northern low albedo regions displaying higher than normal concentrations of sheet silicates and high-silicon glass. Parts of the southern highlands include detectable amounts of high-calcium pyroxenes. Localized concentrations of hematite and olivine have been found. Much of the surface is deeply covered by finely grained iron(III) oxide dust.

Although Mars has no evidence of a structured global magnetic field, observations show that parts of the planet's crust have been magnetized, suggesting that alternating polarity reversals of its dipole field have occurred in the past. This paleomagnetism of magnetically susceptible minerals is similar to the alternating bands found on Earth's ocean floors. One theory, published in 1999 and re-examined in October 2005 (with the help of the *Mars Global Surveyor*), is that these bands suggest plate tectonic activity on Mars four billion years ago, before the planetary dynamo ceased to function and the planet's magnetic field faded.

It is thought that, during the Solar System's formation, Mars was created as the result of a stochastic process of run-away accretion of material from the protoplanetary disk that orbited the Sun. Mars has many distinctive chemical features caused by its position in the Solar System. Elements with comparatively low boiling points, such as chlorine, phosphorus, and sulphur, are much more common on Mars than Earth; these elements were probably pushed outward by the young Sun's energetic solar wind.

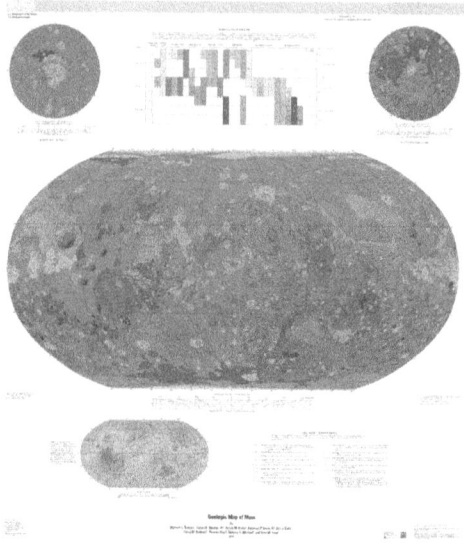

Figure 1: *Geologic map of Mars (USGS, 2014)*

After the formation of the planets, all were subjected to the so-called "Late Heavy Bombardment". About 60% of the surface of Mars shows a record of impacts from that era, whereas much of the remaining surface is probably underlain by immense impact basins caused by those events. There is evidence of an enormous impact basin in the northern hemisphere of Mars, spanning 10,600 by 8,500 km (6,600 by 5,300 mi), or roughly four times the size of the Moon's South Pole – Aitken basin, the largest impact basin yet discovered. This theory suggests that Mars was struck by a Pluto-sized body about four billion years ago. The event, thought to be the cause of the Martian hemispheric dichotomy, created the smooth Borealis basin that covers 40% of the planet.

The geological history of Mars can be split into many periods, but the following are the three primary periods:

- **Noachian period** (named after Noachis Terra): Formation of the oldest extant surfaces of Mars, 4.5 to 3.5 billion years ago. Noachian age surfaces are scarred by many large impact craters. The Tharsis bulge, a volcanic upland, is thought to have formed during this period, with extensive flooding by liquid water late in the period.
- **Hesperian period** (named after Hesperia Planum): 3.5 to between 3.3 and 2.9 billion years ago. The Hesperian period is marked by the formation of extensive lava plains.

Figure 2: *Artist's impression of how Mars may have looked four billion years ago*

- **Amazonian period** (named after Amazonis Planitia): between 3.3 and 2.9 billion years ago to the present. Amazonian regions have few meteorite impact craters, but are otherwise quite varied. Olympus Mons formed during this period, with lava flows elsewhere on Mars.

Geological activity is still taking place on Mars. The Athabasca Valles is home to sheet-like lava flows created about 200 Mya. Water flows in the grabens called the Cerberus Fossae occurred less than 20 Mya, indicating equally recent volcanic intrusions. On February 19, 2008, images from the *Mars Reconnaissance Orbiter* showed evidence of an avalanche from a 700-metre-high (2,300 ft) cliff.

Soil

The *Phoenix* lander returned data showing Martian soil to be slightly alkaline and containing elements such as magnesium, sodium, potassium and chlorine. These nutrients are found in soils on Earth, and they are necessary for growth of plants. Experiments performed by the lander showed that the Martian soil has a basic pH of 7.7, and contains 0.6% of the salt perchlorate.

Streaks are common across Mars and new ones appear frequently on steep slopes of craters, troughs, and valleys. The streaks are dark at first and get lighter with age. The streaks can start in a tiny area, then spread out for hundreds of metres. They have been seen to follow the edges of boulders and other obstacles in their path. The commonly accepted theories include that they are dark underlying layers of soil revealed after avalanches of bright dust or dust

Figure 3: *Exposure of silica-rich dust uncovered by the Spirit rover*

devils. Several other explanations have been put forward, including those that involve water or even the growth of organisms.

Hydrology

Liquid water cannot exist on the surface of Mars due to low atmospheric pressure, which is less than 1% that of Earth's, except at the lowest elevations for short periods. The two polar ice caps appear to be made largely of water. The volume of water ice in the south polar ice cap, if melted, would be sufficient to cover the entire planetary surface to a depth of 11 meters (36 ft). A permafrost mantle stretches from the pole to latitudes of about 60°. Large quantities of water ice are thought to be trapped within the thick cryosphere of Mars. Radar data from *Mars Express* and the *Mars Reconnaissance Orbiter* show large quantities of water ice at both poles (July 2005) and at middle latitudes (November 2008). The Phoenix lander directly sampled water ice in shallow Martian soil on July 31, 2008.

Landforms visible on Mars strongly suggest that liquid water has existed on the planet's surface. Huge linear swathes of scoured ground, known as outflow channels, cut across the surface in about 25 places. These are thought to be a record of erosion caused by the catastrophic release of water from subsurface aquifers, though some of these structures have been hypothesized to result from

Figure 4: *Photomicrograph by Opportunity showing a gray hematite concre-tion, nicknamed "blueberries", indicative of the past existence of liquid water*

the action of glaciers or lava. One of the larger examples, Ma'adim Vallis is 700 km (430 mi) long, much greater than the Grand Canyon, with a width of 20 km (12 mi) and a depth of 2 km (1.2 mi) in places. It is thought to have been carved by flowing water early in Mars's history. The youngest of these channels are thought to have formed as recently as only a few million years ago. Elsewhere, particularly on the oldest areas of the Martian surface, finer-scale, dendritic networks of valleys are spread across significant proportions of the landscape. Features of these valleys and their distribution strongly imply that they were carved by runoff resulting from precipitation in early Mars history. Subsurface water flow and groundwater sapping may play important subsidiary roles in some networks, but precipitation was probably the root cause of the incision in almost all cases.

Along crater and canyon walls, there are thousands of features that appear sim-ilar to terrestrial gullies. The gullies tend to be in the highlands of the south-ern hemisphere and to face the Equator; all are poleward of 30° latitude. A number of authors have suggested that their formation process involves liquid water, probably from melting ice, although others have argued for formation mechanisms involving carbon dioxide frost or the movement of dry dust. No partially degraded gullies have formed by weathering and no superimposed impact craters have been observed, indicating that these are young features,

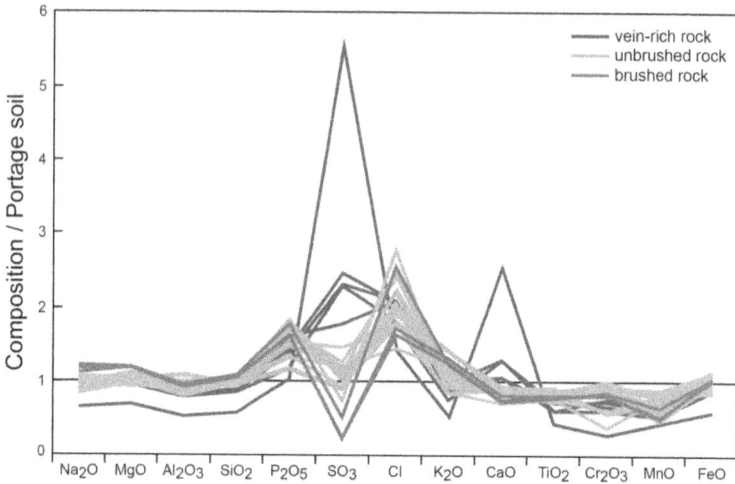

Figure 5: *Composition of "Yellowknife Bay" rocks. Rock veins are higher in calcium and sulfur than "portage" soil (Curiosity, APXS, 2013).*

possibly still active. Other geological features, such as deltas and alluvial fans preserved in craters, are further evidence for warmer, wetter conditions at an interval or intervals in earlier Mars history. Such conditions necessarily require the widespread presence of crater lakes across a large proportion of the surface, for which there is independent mineralogical, sedimentological and geomorphological evidence.

Further evidence that liquid water once existed on the surface of Mars comes from the detection of specific minerals such as hematite and goethite, both of which sometimes form in the presence of water. In 2004, *Opportunity* detected the mineral jarosite. This forms only in the presence of acidic water, which demonstrates that water once existed on Mars. More recent evidence for liquid water comes from the finding of the mineral gypsum on the surface by NASA's Mars rover Opportunity in December 2011. It is believed that the amount of water in the upper mantle of Mars, represented by hydroxyl ions contained within the minerals of Mars's geology, is equal to or greater than that of Earth at 50–300 parts per million of water, which is enough to cover the entire planet to a depth of 200–1,000 m (660–3,280 ft).

In 2005, radar data revealed the presence of large quantities of water ice at the poles and at mid-latitudes. The Mars rover *Spirit* sampled chemical compounds containing water molecules in March 2007. The *Phoenix* lander directly sampled water ice in shallow Martian soil on July 31, 2008.

On March 18, 2013, NASA reported evidence from instruments on the *Curiosity* rover of mineral hydration, likely hydrated calcium sulfate, in several rock samples including the broken fragments of "Tintina" rock and "Sutton Inlier" rock as well as in veins and nodules in other rocks like "Knorr" rock and "Wernicke" rock. Analysis using the rover's DAN instrument provided evidence of subsurface water, amounting to as much as 4% water content, down to a depth of 60 cm (24 in), during the rover's traverse from the *Bradbury Landing* site to the *Yellowknife Bay* area in the *Glenelg* terrain. In September 2015, NASA announced that they had found conclusive evidence of hydrated brine flows on recurring slope lineae, based on spectrometer readings of the darkened areas of slopes. These observations provided confirmation of earlier hypotheses based on timing of formation and their rate of growth, that these dark streaks resulted from water flowing in the very shallow subsurface. The streaks contain hydrated salts, perchlorates, which have water molecules in their crystal structure. The streaks flow downhill in Martian summer, when the temperature is above –23 degrees Celsius, and freeze at lower temperatures. On September 28, 2015, NASA announced the presence of briny flowing salt water on the Martian surface.

Researchers believe that much of the low northern plains of the planet were covered with an ocean hundreds of meters deep, though this remains controversial. In March 2015, scientists stated that such an ocean might have been the size of Earth's Arctic Ocean. This finding was derived from the ratio of water to deuterium in the modern Martian atmosphere compared to that ratio on Earth. The amount of Martian deuterium is eight times the amount that exists on Earth, suggesting that ancient Mars had significantly higher levels of water. Results from the *Curiosity* rover had previously found a high ratio of deuterium in Gale Crater, though not significantly high enough to suggest the former presence of an ocean. Other scientists caution that these results have not been confirmed, and point out that Martian climate models have not yet shown that the planet was warm enough in the past to support bodies of liquid water.

Polar caps

<templatestyles src="Multiple_image/styles.css" />

North polar early summer ice cap (1999)

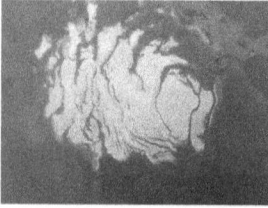

South polar midsummer ice cap (2000)

Mars has two permanent polar ice caps. During a pole's winter, it lies in continuous darkness, chilling the surface and causing the deposition of 25–30% of the atmosphere into slabs of CO_2 ice (dry ice). When the poles are again exposed to sunlight, the frozen CO_2 sublimes. These seasonal actions transport large amounts of dust and water vapor, giving rise to Earth-like frost and large cirrus clouds. Clouds of water-ice were photographed by the *Opportunity* rover in 2004.

The caps at both poles consist primarily (70%) of water ice. Frozen carbon dioxide accumulates as a comparatively thin layer about one metre thick on the north cap in the northern winter only, whereas the south cap has a permanent dry ice cover about eight metres thick. This permanent dry ice cover at the south pole is peppered by flat floored, shallow, roughly circular pits, which repeat imaging shows are expanding by meters per year; this suggests that the permanent CO_2 cover over the south pole water ice is degrading over time. The northern polar cap has a diameter of about 1,000 km (620 mi) during the northern Mars summer, and contains about 1.6 million cubic kilometres (380,000 cu mi) of ice, which, if spread evenly on the cap, would be 2 km (1.2 mi) thick. (This compares to a volume of 2.85 million cubic kilometres (680,000 cu mi) for the Greenland ice sheet.) The southern polar cap has a diameter of 350 km (220 mi) and a thickness of 3 km (1.9 mi). The total volume of ice in the south polar cap plus the adjacent layered deposits has been estimated at 1.6 million cubic km. Both polar caps show spiral troughs, which recent analysis of SHARAD ice penetrating radar has shown are a result of katabatic winds that spiral due to the Coriolis Effect.

The seasonal frosting of areas near the southern ice cap results in the formation of transparent 1-metre-thick slabs of dry ice above the ground. With the arrival of spring, sunlight warms the subsurface and pressure from subliming CO_2 builds up under a slab, elevating and ultimately rupturing it. This leads to geyser-like eruptions of CO_2 gas mixed with dark basaltic sand or dust. This process is rapid, observed happening in the space of a few days, weeks or months, a rate of change rather unusual in geology – especially for Mars. The

Figure 6: *A MOLA-based topographic map showing highlands (red and orange) dominating the southern hemisphere of Mars, lowlands (blue) the northern. Volcanic plateaus delimit regions of the northern plains, whereas the highlands are punctuated by several large impact basins.*

gas rushing underneath a slab to the site of a geyser carves a spiderweb-like pattern of radial channels under the ice, the process being the inverted equivalent of an erosion network formed by water draining through a single plughole.

Geography and naming of surface features

Although better remembered for mapping the Moon, Johann Heinrich Mädler and Wilhelm Beer were the first "areographers". They began by establishing that most of Mars's surface features were permanent and by more precisely determining the planet's rotation period. In 1840, Mädler combined ten years of observations and drew the first map of Mars. Rather than giving names to the various markings, Beer and Mädler simply designated them with letters; Meridian Bay (Sinus Meridiani) was thus feature "*a*".

Today, features on Mars are named from a variety of sources. Albedo features are named for classical mythology. Craters larger than 60 km are named for deceased scientists and writers and others who have contributed to the study of Mars. Craters smaller than 60 km are named for towns and villages of the world with populations of less than 100,000. Large valleys are named for the word "Mars" or "star" in various languages; small valleys are named for rivers.

Large albedo features retain many of the older names, but are often updated to reflect new knowledge of the nature of the features. For example, *Nix Olympica* (the snows of Olympus) has become *Olympus Mons* (Mount Olympus). The surface of Mars as seen from Earth is divided into two kinds of areas, with differing albedo. The paler plains covered with dust and sand rich in reddish iron oxides were once thought of as Martian "continents" and given

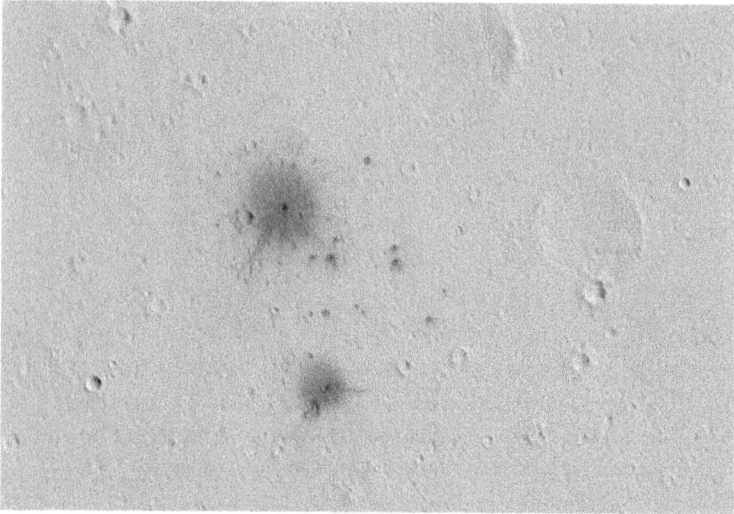

Figure 7: *These new impact craters on Mars occurred sometime between 2008 and 2014, as detected from orbit*

names like Arabia Terra (*land of Arabia*) or Amazonis Planitia (*Amazonian plain*). The dark features were thought to be seas, hence their names Mare Erythraeum, Mare Sirenum and Aurorae Sinus. The largest dark feature seen from Earth is Syrtis Major Planum. The permanent northern polar ice cap is named Planum Boreum, whereas the southern cap is called Planum Australe.

Mars's equator is defined by its rotation, but the location of its Prime Meridian was specified, as was Earth's (at Greenwich), by choice of an arbitrary point; Mädler and Beer selected a line for their first maps of Mars in 1830. After the spacecraft Mariner 9 provided extensive imagery of Mars in 1972, a small crater (later called Airy-0), located in the Sinus Meridiani ("Middle Bay" or "Meridian Bay"), was chosen for the definition of 0.0° longitude to coincide with the original selection.

Because Mars has no oceans and hence no "sea level", a zero-elevation surface had to be selected as a reference level; this is called the *areoid* of Mars, analogous to the terrestrial geoid. Zero altitude was defined by the height at which there is 610.5 Pa (6.105 mbar) of atmospheric pressure. This pressure corresponds to the triple point of water, and it is about 0.6% of the sea level surface pressure on Earth (0.006 atm). In practice, today this surface is defined directly from satellite gravity measurements.

Figure 8: *Bonneville crater and Spirit rover's lander*

Map of quadrangles

For mapping purposes, the United States Geological Survey divides the surface of Mars into thirty "quadrangles", each named for a prominent physiographic feature within that quadrangle. The quadrangles can be seen and explored via the interactive image map below.

File:MGS_MOC_Wide_Angle_Map_of_Mars_PIA03467.jpg

The thirty cartographic quadrangles of Mars, defined by the United States Geological Survey. The quadrangles are numbered with the prefix "MC" for "Mars Chart." Click on a quadrangle name link and you will be taken to the corresponding article. North is at the top; 0°N 180°W[3] is at the far left on the equator. The map images were taken by the Mars Global Surveyor.

[

- view
- talk

]

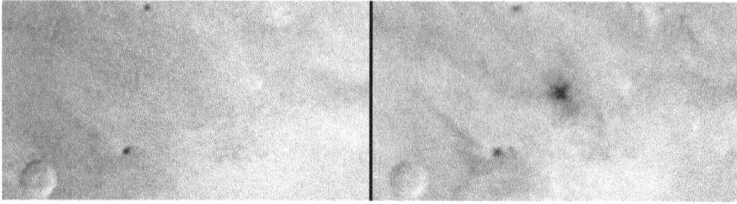

Figure 9: *Fresh asteroid impact on Mars at 3.34°N 219.38°E*[4].
*These before and after images of the same site were taken on the
Martian afternoons of March 27 and 28, 2012 respectively (MRO)*

Impact topography

The dichotomy of Martian topography is striking: northern plains flattened
by lava flows contrast with the southern highlands, pitted and cratered by ancient impacts. Research in 2008 has presented evidence regarding a theory
proposed in 1980 postulating that, four billion years ago, the northern hemisphere of Mars was struck by an object one-tenth to two-thirds the size of
Earth's Moon. If validated, this would make the northern hemisphere of Mars
the site of an impact crater 10,600 by 8,500 km (6,600 by 5,300 mi) in size,
or roughly the area of Europe, Asia, and Australia combined, surpassing the
South Pole–Aitken basin as the largest impact crater in the Solar System.

Mars is scarred by a number of impact craters: a total of 43,000 craters with a
diameter of 5 km (3.1 mi) or greater have been found. The largest confirmed
of these is the Hellas impact basin, a light albedo feature clearly visible from
Earth. Due to the smaller mass of Mars, the probability of an object colliding
with the planet is about half that of Earth. Mars is located closer to the asteroid belt, so it has an increased chance of being struck by materials from that
source. Mars is more likely to be struck by short-period comets, *i.e.*, those
that lie within the orbit of Jupiter. In spite of this, there are far fewer craters
on Mars compared with the Moon, because the atmosphere of Mars provides
protection against small meteors and surface modifying processes have erased
some craters.

Martian craters can have a morphology that suggests the ground became wet
after the meteor impacted.

Volcanoes

The shield volcano Olympus Mons (*Mount Olympus*) is an extinct volcano in
the vast upland region Tharsis, which contains several other large volcanoes.
Olympus Mons is roughly three times the height of Mount Everest, which in

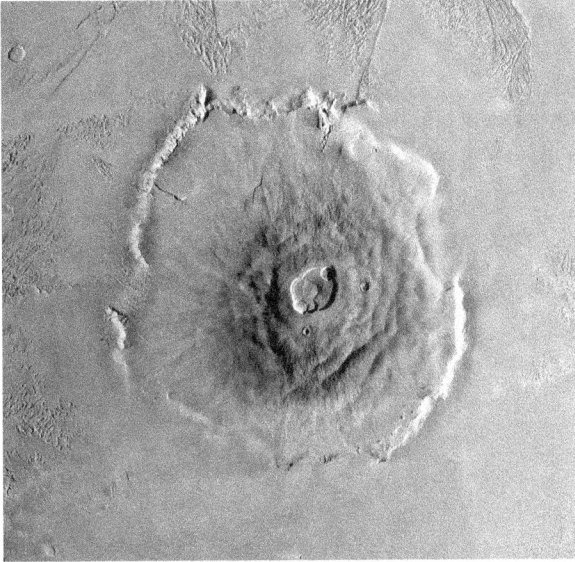

Figure 10: *Viking 1 image of Olympus Mons. The volcano and related terrain are approximately 550 km (340 mi) across.*

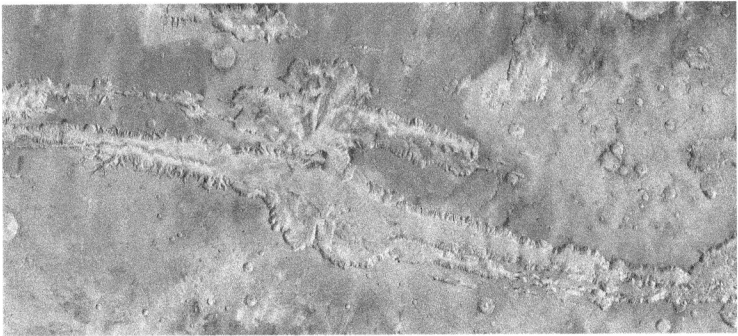

Figure 11: *Valles Marineris (2001 Mars Odyssey)*

comparison stands at just over 8.8 km (5.5 mi). It is either the tallest or second-tallest mountain in the Solar System, depending on how it is measured, with various sources giving figures ranging from about 21 to 27 km (13 to 17 mi) high.

Tectonic sites

The large canyon, Valles Marineris (Latin for "Mariner Valleys", also known as Agathadaemon in the old canal maps), has a length of 4,000 km (2,500 mi) and a depth of up to 7 km (4.3 mi). The length of Valles Marineris is equivalent to the length of Europe and extends across one-fifth the circumference of Mars. By comparison, the Grand Canyon on Earth is only 446 km (277 mi) long and nearly 2 km (1.2 mi) deep. Valles Marineris was formed due to the swelling of the Tharsis area, which caused the crust in the area of Valles Marineris to collapse. In 2012, it was proposed that Valles Marineris is not just a graben, but a plate boundary where 150 km (93 mi) of transverse motion has occurred, making Mars a planet with possibly a two-tectonic plate arrangement.

Holes

Images from the Thermal Emission Imaging System (THEMIS) aboard NASA's Mars Odyssey orbiter have revealed seven possible cave entrances on the flanks of the volcano Arsia Mons. The caves, named after loved ones of their discoverers, are collectively known as the "seven sisters". Cave entrances measure from 100 to 252 m (328 to 827 ft) wide and they are estimated to be at least 73 to 96 m (240 to 315 ft) deep. Because light does not reach the floor of most of the caves, it is possible that they extend much deeper than these lower estimates and widen below the surface. "Dena" is the only exception; its floor is visible and was measured to be 130 m (430 ft) deep. The interiors of these caverns may be protected from micrometeoroids, UV radiation, solar flares and high energy particles that bombard the planet's surface.

Atmosphere

Mars lost its magnetosphere 4 billion years ago, possibly because of numerous asteroid strikes, so the solar wind interacts directly with the Martian iono-sphere, lowering the atmospheric density by stripping away atoms from the outer layer. Both Mars Global Surveyor and Mars Express have detected ionised atmospheric particles trailing off into space behind Mars, and this at-mospheric loss is being studied by the MAVEN orbiter. Compared to Earth, the atmosphere of Mars is quite rarefied. Atmospheric pressure on the sur-face today ranges from a low of 30 Pa (0.030 kPa) on Olympus Mons to over 1,155 Pa (1.155 kPa) in Hellas Planitia, with a mean pressure at the surface level of 600 Pa (0.60 kPa). The highest atmospheric density on Mars is equal to that found 35 km (22 mi) above Earth's surface. The resulting mean sur-face pressure is only 0.6% of that of Earth (101.3 kPa). The scale height of the atmosphere is about 10.8 km (6.7 mi), which is higher than Earth's, 6 km (3.7 mi), because the surface gravity of Mars is only about 38% of Earth's, an

Figure 12: *The tenuous atmosphere of Mars visible on the horizon*

effect offset by both the lower temperature and 50% higher average molecular weight of the atmosphere of Mars.

The atmosphere of Mars consists of about 96% carbon dioxide, 1.93% argon and 1.89% nitrogen along with traces of oxygen and water. The atmosphere is quite dusty, containing particulates about 1.5 μm in diameter which give the Martian sky a tawny color when seen from the surface. It may take on a pink hue due to iron oxide particles suspended in it.

Methane has been detected in the Martian atmosphere; it occurs in extended plumes, and the profiles imply that the methane is released from discrete regions. The concentration of methane fluctuates from about 0.24 ppb during the northern winter to about 0.65 ppb during the summer. In northern midsummer 2003, the principal plume contained 19,000 metric tons of methane, with an estimated source strength of 0.6 kilograms per second. The profiles suggest that there may be two local source regions, the first centered near 30°N 260°W[5] and the second near 0°N 310°W[6]. It is estimated that Mars must produce 270 tonnes per year of methane.

Methane can exist in the Martian atmosphere for only a limited period before it is destroyed—estimates of its lifetime range from 0.6–4 years. Its presence despite this short lifetime indicates that an active source of the gas must be present. Volcanic activity, cometary impacts, and the presence of

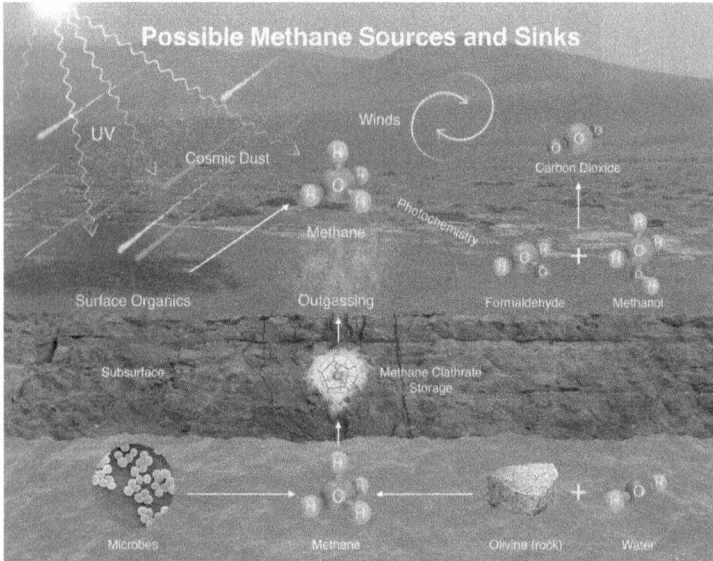

Figure 13: *Potential sources and sinks of methane (CH₄) on Mars*

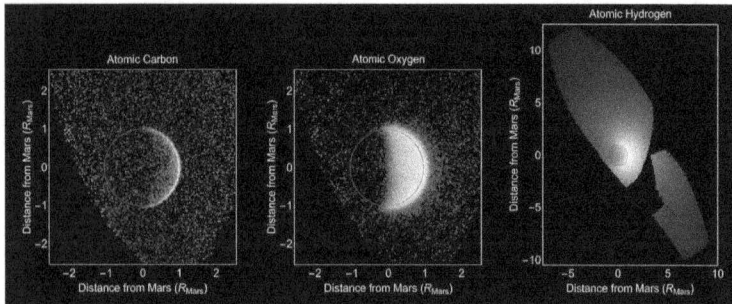

Figure 14: *Escaping atmosphere on Mars (carbon, oxygen, and hydrogen) by MAVEN in UV*

methanogenic microbial life forms are among possible sources. Methane could be produced by a non-biological process called *serpentinization* involving water, carbon dioxide, and the mineral olivine, which is known to be common on Mars.

The *Curiosity* rover, which landed on Mars in August 2012, is able to make measurements that distinguish between different isotopologues of methane, but even if the mission is to determine that microscopic Martian life is the source of the methane, the life forms likely reside far below the surface, outside of the rover's reach. The first measurements with the Tunable Laser Spectrometer (TLS) indicated that there is less than 5 ppb of methane at the landing site at the point of the measurement. On September 19, 2013, NASA scientists, from further measurements by *Curiosity*, reported no detection of atmospheric methane with a measured value of 0.18 ± 0.67 ppbv corresponding to an upper limit of only 1.3 ppbv (95% confidence limit) and, as a result, conclude that the probability of current methanogenic microbial activity on Mars is reduced.

The Mars Orbiter Mission by India is searching for methane in the atmosphere, while the ExoMars Trace Gas Orbiter, launched in 2016, would further study the methane as well as its decomposition products, such as formaldehyde and methanol.[7]

On December 16, 2014, NASA reported the *Curiosity* rover detected a "tenfold spike", likely localized, in the amount of methane in the Martian atmosphere. Sample measurements taken "a dozen times over 20 months" showed increases in late 2013 and early 2014, averaging "7 parts of methane per billion in the atmosphere." Before and after that, readings averaged around one-tenth that level.

Ammonia was tentatively detected on Mars by the Mars Express satellite, but with its relatively short lifetime, it is not clear what produced it. Ammonia is not stable in the Martian atmosphere and breaks down after a few hours. One possible source is volcanic activity.

In September 2017, NASA reported radiation levels on the surface of the planet Mars were temporarily doubled, and were associated with an aurora 25 times brighter than any observed earlier, due to a massive, and unexpected, solar storm in the middle of the month.

Aurora

In 1994, the European Space Agency's Mars Express found an ultraviolet glow coming from "magnetic umbrellas" in the southern hemisphere. Mars does not have a global magnetic field which guides charged particles entering the atmosphere. Mars has multiple umbrella-shaped magnetic fields mainly in the southern hemisphere, which are remnants of a global field that decayed billions of years ago.

In late December 2014, NASA's MAVEN spacecraft detected evidence of widespread auroras in Mars's northern hemisphere and descended to approximately 20–30 degrees North latitude of Mars's equator. The particles causing the aurora penetrated into the Martian atmosphere, creating auroras below 100 km above the surface, Earth's auroras range from 100 km to 500 km above the surface. Magnetic fields in the solar wind drape over Mars, into the atmosphere, and the charged particles follow the solar wind magnetic field lines into the atmosphere, causing auroras to occur outside the magnetic umbrellas.

On March 18, 2015, NASA reported the detection of an aurora that is not fully understood and an unexplained dust cloud in the atmosphere of Mars.

Climate

Of all the planets in the Solar System, the seasons of Mars are the most Earth-like, due to the similar tilts of the two planets' rotational axes. The lengths of the Martian seasons are about twice those of Earth's because Mars's greater distance from the Sun leads to the Martian year being about two Earth years long. Martian surface temperatures vary from lows of about −143 °C (−225 °F) at the winter polar caps to highs of up to 35 °C (95 °F) in equatorial summer. The wide range in temperatures is due to the thin atmosphere which cannot store much solar heat, the low atmospheric pressure, and the low thermal inertia of Martian soil. The planet is 1.52 times as far from the Sun as Earth, resulting in just 43% of the amount of sunlight.

If Mars had an Earth-like orbit, its seasons would be similar to Earth's because its axial tilt is similar to Earth's. The comparatively large eccentricity of the Martian orbit has a significant effect. Mars is near perihelion when it is summer in the southern hemisphere and winter in the north, and near aphelion when it is winter in the southern hemisphere and summer in the north. As a result, the seasons in the southern hemisphere are more extreme and the seasons in the northern are milder than would otherwise be the case. The summer temperatures in the south can be warmer than the equivalent summer temperatures in the north by up to 30 °C (54 °F).

Mars has the largest dust storms in the Solar System, reaching speeds of over 160 km/h (100 mph). These can vary from a storm over a small area, to gigantic storms that cover the entire planet. They tend to occur when Mars is closest to the Sun, and have been shown to increase the global temperature.

<templatestyles src="Multiple_image/styles.css" />

Dust storms on Mars

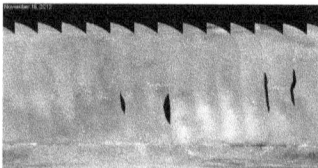

November 18, 2012

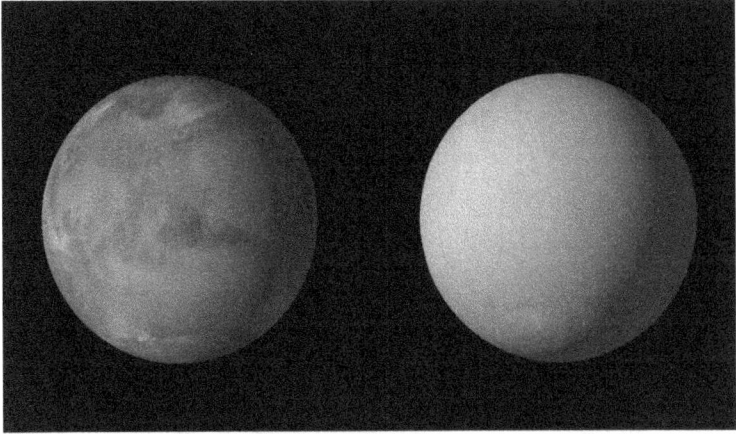

Figure 15:
Mars (before/after) dust storm (July 2018)

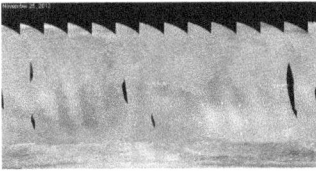

November 25, 2012

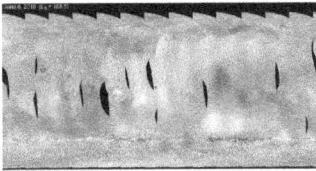

June 6, 2018

Locations of the *Opportunity* and *Curiosity* rovers are noted

Orbit and rotation

Mars's average distance from the Sun is roughly 230 million km (143 million mi), and its orbital period is 687 (Earth) days. The solar day (or sol) on Mars is only slightly longer than an Earth day: 24 hours, 39 minutes, and 35.244 seconds. A Martian year is equal to 1.8809 Earth years, or 1 year, 320 days, and 18.2 hours.

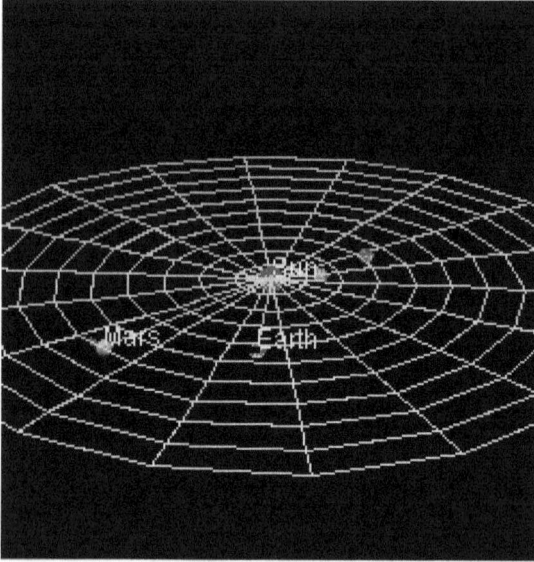

Figure 16: *Mars is about 230 million km (143 million mi) from the Sun; its orbital period is 687 (Earth) days, depicted in red. Earth's orbit is in blue.*

The axial tilt of Mars is 25.19 degrees relative to its orbital plane, which is similar to the axial tilt of Earth. As a result, Mars has seasons like Earth, though on Mars they are nearly twice as long because its orbital period is that much longer. In the present day epoch, the orientation of the north pole of Mars is close to the star Deneb.

Mars has a relatively pronounced orbital eccentricity of about 0.09; of the seven other planets in the Solar System, only Mercury has a larger orbital eccentricity. It is known that in the past, Mars has had a much more circular orbit. At one point, 1.35 million Earth years ago, Mars had an eccentricity of roughly 0.002, much less than that of Earth today. Mars's cycle of eccentricity is 96,000 Earth years compared to Earth's cycle of 100,000 years. Mars has a much longer cycle of eccentricity, with a period of 2.2 million Earth years, and this overshadows the 96,000-year cycle in the eccentricity graphs. For the last 35,000 years, the orbit of Mars has been getting slightly more eccentric because of the gravitational effects of the other planets. The closest distance between Earth and Mars will continue to mildly decrease for the next 25,000 years.

Figure 17: *Viking 1 lander's sampling arm scooped up soil samples for tests (Chryse Planitia)*

Habitability and search for life

The current understanding of planetary habitability—the ability of a world to develop environmental conditions favorable to the emergence of life—favors planets that have liquid water on their surface. Most often this requires the orbit of a planet to lie within the habitable zone, which for the Sun extends from just beyond Venus to about the semi-major axis of Mars. During perihelion, Mars dips inside this region, but Mars's thin (low-pressure) atmosphere prevents liquid water from existing over large regions for extended periods. The past flow of liquid water demonstrates the planet's potential for habitability. Recent evidence has suggested that any water on the Martian surface may have been too salty and acidic to support regular terrestrial life.

The lack of a magnetosphere and the extremely thin atmosphere of Mars are a challenge: the planet has little heat transfer across its surface, poor insulation against bombardment of the solar wind and insufficient atmospheric pressure to retain water in a liquid form (water instead sublimes to a gaseous state). Mars is nearly, or perhaps totally, geologically dead; the end of volcanic activity has apparently stopped the recycling of chemicals and minerals between the surface and interior of the planet.

In situ investigations have been performed on Mars by the *Viking* landers, *Spirit* and *Opportunity* rovers, *Phoenix* lander, and *Curiosity* rover. Evidence suggests that the planet was once significantly more habitable than it is today, but whether living organisms ever existed there remains unknown. The Viking probes of the mid-1970s carried experiments designed to detect microorganisms in Martian soil at their respective landing sites and had positive results,

Figure 18: *Detection of impact glass deposits (green spots)
at Alga crater, a possible site for preserved ancient life*

including a temporary increase of CO

2 production on exposure to water and nutrients. This sign of life was later disputed by scientists, resulting in a continuing debate, with NASA scientist Gilbert Levin asserting that Viking may have found life. A re-analysis of the Viking data, in light of modern knowledge of extremophile forms of life, has suggested that the Viking tests were not sophisticated enough to detect these forms of life. The tests could even have killed a (hypothetical) life form. Tests conducted by the Phoenix Mars lander have shown that the soil has a alkaline pH and it contains magnesium, sodium, potassium and chloride. The soil nutrients may be able to support life, but life would still have to be shielded from the intense ultraviolet light. A recent analysis of martian meteorite EETA79001 found 0.6 ppm ClO−

4, 1.4 ppm ClO−

3, and 16 ppm NO−

3, most likely of Martian origin. The ClO−

3 suggests the presence of other highly oxidizing oxychlorines, such as ClO−

2 or ClO, produced both by UV oxidation of Cl and X-ray radiolysis of ClO−

4. Thus, only highly refractory and/or well-protected (sub-surface) organics or life forms are likely to survive.

A 2014 analysis of the Phoenix WCL showed that the Ca(ClO

4)

2 in the Phoenix soil has not interacted with liquid water of any form, perhaps for as long as 600 Myr. If it had, the highly soluble Ca(ClO

Figure 19: *This image from Gale crater in 2018 prompted speculation that some shapes were worm-like fossils, but they were geological formations probably formed under water.*

4)
2 in contact with liquid water would have formed only CaSO
4. This suggests a severely arid environment, with minimal or no liquid water interaction.

Scientists have proposed that carbonate globules found in meteorite ALH84001, which is thought to have originated from Mars, could be fossilized microbes extant on Mars when the meteorite was blasted from the Martian surface by a meteor strike some 15 million years ago. This proposal has been met with skepticism, and an exclusively inorganic origin for the shapes has been proposed.

Small quantities of methane and formaldehyde detected by Mars orbiters are both claimed to be possible evidence for life, as these chemical compounds would quickly break down in the Martian atmosphere. Alternatively, these compounds may instead be replenished by volcanic or other geological means, such as serpentinization.

Impact glass, formed by the impact of meteors, which on Earth can preserve signs of life, has been found on the surface of the impact craters on Mars.

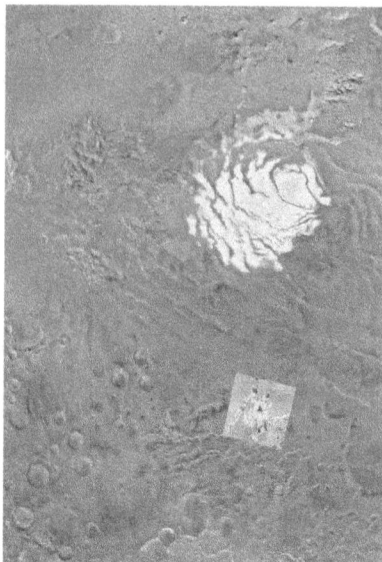

Figure 20: *Location of subsurface water in Planum Australe*

Likewise, the glass in impact craters on Mars could have preserved signs of life if life existed at the site.

In May 2017, evidence of the earliest known life on land on Earth may have been found in 3.48-billion-year-old geyserite and other related mineral deposits (often found around hot springs and geysers) uncovered in the Pilbara Craton of Western Australia. These findings may be helpful in deciding where best to search for early signs of life on the planet Mars.

In early 2018, media reports speculated that certain rock features at a site called Jura looked like a type of fossil, but project scientists say the formations likely resulted from a geological process at the bottom of an ancient drying lakebed, and are related to mineral veins in the area similar to gypsum crystals.

On June 7, 2018, NASA announced that the *Curiosity* rover had discovered organic compounds in sedimentary rocks dating to three billion years old, indicating that some of the building blocks for life were present.

In July 2018, scientists reported the discovery of a subglacial lake on Mars, the first known stable body of water on the planet. It sits 1.5 km (0.9 mi) below the surface at the base of the southern polar ice cap and is about 20 km (12 mi) wide. The lake was discovered using the MARSIS radar on board the *Mars Express* orbiter, and the profiles were collected between May 2012 and December 2015. The lake is centered at 193°E, 81°S, a flat area that does not exhibit any peculiar topographic characteristics. It is mostly surrounded by higher ground except on its eastern side, where there is a depression.

Moons

<templatestyles src="Multiple_image/styles.css" />

Enhanced-color HiRISE image of Phobos, showing a series of mostly parallel grooves and crater chains, with Stickney crater at right

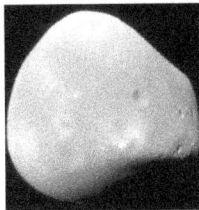

Enhanced-color HiRISE image of Deimos (not to scale), showing its smooth blanket of regolith

Mars has two relatively small (compared to Earth's) natural moons, Phobos (about 22 km (14 mi) in diameter) and Deimos (about 12 km (7.5 mi) in diameter), which orbit close to the planet. Asteroid capture is a long-favored theory, but their origin remains uncertain. Both satellites were discovered in 1877 by Asaph Hall; they are named after the characters Phobos (panic/fear) and Deimos (terror/dread), who, in Greek mythology, accompanied their father Ares, god of war, into battle. Mars was the Roman counterpart of Ares. In modern Greek, though, the planet retains its ancient name *Ares* (Aris: Ἄρης).

From the surface of Mars, the motions of Phobos and Deimos appear different from that of the Moon. Phobos rises in the west, sets in the east, and rises again in just 11 hours. Deimos, being only just outside synchronous orbit – where the orbital period would match the planet's period of rotation – rises as expected in the east but slowly. Despite the 30-hour orbit of Deimos, 2.7 days elapse between its rise and set for an equatorial observer, as it slowly falls behind the rotation of Mars.

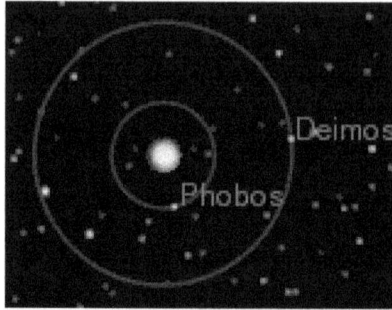

Figure 21: *Orbits of Phobos and Deimos (to scale)*

Because the orbit of Phobos is below synchronous altitude, the tidal forces from the planet Mars are gradually lowering its orbit. In about 50 million years, it could either crash into Mars's surface or break up into a ring structure around the planet.

The origin of the two moons is not well understood. Their low albedo and carbonaceous chondrite composition have been regarded as similar to asteroids, supporting the capture theory. The unstable orbit of Phobos would seem to point towards a relatively recent capture. But both have circular orbits, near the equator, which is unusual for captured objects and the required capture dynamics are complex. Accretion early in the history of Mars is plausible, but would not account for a composition resembling asteroids rather than Mars itself, if that is confirmed.

A third possibility is the involvement of a third body or a type of impact disruption. More-recent lines of evidence for Phobos having a highly porous interior, and suggesting a composition containing mainly phyllosilicates and other minerals known from Mars, point toward an origin of Phobos from material ejected by an impact on Mars that reaccreted in Martian orbit, similar to the prevailing theory for the origin of Earth's moon. Although the VNIR spectra of the moons of Mars resemble those of outer-belt asteroids, the thermal infrared spectra of Phobos are reported to be inconsistent with chondrites of any class.

Mars may have moons smaller than 50 to 100 metres (160 to 330 ft) in diameter, and a dust ring is predicted to exist between Phobos and Deimos.

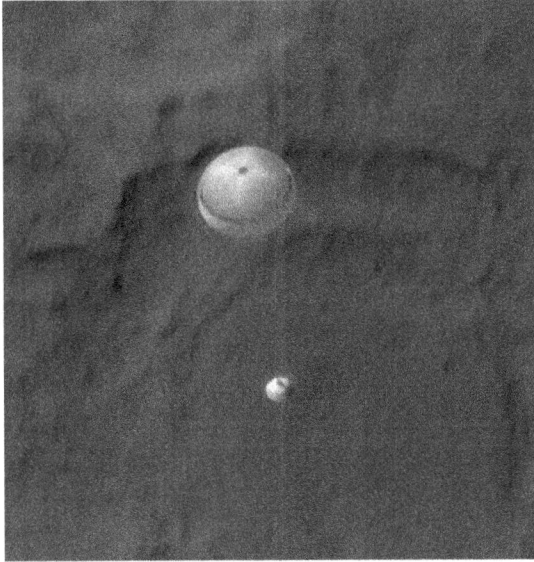

Figure 22: *Mars Science Laboratory under parachute during its atmospheric entry at Mars*

Exploration

Dozens of crewless spacecraft, including orbiters, landers, and rovers, have been sent to Mars by the Soviet Union, the United States, Europe, and India to study the planet's surface, climate, and geology.

As of 2018[8], Mars is host to eight functioning spacecraft: six in orbit—*2001 Mars Odyssey*, *Mars Express*, *Mars Reconnaissance Orbiter*, MAVEN, Mars Orbiter Mission and ExoMars Trace Gas Orbiter—and two on the surface—Mars Exploration Rover *Opportunity* and the Mars Science Laboratory *Curiosity*. The public can request images of Mars via the *Mars Reconnaissance Orbiter*'s HiWish program.

The Mars Science Laboratory, named *Curiosity*, launched on November 26, 2011, and reached Mars on August 6, 2012 UTC. It is larger and more advanced than the Mars Exploration Rovers, with a movement rate up to 90 m (300 ft) per hour. Experiments include a laser chemical sampler that can deduce the make-up of rocks at a distance of 7 m (23 ft). On February 10, 2013, the *Curiosity* rover obtained the first deep rock samples ever taken from another planetary body, using its on-board drill. The same year, it discovered that Mars's soil contains between 1.5% and 3% water by mass (albeit attached to other compounds and thus not freely accessible). Observations by the *Mars*

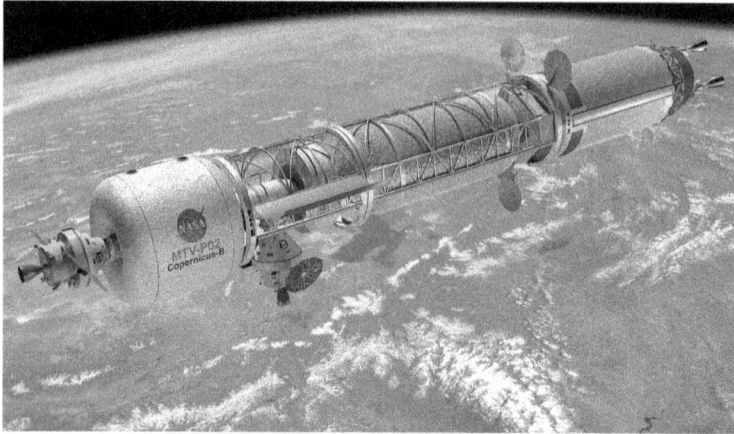

Figure 23: *Concept for a Bimodal Nuclear
Thermal Transfer Vehicle in low Earth orbit*

Reconnaissance Orbiter had previously revealed the possibility of flowing water during the warmest months on Mars.

On September 24, 2014, Mars Orbiter Mission (MOM), launched by the Indian Space Research Organisation, reached Mars orbit. ISRO launched MOM on November 5, 2013, with the aim of analyzing the Martian atmosphere and topography. The Mars Orbiter Mission used a Hohmann transfer orbit to escape Earth's gravitational influence and catapult into a nine-month-long voyage to Mars. The mission is the first successful Asian interplanetary mission.

The European Space Agency, in collaboration with Roscosmos, launched the ExoMars Trace Gas Orbiter and *Schiaparelli* lander on March 14, 2016. While the Trace Gas Orbiter successfully entered Mars orbit on October 19, 2016, *Schiaparelli* crashed during its landing attempt.

Future

In May 2018 NASA's *InSight* lander was launched, along with the twin MarCO CubeSats that will fly by Mars and provide a telemetry relay for the landing. The mission is expected to arrive at Mars in November 2018. NASA plans to launch its Mars 2020 astrobiology rover in July or August 2020.

The European Space Agency will launch the ExoMars rover and surface platform in July 2020.

The United Arab Emirates' *Mars Hope* orbiter is planned for launch in 2020, reaching Mars orbit in 2021. The probe will make a global study of the Martian atmosphere.

Several plans for a human mission to Mars have been proposed throughout the 20th century and into the 21st century, but no active plan has an arrival date sooner than the 2020s. SpaceX founder Elon Musk presented a plan in September 2016 to, optimistically, launch space tourists to Mars in 2024 at an estimated development cost of US$10 billion. In October 2016, President Barack Obama renewed U.S. policy to pursue the goal of sending humans to Mars in the 2030s, and to continue using the International Space Station as a technology incubator in that pursuit. The NASA Authorization Act of 2017 directed NASA to get humans near or on the surface of Mars by the early 2030s.

Astronomy on Mars

With the presence of various orbiters, landers, and rovers, it is possible to practice astronomy from Mars. Although Mars's moon Phobos appears about one-third the angular diameter of the full moon on Earth, Deimos appears more or less star-like, looking only slightly brighter than Venus does from Earth.

Various phenomena seen from Earth have also been observed from Mars, such as meteors and auroras. The apparent sizes of the moons Phobos and Deimos are sufficiently smaller than that of the Sun; thus, their partial "eclipses" of the Sun are best considered transits (see transit of Deimos and Phobos from Mars). Transits of Mercury and Venus have been observed from Mars. A transit of Earth will be seen from Mars on November 10, 2084.

On October 19, 2014, Comet Siding Spring passed extremely close to Mars, so close that the coma may have enveloped Mars.

<templatestyles src="Multiple_image/styles.css" />

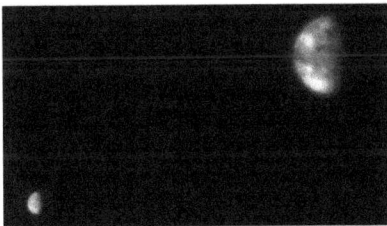

Earth and the Moon (*MRO* HiRISE, November 2016)

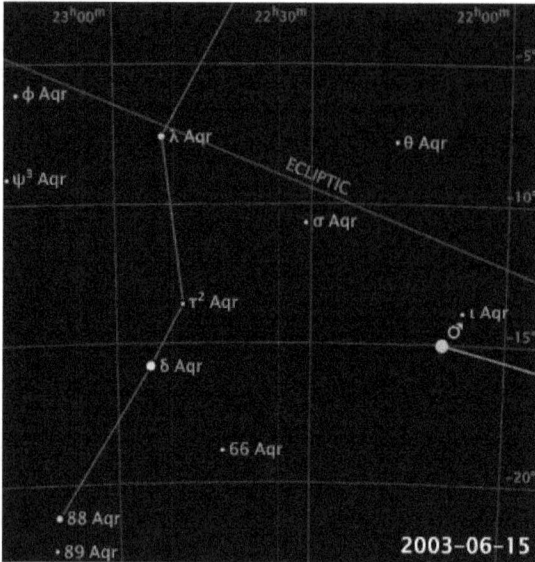

Figure 24: *Animation of the apparent retro-grade motion of Mars in 2003 as seen from Earth*

Phobos transits the Sun (*Opportunity*, March 10, 2004)

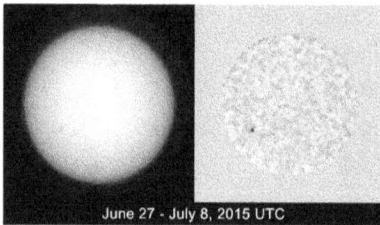

Tracking sunspots from Mars

Viewing

Because the orbit of Mars is eccentric, its apparent magnitude at opposition from the Sun can range from –3.0 to –1.4. The minimum brightness is magnitude +1.6 when the planet is in conjunction with the Sun. Mars usually appears distinctly yellow, orange, or red; the actual color of Mars is closer to butterscotch, and the redness seen is just dust in the planet's atmosphere. NASA's *Spirit* rover has taken pictures of a greenish-brown, mud-colored landscape with blue-grey rocks and patches of light red sand. When farthest away from

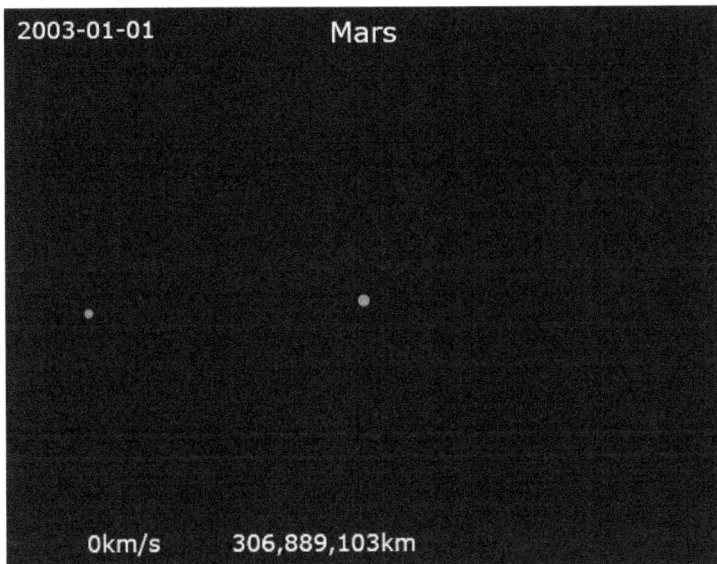

Figure 25: *Geocentric animation of Mars's orbit relative to Earth from January 2003 to January 2019*

Mars · Earth

Earth, it is more than seven times farther away than when it is closest. When least favorably positioned, it can be lost in the Sun's glare for months at a time. At its most favorable times—at 15- or 17-year intervals, and always between late July and late September—a lot of surface detail can be seen with a telescope. Especially noticeable, even at low magnification, are the polar ice caps.

As Mars approaches opposition, it begins a period of retrograde motion, which means it will appear to move backwards in a looping motion with respect to the background stars. The duration of this retrograde motion lasts for about 72 days, and Mars reaches its peak luminosity in the middle of this motion.

Closest approaches

Relative

The point at which Mars's geocentric longitude is 180° different from the Sun's is known as opposition, which is near the time of closest approach to Earth. The time of opposition can occur as much as 8.5 days away from the closest approach. The distance at close approach varies between about 54 and 103 million km (34 and 64 million mi) due to the planets' elliptical orbits, which causes comparable variation in angular size. The last Mars opposition occurred on July 27, 2018, at a distance of about 58 million km (36 million mi). The next Mars opposition occurs on October 13, 2020, at a distance of about 63 million km (39 million mi). The average time between the successive oppositions of Mars, its synodic period, is 780 days; but the number of days between the dates of successive oppositions can range from 764 to 812.

As Mars approaches opposition it begins a period of retrograde motion, which makes it appear to move backwards in a looping motion relative to the background stars. The duration of this retrograde motion is about 72 days.

Absolute, around the present time

Mars made its closest approach to Earth and maximum apparent brightness in nearly 60,000 years, 55,758,006 km (0.37271925 AU; 34,646,419 mi), magnitude –2.88, on August 27, 2003, at 9:51:13 UTC. This occurred when Mars was one day from opposition and about three days from its perihelion, making it particularly easy to see from Earth. The last time it came so close is estimated to have been on September 12, 57,617 BC, the next time being in 2287. This record approach was only slightly closer than other recent close approaches. For instance, the minimum distance on August 22, 1924, was 0.37285 AU, and the minimum distance on August 24, 2208, will be 0.37279 AU.

Historical observations

The history of observations of Mars is marked by the oppositions of Mars, when the planet is closest to Earth and hence is most easily visible, which occur every couple of years. Even more notable are the perihelic oppositions of Mars, which occur every 15 or 17 years and are distinguished because Mars is close to perihelion, making it even closer to Earth.

Ancient and medieval observations

The ancient Sumerians believed that Mars was Nergal, the god of war and plague. During Sumerian times, Nergal was a minor deity of little significance, but, during later times, his main cult center was the city of Nineveh. In Mesopotamian texts, Mars is referred to as the "star of judgement of the fate of the dead". The existence of Mars as a wandering object in the night sky was recorded by the ancient Egyptian astronomers and, by 1534 BCE, they were familiar with the retrograde motion of the planet. By the period of the Neo-Babylonian Empire, the Babylonian astronomers were making regular records of the positions of the planets and systematic observations of their behavior. For Mars, they knew that the planet made 37 synodic periods, or 42 circuits

Figure 26: *Galileo Galilei, first person to see Mars via telescope in 1610.*

of the zodiac, every 79 years. They invented arithmetic methods for making minor corrections to the predicted positions of the planets.

In the fourth century BCE, Aristotle noted that Mars disappeared behind the Moon during an occultation, indicating that the planet was farther away. Ptolemy, a Greek living in Alexandria, attempted to address the problem of the orbital motion of Mars. Ptolemy's model and his collective work on astronomy was presented in the multi-volume collection *Almagest*, which became the authoritative treatise on Western astronomy for the next fourteen centuries. Literature from ancient China confirms that Mars was known by Chinese astronomers by no later than the fourth century BCE. In the fifth century CE, the Indian astronomical text *Surya Siddhanta* estimated the diameter of Mars. In the East Asian cultures, Mars is traditionally referred to as the "fire star" (Chinese: 火星), based on the Five elements.

During the seventeenth century, Tycho Brahe measured the diurnal parallax of Mars that Johannes Kepler used to make a preliminary calculation of the relative distance to the planet. When the telescope became available, the diurnal parallax of Mars was again measured in an effort to determine the Sun-Earth distance. This was first performed by Giovanni Domenico Cassini in 1672. The early parallax measurements were hampered by the quality of the instruments. The only occultation of Mars by Venus observed was that of October

13, 1590, seen by Michael Maestlin at Heidelberg. In 1610, Mars was viewed by Galileo Galilei, who was first to see it via telescope. The first person to draw a map of Mars that displayed any terrain features was the Dutch astronomer Christiaan Huygens.

Martian "canals"

<templatestyles src="Multiple_image/styles.css" />

Map of Mars by Giovanni Schiaparelli

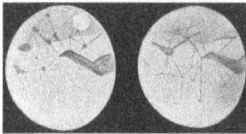

Mars sketched as observed by Lowell before 1914 (south on top)

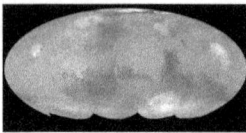

Map of Mars from the *Hubble Space Telescope* as seen near the 1999 opposition (north on top)

By the 19th century, the resolution of telescopes reached a level sufficient for surface features to be identified. A perihelic opposition of Mars occurred on September 5, 1877. In that year, the Italian astronomer Giovanni Schiaparelli used a 22 cm (8.7 in) telescope in Milan to help produce the first detailed map of Mars. These maps notably contained features he called *canali*, which were later shown to be an optical illusion. These *canali* were supposedly long, straight lines on the surface of Mars, to which he gave names of famous rivers on Earth. His term, which means "channels" or "grooves", was popularly mistranslated in English as "canals".

Influenced by the observations, the orientalist Percival Lowell founded an observatory which had 30 and 45 cm (12 and 18 in) telescopes. The observatory was used for the exploration of Mars during the last good opportunity in 1894 and the following less favorable oppositions. He published several books on Mars and life on the planet, which had a great influence on the public. The

♂

canali were independently found by other astronomers, like Henri Joseph Perrotin and Louis Thollon in Nice, using one of the largest telescopes of that time.

The seasonal changes (consisting of the diminishing of the polar caps and the dark areas formed during Martian summer) in combination with the canals led to speculation about life on Mars, and it was a long-held belief that Mars contained vast seas and vegetation. The telescope never reached the resolution required to give proof to any speculations. As bigger telescopes were used, fewer long, straight *canali* were observed. During an observation in 1909 by Flammarion with an 84 cm (33 in) telescope, irregular patterns were observed, but no *canali* were seen.

Even in the 1960s articles were published on Martian biology, putting aside explanations other than life for the seasonal changes on Mars. Detailed scenarios for the metabolism and chemical cycles for a functional ecosystem have been published.

Spacecraft visitation

Once spacecraft visited the planet during NASA's Mariner missions in the 1960s and 70s, these concepts were radically broken. The results of the Viking life-detection experiments aided an intermission in which the hypothesis of a hostile, dead planet was generally accepted.

Mariner 9 and Viking allowed better maps of Mars to be made using the data from these missions, and another major leap forward was the Mars Global Surveyor mission, launched in 1996 and operated until late 2006, that allowed complete, extremely detailed maps of the Martian topography, magnetic field and surface minerals to be obtained. These maps are available online; for example, at Google Mars. Mars Reconnaissance Orbiter and Mars Express continued exploring with new instruments, and supporting lander missions. NASA provides two online tools: Mars Trek, which provides visualizations of the planet using data from 50 years of exploration, and Experience Curiosity, which simulates traveling on Mars in 3-D with Curiosity.

In culture

Mars is named after the Roman god of war. In different cultures, Mars represents masculinity and youth. Its symbol, a circle with an arrow pointing out to the upper right, is used as a symbol for the male gender.

Figure 27: *An 1893 soap ad playing on the popular idea that Mars was populated*

The many failures in Mars exploration probes resulted in a satirical counter-culture blaming the failures on an Earth-Mars "Bermuda Triangle", a "Mars Curse", or a "Great Galactic Ghoul" that feeds on Martian spacecraft.

Intelligent "Martians"

The fashionable idea that Mars was populated by intelligent Martians exploded in the late 19th century. Schiaparelli's "canali" observations combined with Percival Lowell's books on the subject put forward the standard notion of a planet that was a drying, cooling, dying world with ancient civilizations constructing irrigation works.

Many other observations and proclamations by notable personalities added to what has been termed "Mars Fever". In 1899, while investigating atmospheric radio noise using his receivers in his Colorado Springs lab, inventor Nikola Tesla observed repetitive signals that he later surmised might have been radio communications coming from another planet, possibly Mars. In a 1901 interview Tesla said:

It was some time afterward when the thought flashed upon my mind that the disturbances I had observed might be due to an intelligent control. Although I could not decipher their meaning, it was impossible for me to think of them as having been entirely accidental. The feeling is constantly growing on me that I had been the first to hear the greeting of one planet to another.

Tesla's theories gained support from Lord Kelvin who, while visiting the United States in 1902, was reported to have said that he thought Tesla had picked up Martian signals being sent to the United States. Kelvin "emphatically" denied this report shortly before leaving: "What I really said was that the inhabitants of Mars, if there are any, were doubtless able to see New York, particularly the glare of the electricity."

In a *New York Times* article in 1901, Edward Charles Pickering, director of the Harvard College Observatory, said that they had received a telegram from Lowell Observatory in Arizona that seemed to confirm that Mars was trying to communicate with Earth.

Early in December 1900, we received from Lowell Observatory in Arizona a telegram that a shaft of light had been seen to project from Mars (the Lowell observatory makes a specialty of Mars) lasting seventy minutes. I wired these facts to Europe and sent out neostyle copies through this country. The observer there is a careful, reliable man and there is no reason to doubt that the light existed. It was given as from a well-known geographical point on Mars. That was all. Now the story has gone the world over. In Europe it is stated that I have been in communication with Mars, and all sorts of exaggerations have spring up. Whatever the light was, we have no means of knowing. Whether it had intelligence or not, no one can say. It is absolutely inexplicable.

Pickering later proposed creating a set of mirrors in Texas, intended to signal Martians.

In recent decades, the high-resolution mapping of the surface of Mars, culminating in Mars Global Surveyor, revealed no artifacts of habitation by "intelligent" life, but pseudoscientific speculation about intelligent life on Mars continues from commentators such as Richard C. Hoagland. Reminiscent of the *canali* controversy, these speculations are based on small scale features perceived in the spacecraft images, such as "pyramids" and the "Face on Mars". Planetary astronomer Carl Sagan wrote:

Mars has become a kind of mythic arena onto which we have projected our Earthly hopes and fears.

The depiction of Mars in fiction has been stimulated by its dramatic red color and by nineteenth century scientific speculations that its surface conditions might support not just life but intelligent life. Thus originated a large number of science fiction scenarios, among which is H. G. Wells' *The War of the Worlds*, published in 1898, in which Martians seek to escape their dying planet by invading Earth.

Figure 28: *Martian tripod illustration from the 1906 French edition of The War of the Worlds by H. G. Wells*

Influential works included Ray Bradbury's *The Martian Chronicles*, in which human explorers accidentally destroy a Martian civilization, Edgar Rice Burroughs' *Barsoom* series, C. S. Lewis' novel *Out of the Silent Planet* (1938), and a number of Robert A. Heinlein stories before the mid-sixties.

Jonathan Swift made reference to the moons of Mars, about 150 years before their actual discovery by Asaph Hall, detailing reasonably accurate descriptions of their orbits, in the 19th chapter of his novel *Gulliver's Travels*.

A comic figure of an intelligent Martian, Marvin the Martian, appeared in *Haredevil Hare* (1948) as a character in the Looney Tunes animated cartoons of Warner Brothers, and has continued as part of popular culture to the present.

After the Mariner and Viking spacecraft had returned pictures of Mars as it really is, an apparently lifeless and canal-less world, these ideas about Mars had to be abandoned, and a vogue for accurate, realist depictions of human colonies on Mars developed, the best known of which may be Kim Stanley Robinson's *Mars* trilogy. Pseudo-scientific speculations about the Face on Mars and other enigmatic landmarks spotted by space probes have meant that ancient civilizations continue to be a popular theme in science fiction, especially in film.

Notes

References

External links

- Mars[9] at Curlie (based on DMOZ)
- Mars Exploration Program[10] at NASA.gov
- Google Mars[11] and Google Mars 3D[12], interactive maps of the planet
- Geody Mars[13], mapping site that supports NASA World Wind, Celestia, and other applications

Images

- Mars images[14] by NASA's Planetary Photojournal
- Mars images[15] by NASA's Mars Exploration Program
- Mars images[16] by Malin Space Science Systems
- HiRISE image catalog[17] by the University of Arizona

Videos

- Rotating color globe of Mars[18] by the National Oceanic and Atmospheric Administration
- Rotating geological globe of Mars[19] by the United States Geological Survey
- NASA's *Curiosity* Finds Ancient Streambed – First Evidence of Water on Mars[20] on YouTube by The Science Channel (2012, 4:31)
- Flight Into Mariner Valley[21] by Arizona State University
- High resolution video[22] simulation of rotating Mars by Seán Doran, showing Arabia Terra, Valles Marineris and Tharsis (see album[23] for more)

Cartographic resources

- Mars nomenclature[24] and quadrangle maps with feature names[25] by the United States Geological Survey
- Geological map of Mars[26] by the United States Geological Survey
- Viking orbiter photomap[27] by Eötvös Loránd University
- Mars Global Surveyor topographical map[28] by Eötvös Loránd University

<indicator name="featured-star"> ⭐ </indicator>

Geology of Mars

Geology of Mars

The **geology of Mars** is the scientific study of the surface, crust, and interior of the planet Mars. It emphasizes the composition, structure, history, and physical processes that shape the planet. It is analogous to the field of terrestrial geology. In planetary science, the term *geology* is used in its broadest sense to mean the study of the solid parts of planets and moons. The term incorporates aspects of geophysics, geochemistry, mineralogy, geodesy, and cartography. A neologism, **areology**, from the Greek word *Arēs* (Mars), sometimes appears as a synonym for Mars's geology in the popular media and works of science fiction (e.g. Kim Stanley Robinson's Mars trilogy).

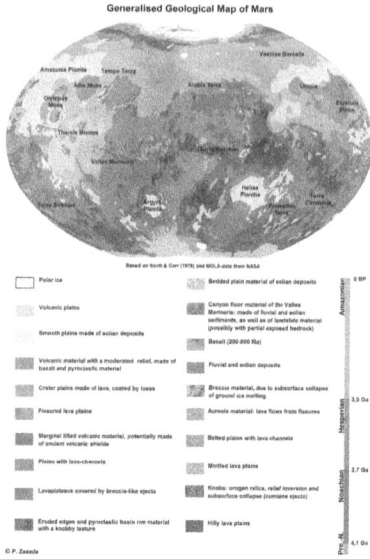

Figure 29: *Generalised geological map of Mars*[29]

Figure 30: *Mars as seen by the Hubble Space Telescope*

Geological map of Mars (2014)

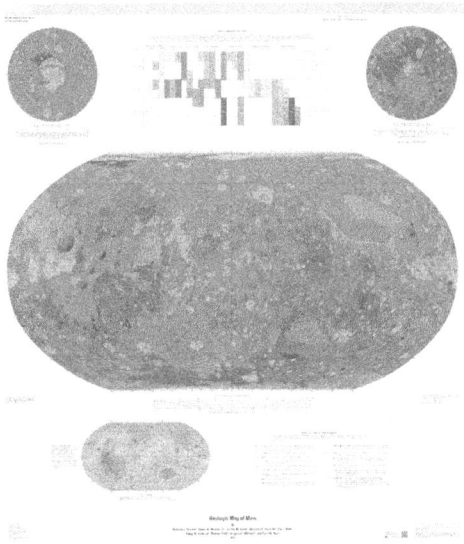

File:USGS-MarsMap-sim3292-20140714-crop.png

Mars - geologic map (USGS; July 14, 2014) (full image)

Global Martian topography and large-scale features

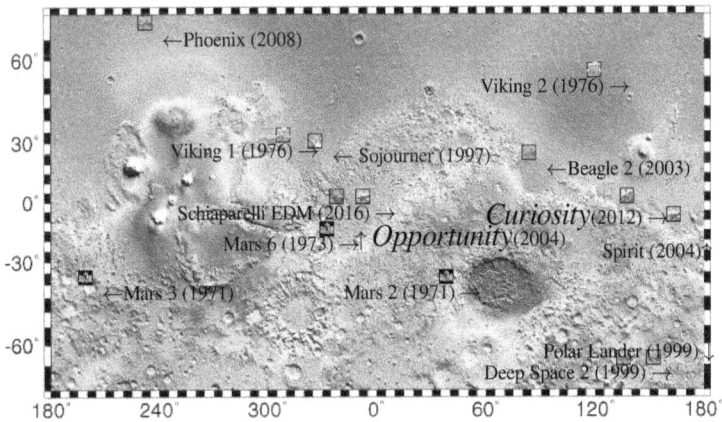

Composition of Mars

Mars is a differentiated, terrestrial planet.

Global physiography

Most of our current knowledge about the geology of Mars comes from studying landforms and relief features (terrain) seen in images taken by orbiting
spacecraft. Mars has a number of distinct, large-scale surface features that indicate the types of geological processes that have operated on the planet over
time. This section introduces several of the larger physiographic regions of
Mars. Together, these regions illustrate how geologic processes involving volcanism, tectonism, water, ice, and impacts have shaped the planet on a global
scale.

Hemispheric dichotomy

The northern and southern hemispheres of Mars are strikingly different from
each other in topography and physiography. This dichotomy is a fundamental
global geologic feature of the planet. Simply stated, the northern part of the
planet is an enormous topographic depression. About one-third of the planet's
surface (mostly in the northern hemisphere) lies 3–6 km lower in elevation
than the southern two-thirds. This is a first-order relief feature on par with
the elevation difference between Earth's continents and ocean basins. The dichotomy is also expressed in two other ways: as a difference in impact crater
density and crustal thickness between the two hemispheres. The hemisphere
south of the dichotomy boundary (often called the southern highlands or up-
lands) is very heavily cratered and ancient, characterized by rugged surfaces
that date back to the period of heavy bombardment. In contrast, the lowlands
north of the dichotomy boundary have few large craters, are very smooth and
flat, and have other features indicating that extensive resurfacing has occurred
since the southern highlands formed. The third distinction between the two
hemispheres is in crustal thickness. Topographic and geophysical gravity data
indicate that the crust in the southern highlands has a maximum thickness of
about 58 km (36 mi), whereas crust in the northern lowlands "peaks" at around
32 km (20 mi) in thickness. The location of the dichotomy boundary varies in
latitude across Mars and depends on which of the three physical expressions
of the dichotomy is being considered.

The origin and age of the hemispheric dichotomy are still debated. Hypotheses
of origin generally fall into two categories: one, the dichotomy was produced
by a mega-impact event or several large impacts early in the planet's history
(exogenic theories) or two, the dichotomy was produced by crustal thinning in

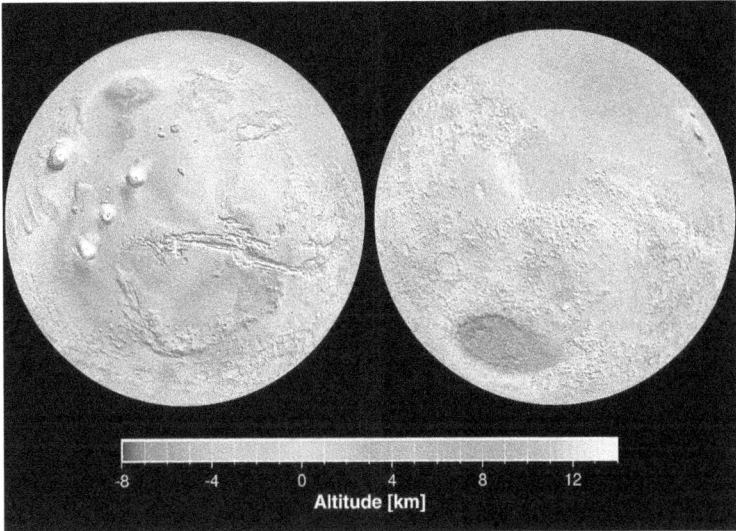

Figure 31: *Mars Orbital Laser Altimeter (MOLA) colorized shaded-relief maps showing elevations in the western and eastern hemispheres of Mars. (Left): The western hemisphere is dominated by the Tharsis region (red and brown). Tall volcanoes appear white. Valles Marineris (blue) is the long gash-like feature to the right. (Right): Eastern hemisphere shows the cratered highlands (yellow to red) with the Hellas basin (deep blue/ purple) at lower left. The Elysium province is at the upper right edge. Areas north of the dichotomy boundary appear as shades of blue on both maps.*

the northern hemisphere by mantle convection, overturning, or other chemical and thermal processes in the planet's interior (endogenic theories). One endogenic model proposes an early episode of plate tectonics producing a thinner crust in the north, similar to what is occurring at spreading plate boundaries on Earth. Whatever its origin, the Martian dichotomy appears to be extremely old. A new theory based on the Southern Polar Giant Impact and validated by the discovery of twelve hemispherical alignments shows that exogenic theories appear to be stronger than endogenic theories and that Mars never had plate tectonics that could modify the dichotomy. Laser altimeter and radar sounding data from orbiting spacecraft have identified a large number of basin-sized structures previously hidden in visual images. Called quasi-circular depressions (QCDs), these features likely represent derelict impact craters from the period of heavy bombardment that are now covered by a veneer of younger deposits. Crater counting studies of QCDs suggest that the underlying surface in the northern hemisphere is at least as old as the oldest exposed crust in the southern highlands. The ancient age of the dichotomy places a significant constraint on theories of its origin.

Tharsis and Elysium volcanic provinces

Straddling the dichotomy boundary in Mars's western hemisphere is a massive volcano-tectonic province known as the Tharsis region or the Tharsis bulge. This immense, elevated structure is thousands of kilometers in diameter and covers up to 25% of the planet's surface. Averaging 7–10 km above datum (Martian "sea" level), Tharsis contains the highest elevations on the planet and the largest known volcanoes in the Solar System. Three enormous volcanoes, Ascraeus Mons, Pavonis Mons, and Arsia Mons (collectively known as the Tharsis Montes), sit aligned NE-SW along the crest of the bulge. The vast Alba Mons (formerly Alba Patera) occupies the northern part of the region. The huge shield volcano Olympus Mons lies off the main bulge, at the western edge of the province. The extreme massiveness of Tharsis has placed tremendous stresses on the planet's lithosphere. As a result, immense extensional fractures (grabens and rift valleys) radiate outward from Tharsis, extending halfway around the planet.[30]

A smaller volcanic center lies several thousand kilometers west of Tharsis in Elysium. The Elysium volcanic complex is about 2,000 kilometers in diameter and consists of three main volcanoes, Elysium Mons, Hecates Tholus, and Albor Tholus. The Elysium group of volcanoes is thought to be somewhat different from the Tharsis Montes, in that development of the former involved both lavas and pyroclastics.

Large impact basins

Several enormous, circular impact basins are present on Mars. The largest one that is readily visible is the Hellas basin located in the southern hemisphere. It is the second largest confirmed impact structure on the planet, centered at about 64°E longitude and 40°S latitude. The central part of the basin (Hellas Planitia) is 1,800 km in diameter[31] and surrounded by a broad, heavily eroded annular rim structure characterized by closely spaced rugged irregular mountains (massifs), which probably represent uplifted, jostled blocks of old pre-basin crust.[32] (See Anseris Mons, for example.) Ancient, low-relief volcanic constructs (highland paterae) are located on the northeastern and southwestern portions of the rim. The basin floor contains thick, structurally complex sedimentary deposits that have a long geologic history of deposition, erosion, and internal deformation. The lowest elevations on the planet are located within the Hellas basin, with some areas of the basin floor lying over 8 km below datum.

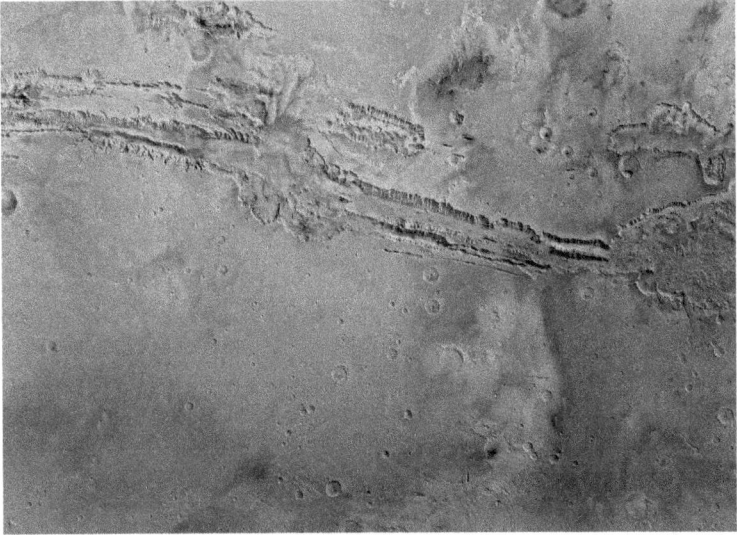

Figure 32: *Viking Orbiter 1 view image of Valles Marineris.*

The two other large impact structures on the planet are the Argyre and Isidis basins. Like Hellas, Argyre (800 km in diameter) is located in the southern highlands and is surrounded by a broad ring of mountains. The mountains in the southern portion of the rim, Charitum Montes, may have been eroded by valley glaciers and ice sheets at some point in Mars's history. The Isidis basin (roughly 1,000 km in diameter) lies on the dichotomy boundary at about 87°E longitude. The northeastern portion of the basin rim has been eroded and is now buried by northern plains deposits, giving the basin a semicircular outline. The northwestern rim of the basin is characterized by arcuate grabens (Nili Fossae) that are circumferential to the basin. One additional large basin, Utopia, is completely buried by northern plains deposits. Its outline is clearly discernable only from altimetry data. All of the large basins on Mars are extremely old, dating back to the late heavy bombardment. They are thought to be comparable in age to the Imbrium and Orientale basins on the Moon.

Equatorial canyon system

Near the equator in the western hemisphere lies an immense system of deep, interconnected canyons and troughs collectively known as the Valles Marineris. The canyon system extends eastward from Tharsis for a length of over 4,000 km, nearly a quarter of the planet's circumference. If placed on Earth, Valles Marineris would span the width of North America.[33] In places, the canyons are up to 300 km wide and 10 km deep. Often compared to Earth's Grand Canyon, the Valles Marineris has a very different origin than its tinier, so-called counterpart on Earth. The Grand Canyon is largely a product of water erosion. The Martian equatorial canyons were of tectonic origin, i.e. they were formed mostly by faulting. They could be similar to the East African Rift valleys. The canyons represent the surface expression of powerful extensional strain in the Martian crust, probably due to loading from the Tharsis bulge.

Chaotic terrain and outflow channels

The terrain at the eastern end of the Valles Marineris grades into dense jumbles of low rounded hills that seem to have formed by the collapse of upland surfaces to form broad, rubble-filled hollows. Called chaotic terrain, these areas mark the heads of huge outflow channels that emerge full size from the chaotic terrain and empty (debouch) northward into Chryse Planitia. The presence of streamlined islands and other geomorphic features indicate that the channels were most likely formed by catastrophic releases of water from aquifers or the melting of subsurface ice. However, these features could also be formed by abundant volcanic lava flows coming from Tharsis. The channels, which include Ares, Shalbatana, Simud, and Tiu Valles, are enormous by terrestrial standards, and the flows that formed them correspondingly immense. For example, the peak discharge required to carve the 28-km-wide Ares Vallis is estimated to have been 14 million cubic metres (500 million cu ft) per second, over ten thousand times the average discharge of the Mississippi River.

Ice caps

The polar ice caps are well-known telescopic features of Mars, first identified by Christiaan Huygens in 1672.[34] Since the 1960s, we have known that the seasonal caps (those seen in the telescope to grow and wane seasonally) are composed of carbon dioxide (CO_2) ice that condenses out of the atmosphere as temperatures fall to 148 K, the frost point of CO_2, during the polar wintertime. In the north, the CO_2 ice completely dissipates (sublimes) in summer, leaving behind a residual cap of water (H_2O) ice. At the south pole, a small residual cap of CO_2 ice remains in summer.

Both residual ice caps overlie thick layered deposits of interbedded ice and dust. In the north, the layered deposits form a 3 km-high, 1,000 km-diameter plateau called Planum Boreum. A similar kilometers-thick plateau, Planum Australe, lies in the south. Both plana (the Latin plural of planum) are sometimes treated as synonymous with the polar ice caps, but the permanent ice (seen as the high albedo, white surfaces in images) forms only a relatively thin mantle on top of the layered deposits. The layered deposits probably represent alternating cycles of dust and ice deposition caused by climate changes related to variations in the

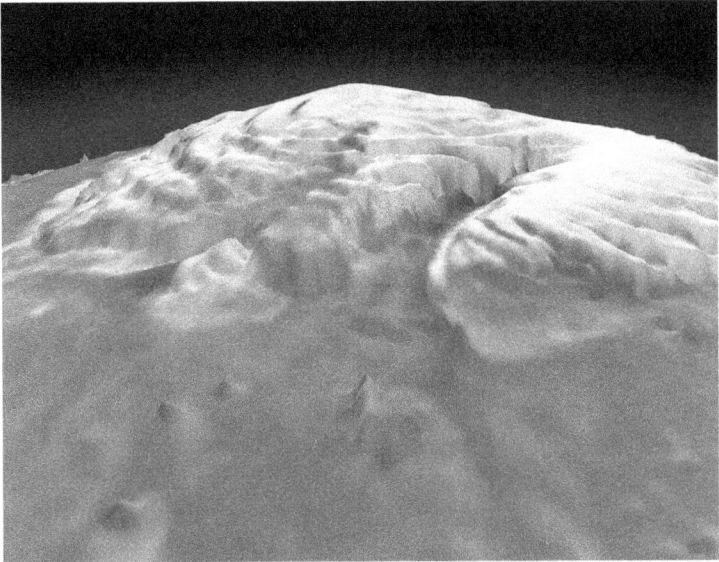

Figure 33: *Mars Orbital Laser Altimeter (MOLA) derived image of Planum Boreum. Vertical exaggeration is extreme. Note that residual ice cap is only the thin veneer (shown in white) on top of the plateau.*

planet's orbital parameters over time (see also Milankovitch cycles). The polar layered deposits are some of the youngest geologic units on Mars.

Albedo features

No topography is visible on Mars from Earth. The bright areas and dark markings seen through a telescope are albedo features. The bright, red-ochre areas are locations where fine dust covers the surface. Bright areas (excluding the polar caps and clouds) include Hellas, Tharsis, and Arabia Terra. The dark gray markings represent areas that the wind has swept clean of dust, leaving behind the lower layer of dark, rocky material. Dark markings are most distinct in a broad belt from 0° to 40° S latitude. However, the most prominent dark marking, Syrtis Major Planum, is in the northern hemisphere. The classical albedo feature, Mare Acidalium (Acidalia Planitia), is another prominent dark area in the northern hemisphere. A third type of area, intermediate in color and albedo, is also present and thought to represent regions containing a mixture of the material from the bright and dark areas.

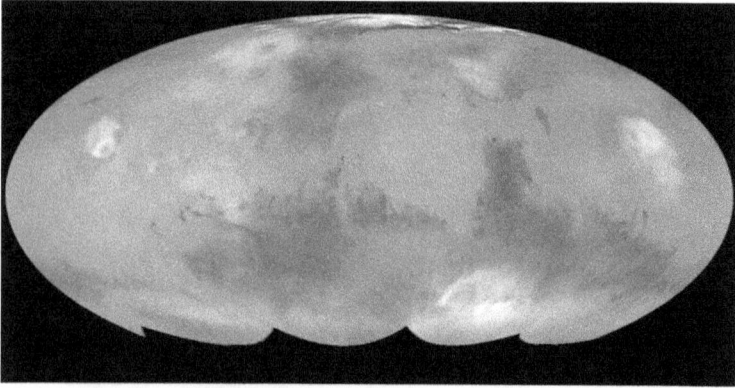

Figure 34: *Mollweide projection of albedo features on Mars from Hubble Space Telescope. Bright ochre areas in left, center, and right are Tharsis, Arabia, and Elysium, respectively. The dark region at top center left is Acidalium Planitia. Syrtis Major is the dark area projecting upward in the center right. Note orographic clouds over Olympus and Elysium Montes (left and right, respectively).*

Impact craters

Impact craters were first identified on Mars by the Mariner 4 spacecraft in 1965. Early observations showed that Martian craters were generally shallower and smoother than lunar craters, indicating that Mars has a more active history of erosion and deposition than the Moon.

In other aspects, Martian craters resemble lunar craters. Both are products of hypervelocity impacts and show a progression of morphology types with increasing size. Martian craters below about 7 km in diameter are called simple craters; they are bowl-shaped with sharp raised rims and have depth/diameter ratios of about 1/5. Martian craters change from simple to more complex types at diameters of roughly 5 to 8 km. Complex craters have central peaks (or peak complexes), relatively flat floors, and terracing or slumping along the inner walls. Complex craters are shallower than simple craters in proportion to their widths, with depth/diameter ratios ranging from 1/5 at the simple-to-complex transition diameter ($\sim$7 km) to about 1/30 for a 100-km diameter crater. Another transition occurs at crater diameters of around 130 km as central peaks turn into concentric rings of hills to form multi-ring basins.

Mars has the greatest diversity of impact crater types of any planet in the Solar System. This is partly because the presence of both rocky and volatile-rich layers in the subsurface produces a range of morphologies even among craters within the same size classes. Mars also has an atmosphere that plays a role in ejecta emplacement and subsequent erosion. Moreover, Mars has a rate of volcanic and tectonic activity low enough that ancient, eroded craters are still preserved, yet high enough to have resurfaced large areas of the planet, producing a diverse range of crater populations of widely differing ages. Over 42,000 impact craters greater than 5 km in diameter have been catalogued on Mars, and the number of smaller craters is probably innumerable. The density of craters on Mars is highest in the southern hemisphere, south of the dichotomy boundary. This is where most of the large craters and basins are located.

Crater morphology provides information about the physical structure and composition of the surface and subsurface at the time of impact. For example, the size of central peaks in Martian craters is larger than comparable craters on Mercury or the Moon.[35] In addition, the central peaks of many large craters on Mars have pit craters at their summits. Central pit craters are rare on the Moon but are very common on Mars and the icy satellites of the outer Solar System. Large central peaks and the abundance of pit craters probably indicate the presence of near-surface ice at the time of impact. Polewards of 30 degrees of latitude, the form of older impact craters is rounded out ("softened") by acceleration of soil creep by ground ice.

The most notable difference between Martian craters and other craters in the Solar System is the presence of lobate (fludized) ejecta blankets. Many craters at equatorial and mid-latitudes on Mars have this form of ejecta morphology, which is thought to arise when the impacting object melts ice in the subsurface. Liquid water in the ejected material forms a muddy slurry that flows along the surface, producing the characteristic lobe shapes. The crater Yuty is a good example of a rampart crater, which is so called because of the rampart-like edge to its ejecta blanket.

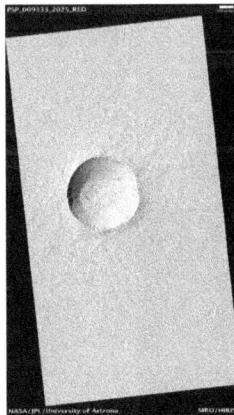

Figure 35: *HiRISE image of simple rayed crater on southeastern flank of Elysium Mons.*

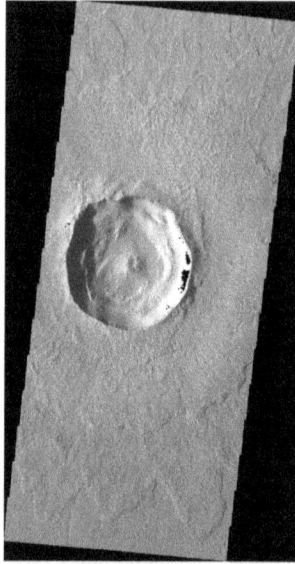

Figure 36: *THEMIS image of complex crater with fluidized ejecta. Note central peak with pit crater.*

Figure 37: *Viking orbiter image of Yuty crater showing lobate ejecta.*

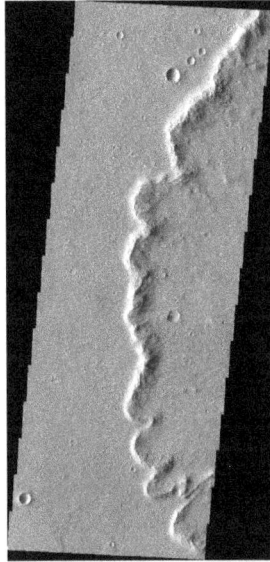

Figure 38: *THEMIS close-up view of ejecta from 17-km di-ameter crater at 21°S, 285°E. Note prominent rampart.*

Martian craters are commonly classified by their ejecta. Craters with one ejecta layer are called single-layer ejecta (SLE) craters. Craters with two superposed ejecta blankets are called double-layer ejecta (DLE) craters, and craters with more than two ejecta layers are called multiple-layered ejecta (MLE) craters. These morphological differences are thought to reflect compositional differences (i.e. interlayered ice, rock, or water) in the subsurface at the time of impact.[36]

Martian craters show a large diversity of preservational states, from extremely fresh to old and eroded. Degraded and infilled impact craters record variations in volcanic, fluvial, and eolian activity over geologic time. Pedestal craters are craters with their ejecta sitting above the surrounding terrain to form raised platforms. They occur because the crater's ejecta forms a resistant layer so that the area nearest the crater erodes more slowly than the rest of the region. Some pedestals are hundreds of meters above the surrounding area, meaning that hundreds of meters of material were eroded away. Pedestal craters were first observed during the Mariner 9 mission in 1972.[37,38]

Figure 39: *Pedestal crater in Amazonis quadrangle as seen by HiRISE.*

Volcanism

Volcanic structures and landforms cover large portions of the Martian surface. The most conspicuous volcanoes on Mars are located in Tharsis and Elysium. Geologists think one of the reasons volcanoes on Mars were able to grow so large is that Mars has fewer tectonic boundaries in comparison to Earth. Lava from a stationary hot spot was able to accumulate at one location on the surface for many hundreds of millions of years.

Scientists have never recorded an active volcano eruption on the surface of Mars. Searches for thermal signatures and surface changes within the last decade have not yielded evidence for active volcanism.

On October 17, 2012, the *Curiosity rover* on the planet Mars at "Rocknest" performed the first X-ray diffraction analysis of Martian soil. The results from the rover's CheMin analyzer revealed the presence of several minerals, including feldspar, pyroxenes and olivine, and suggested that the Martian soil in the sample was similar to the "weathered basaltic soils" of Hawaiian volcanoes. In July 2015, the same rover identified tridymite in a rock sample from Gale Crater, leading scientists to conclude that silicic volcanism might have played a much more prevalent role in the planet's volcanic history than previously thought.

Figure 40: *First X-ray diffraction view of Martian soil - CheMin analysis reveals feldspar, pyroxenes, olivine and more (Curiosity rover at "Rocknest", October 17, 2012).*

Sedimentology

Flowing water appears to have been common on the surface of Mars at various points in its history, and especially on ancient Mars. Many of these flows carved the surface, forming valley networks and producing sediment. This sediment has been redeposited in a wide variety of wet environments, including in alluvial fans, meandering channels, deltas, lakes, and perhaps even oceans.[39,40] The processes of deposition and transportation are associated with gravity. Due to gravity, related differences in water fluxes and flow speeds, inferred from grain size distributions, Martian landscapes were created by different environmental conditions.[41] Nevertheless, there are other ways of estimating the amount of water on ancient Mars (see: Water on Mars). Groundwater has been implicated in the cementation of aeolian sediments and the formation and transport of a wide variety of sedimentary minerals including clays, sulphates and hematite.

When the surface has been dry, wind has been a major geomorphic agent. Wind driven sand bodies like megaripples and dunes are extremely common on the modern Martian surface, and Opportunity has documented abundant

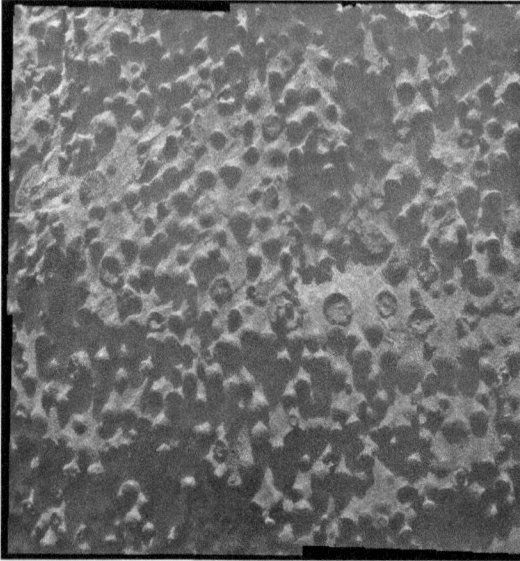

Figure 41: *Collection of spheres, each about 3 mm in diameter as seen by Opportunity rover*

aeolian sandstones on its traverse.[42] Ventifacts, like Jake Matijevic (rock), are an other aeolian landform on the Martian Surface.[43]

A wide variety of other sedimentological facies are also present locally on Mars, including glacial deposits, hot springs, dry mass movement deposits (especially landslides), and cryogenic and periglacial material, amongst many others. Evidence for ancient rivers, a lake, and dune fields have all been observed in the preserved strata by rovers at Meridiani Planum and Gale crater.

Groundwater on Mars

One group of researchers proposed that some of the layers on Mars were caused by groundwater rising to the surface in many places, especially inside of craters. According to the theory, groundwater with dissolved minerals came to the surface, in and later around craters, and helped to form layers by adding minerals (especially sulfate) and cementing sediments. This hypothesis is supported by a groundwater model and by sulfates discovered in a wide area. At first, by examining surface materials with Opportunity Rover, scientists discovered that groundwater had repeatedly risen and deposited sulfates. Later studies with instruments on board the Mars Reconnaissance Orbiter showed that the same kinds of materials exist in a large area that included Arabia.[44]

Interesting geomorphological features

Avalanches

On February 19, 2008, images obtained by the HiRISE camera on the Mars Reconnaissance Orbiter showed a spectacular avalanche, in which debris thought to be fine-grained ice, dust, and large blocks fell from a 700-metre (2,300 ft) high cliff. Evidence of the avalanche included dust clouds rising from the cliff afterwards.[45] Such geological events are theorized to be the cause of geologic patterns known as slope streaks.

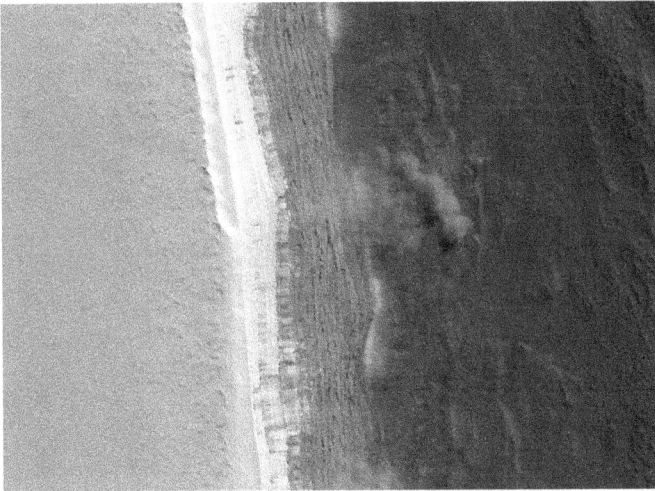

Figure 42: *Image of the February 19, 2008 Mars avalanche captured by the Mars Reconnaissance Orbiter.*

Figure 43: *Closer shot of the avalanche.*

Figure 44: *Dust clouds rise above the deep cliff.*

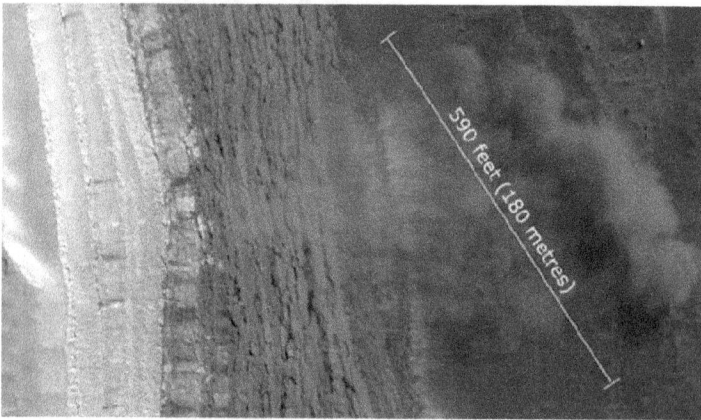

Figure 45: *A photo with scale demonstrates the size of the avalanche.*

Possible Caves

NASA scientists studying pictures from the Odyssey spacecraft have spotted what might be seven caves on the flanks of the Arsia Mons volcano on Mars. The pit entrances measure from 100 to 252 metres (328 to 827 ft) wide and they are believed to be at least 73 to 96 metres (240 to 315 ft) deep. See image below: the pits have been informally named (A) Dena, (B) Chloe, (C) Wendy, (D) Annie, (E) Abby (left) and Nikki, and (F) Jeanne. Because light did not reach the floor of most of the pits, it is likely that they extend much deeper than these lower estimates.Wikipedia:Citation needed Dena's floor was observed and found to be 130m deep. Further investigation suggested that these were not necessarily lava tube "skylights". Review of the images has resulted in yet more discoveries of deep pits.

Figure 46: *A cave on Mars ("Jeanne") as seen by the Mars Reconnaissance Orbiter.*

Figure 47: *HiRISE closeup of Jeanne showing afternoon illumination of the east wall of the shaft.*

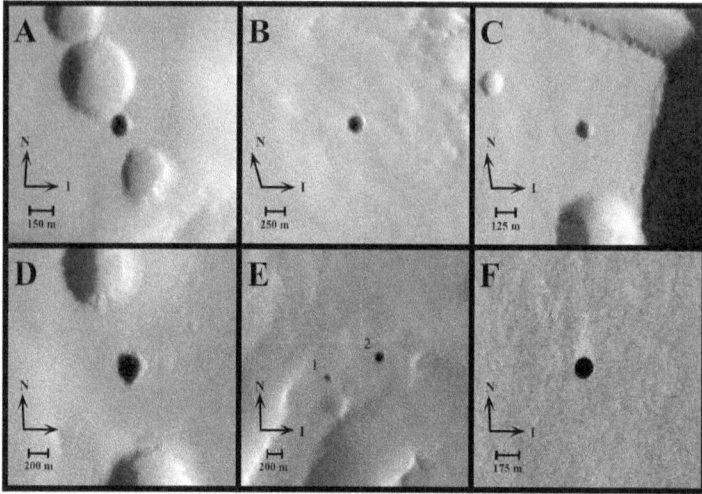

Figure 48: *THEMIS image of cave entrances on Mars.*

It has been suggested that human explorers on Mars could use lava tubes as shelters. The caves may be the only natural structures offering protection from the micrometeoroids, UV radiation, solar flares, and high energy particles that bombard the planet's surface.

Inverted relief

Some areas of Mars show inverted relief, where features that were once depressions, like streams, are now above the surface. It is believed that materials like large rocks were deposited in low-lying areas. Later, wind erosion removed much of the surface layers, but left behind the more resistant deposits. Other ways of making inverted relief might be lava flowing down a stream bed or materials being cemented by minerals dissolved in water. On Earth, materials cemented by silica are highly resistant to all kinds of erosional forces. Examples of inverted channels on Earth are found in the Cedar Mountain Formation near Green River, Utah. Inverted relief in the shape of streams are further evidence of water flowing on the Martian surface in past times. Inverted relief in the form of stream channels suggest that the climate was different—much wetter—when the inverted channels were formed.

In an article published in January 2010, a large group of scientists endorsed the idea of searching for life in Miyamoto Crater because of inverted stream channels and minerals that indicated the past presence of water.

Images of other examples of inverted terrain are shown below from various parts of Mars.

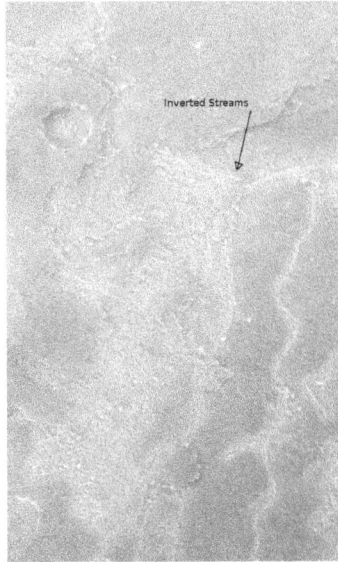

Figure 49: *Inverted Streams near Juventae Chasma, as seen by Mars Global Surveyor. These streams begin at the top of a ridge then run together.*

Figure 50: *Inverted Channel with many branches in Syrtis Major quadrangle.*

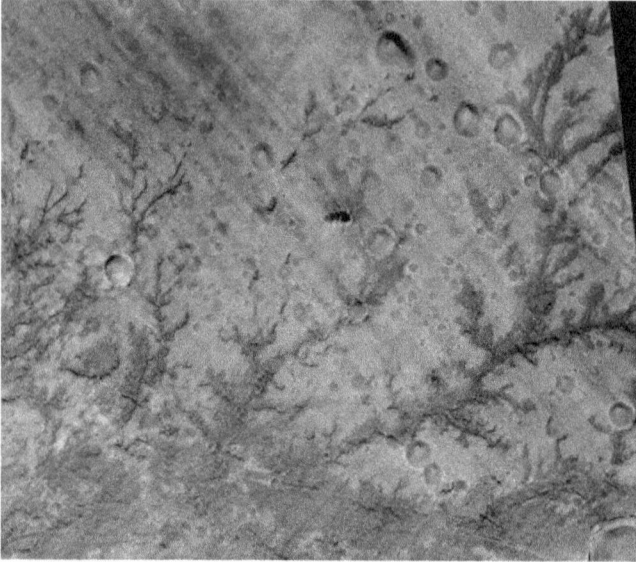

Figure 51: *Inverted Stream Channels in Antoniadi Crater,*
as seen by HiRISE. Image in Syrtis Major quadrangle.

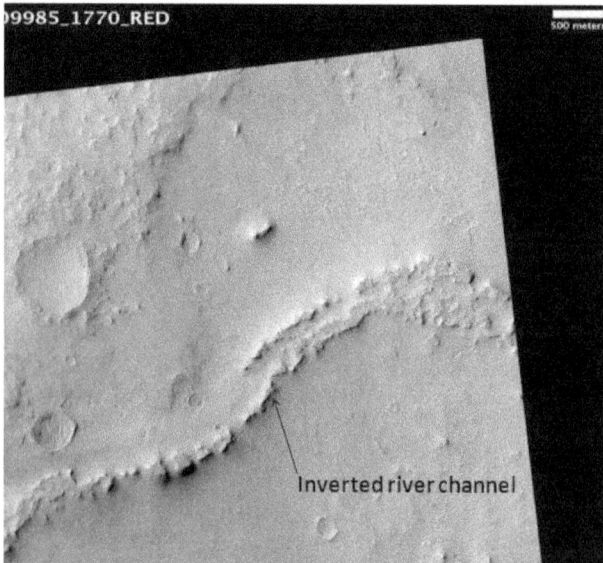

Figure 52: *Inverted Channel in Miyamoto Crater, as seen by HiRISE. Image*
is located in Margaritifer Sinus quadrangle. The scale bar is 500 meters long.

Notable rocks on Mars

Adiron-dack (*Spirit*)	Barnacle Bill (*Sojourner*)	Bathurst Inlet (*Curiosity*)	Big Joe* (*Viking*)	Block Island (*Opportunity*) **M**	Bounce (*Opportunity*)	Corona-tion (*Curiosity*)	El Capitan (*Opportunity*)
Esper-ance* (*Opportunity*)	Goulburn (*Curiosity*)	Heat Shield (*Opportunity*) **M**	Home Plate (*Spirit*)	Hottah (*Curiosity*)	Jake Matijevic (*Curiosity*)	Last Chance (*Opportunity*)	Link (*Curiosity*)
Mackinac Island (*Opportunity*) **M**	Mimi* (*Spirit*)	Oileán Ruaidh (*Opportunity*) **M**	Pot of Gold (*Spirit*)	Rocknest 3 (*Curiosity*)	Shelter Island (*Opportunity*) **M**	Tintina (*Curiosity*)	Yogi (*Sojourner*)

Notes: * = linked article is about the mission that encountered this rock; **M** = Meteorite - (
This box:
- view
- talk
- edit[46]
)

Bibliography

- Carr, Michael (2006). *The surface of Mars*. Cambridge, UK: Cambridge University Press. ISBN 0-521-87201-4.
- Hartmann, W. (2003). *A Traveler's Guide to Mars: The Mysterious Landscapes of the Red Planet*. New York: Workman Publishing. ISBN 978-0-7611-2606-5.

External links

- Mars - Geologic Map[47] (USGS, 2014) (original[48] / crop / full / video (00:56)[49]).
- Mars - Geologic Map[50] (USGS, 1978).
- Animated flights over Mars at 100 meter altitude[51]
- Oblique-impact complex on Mars (Syria Planum and Sinai Planum)[52]
- Presents good images, distances, and elevations/NASA[53]

Martian soil

Martian soil

Martian soil is the fine regolith found on the surface of Mars. Its properties can differ significantly from those of terrestrial soil. The term Martian soil typically refers to the finer fraction of regolith. On Earth, the term "soil" usually includes organic content. In contrast, planetary scientists adopt a functional definition of soil to distinguish it from rocks. Rocks generally refer to 10 cm scale and larger materials (e.g., fragments, breccia, and exposed outcrops) with high thermal inertia, with areal fractions consistent with the Viking Infrared Thermal Mapper (IRTM) data, and immobile under current aeolian conditions. Consequently, rocks classify as grains exceeding the size of cobbles on the Wentworth scale.

This approach enables agreement across Martian remote sensing methods that span the electromagnetic spectrum from gamma to radio waves. "Soil" refers to all other, typically unconsolidated, material including those sufficiently fine-grained to be mobilized by wind. Soil consequently encompasses a variety of regolith components identified at landing sites. Typical examples include: bedform armor, clasts, concretions, drift, dust, rocky fragments, and sand. The functional definition reinforces a recently proposed genetic definition of soil on terrestrial bodies (including asteroids and satellites) as an unconsolidated and chemically weathered surficial layer of fine-grained mineral or organic material exceeding centimeter scale thickness, with or without coarse elements and cemented portions.

Martian dust generally connotes even finer materials than Martian soil, the fraction which is less than 30 micrometres in diameter. Disagreement over the significance of soil's definition arises due to the lack of an integrated concept of soil in the literature. The pragmatic definition "medium for plant growth" has been commonly adopted in the planetary science community but a more complex definition describes soil as "(bio)geochemically/physically altered material at the surface of a planetary body that encompasses surficial extraterrestrial

Figure 53: *Curiosity's view of Martian soil and boulders after crossing the "Dingo Gap" sand dune (February 9, 2014; raw color[54]).*

telluric deposits." This definition emphasizes that soil is a body that retains information about its environmental history and that does not need the presence of life to form.

Observations

<templatestyles src="Multiple_image/styles.css" />

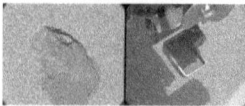

First use of the Curiosity rover scooper as it sifts a load of sand at *"Rocknest"* (October 7, 2012).

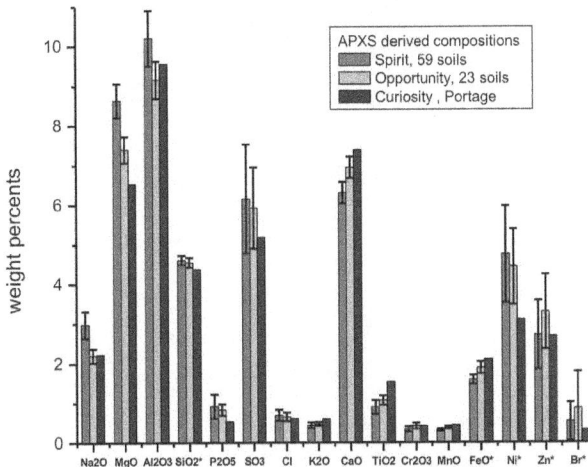

Figure 54: *Comparison of Soils on Mars - Samples by Curiosity rover, Opportunity rover, Spirit rover (December 3, 2012).*

Mars is covered with vast expanses of sand and dust and its surface is littered with rocks and boulders. The dust is occasionally picked up in vast planet-wide dust storms. Mars dust is very fine, and enough remains suspended in the atmosphere to give the sky a reddish hue. The reddish hue is due to rusting iron minerals presumably formed a few billion years ago when Mars was warm and wet, but now that Mars is cold and dry, modern rusting may be due to a superoxide that forms on minerals exposed to ultraviolet rays in sunlight. The sand is believed to move only slowly in the Martian winds due to the very low density of the atmosphere in the present epoch. In the past, liquid water flowing in gullies and river valleys may have shaped the Martian regolith. Mars researchers are studying whether groundwater sapping is shaping the Martian regolith in the present epoch, and whether carbon dioxide hydrates exist on Mars and play a role.

It is believed that large quantities of water and carbon dioxideWikipedia:Citation needed ices remain frozen within the regolith in the equatorial parts of Mars and on its surface at higher latitudes. Water contents of Martian regolith range from <2% by weight to more than ?. The presence of olivine, which is an easily weatherable primary mineral, has been interpreted to mean that physical rather than chemical weathering processes currently dominate on Mars. High concentrations of ice in soils are thought

Figure 55: *First X-ray diffraction view of Martian soil - CheMin analysis reveals feldspar, pyroxenes, olivine and more (Curiosity rover at "Rocknest", October 17, 2012).*

to be the cause of accelerated soil creep, which forms the rounded "softened terrain" characteristic of the Martian midlatitudes.

In June, 2008, the Phoenix Lander returned data showing Martian soil to be slightly alkaline and containing vital nutrients such as magnesium, sodium, potassium and chloride, all of which are necessary for living organisms to grow. Scientists compared the soil near Mars' north pole to that of backyard gardens on Earth, and concluded that it could be suitable for growth of plants. However, in August, 2008, the Phoenix Lander conducted simple chemistry experiments, mixing water from Earth with Martian soil in an attempt to test its pH, and discovered traces of the salt perchlorate, while also confirming many scientists' theories that the Martian surface was considerably basic, measuring at 8.3. The presence of the perchlorate, if confirmed, would make Martian soil more exotic than previously believed. Further testing is necessary to eliminate the possibility of the perchlorate readings being caused by terrestrial sources, which may have migrated from the spacecraft either into samples or the instrumentation.

While our understanding of Martian soils is extremely rudimentary, their diversity may raise the question of how we might compare them with our Earth-based soils. Applying an Earth-based system is largely debatable but a simple

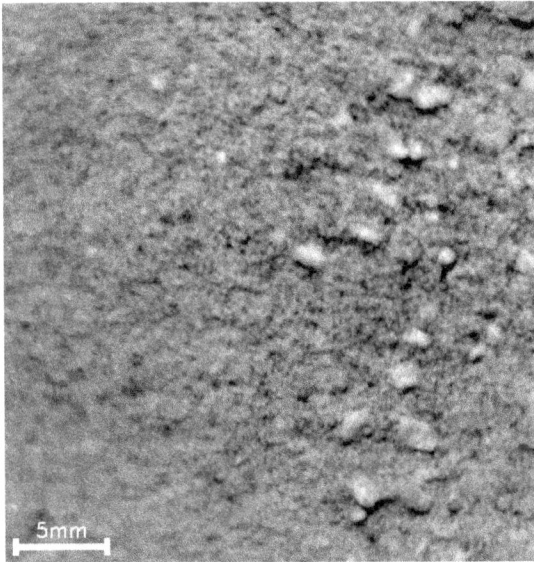

Figure 56: *"Sutton Inlier" soil on Mars - target of ChemCam's laser - Curiosity rover (May 11, 2013).*

option is to distinguish the (largely) biotic Earth from the abiotic Solar System, and include all non-Earth soils in a new World Reference Base for Soil Resources Reference Group or USDA soil taxonomy Order, which might be tentatively called Astrosols.

On October 17, 2012 (Curiosity rover at "Rocknest"), the first X-ray diffraction analysis of Martian soil was performed. The results revealed the presence of several minerals, including feldspar, pyroxenes and olivine, and suggested that the Martian soil in the sample was similar to the "weathered basaltic soils" of Hawaiian volcanoes. Hawaiian volcanic ash has been used as Martian regolith simulant by researchers since 1998.

In December 2012, scientists working on the Mars Science Laboratory mission announced that an extensive soil analysis of Martian soil performed by the Curiosity rover showed evidence of water molecules, sulphur and chlorine, as well as hints of organic compounds. However, terrestrial contamination, as the source of the organic compounds, could not be ruled out.

On September 26, 2013, NASA scientists reported the Mars *Curiosity* rover detected "abundant, easily accessible" water (1.5 to 3 weight percent) in soil samples at the Rocknest region of Aeolis Palus in Gale Crater. In addition,

NASA reported that the *Curiosity* rover found two principal soil types: a fine-grained mafic type and a locally derived, coarse-grained felsic type. The mafic type, similar to other martian soils and martian dust, was associated with hydration of the amorphous phases of the soil. Also, perchlorates, the presence of which may make detection of life-related organic molecules difficult, were found at the Curiosity rover landing site (and earlier at the more polar site of the Phoenix lander) suggesting a "global distribution of these salts". NASA also reported that Jake M rock, a rock encountered by *Curiosity* on the way to Glenelg, was a mugearite and very similar to terrestrial mugearite rocks.

Atmospheric dust

<templatestyles src="Multiple_image/styles.css" />

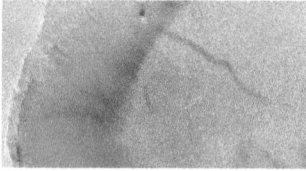

Dust devil on Mars (MGS).

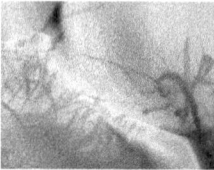

Dust devils cause twisting dark trails on the Martian surface.

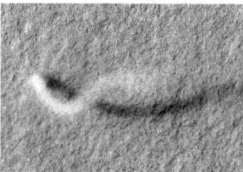

Serpent Dust Devil of Mars (MRO).

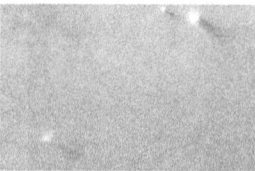

Dust devils in Valles Marineris (MRO).

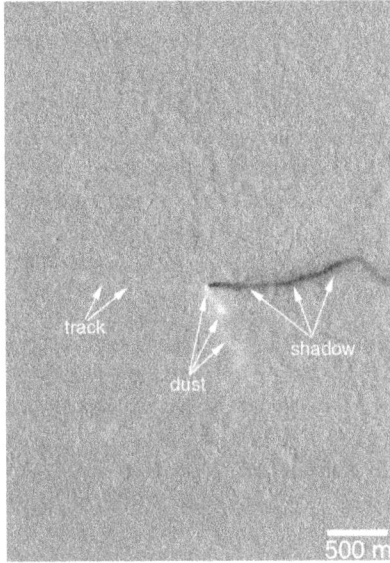

Figure 57: *Martian Dust Devil - in Amazonis Plani-tia (April 10, 2001) (also[55]) (video (02:19)[56]).*

<templatestyles src="Multiple_image/styles.css" />

Dust storms on Mars.

June 6, 2018.

November 25, 2012

November 18, 2012

Locations of *Opportunity* and *Curiosity* rovers are noted (MRO).

Similarly sized **dust** will settle from the thinner Martian atmosphere sooner than it would on Earth. For example, the dust suspended by the 2001 global dust storms on Mars only remained in the Martian atmosphere for 0.6 years, while the dust from Mt. Pinatubo took about 2 years to settle. However, under current Martian conditions, the mass movements involved are generally much smaller than on Earth. Even the 2001 global dust storms on Mars moved only the equivalent of a very thin dust layer – about 3 μm thick if deposited with uniform thickness between 58° north and south of the equator. Dust deposition at the two rover sites has proceeded at a rate of about the thickness of a grain every 100 sols.

The difference in the concentration of dust in Earth's atmosphere and that of Mars stems from a key factor. On Earth, dust that leaves atmospheric suspension usually gets aggregated into larger particles through the action of soil moisture or gets suspended in oceanic waters. It helps that most of earth's surface is covered by liquid water. Neither process occurs on Mars, leaving deposited dust available for suspension back into the Martian atmosphere. In fact, the composition of Martian atmospheric dust – very similar to surface dust – as observed by the Mars Global Surveyor Thermal Emission Spectrometer, may be volumetrically dominated by composites of plagioclase feldspar and zeolite which can be mechanically derived from Martian basaltic rocks without chemical alteration. Observations of the Mars Exploration Rovers' magnetic dust traps suggest that about 45% of the elemental iron in atmospheric dust is maximally (3+) oxidized and that nearly half exists in titanomagnetite,[57] both consistent with mechanical derivation of dust with aqueous alteration limited to just thin films of water. Collectively, these observations support the absence of water-driven dust aggregation processes on Mars. Furthermore, wind activity dominates the surface of Mars at present, and the abundant dune fields of Mars can easily yield particles into atmospheric suspension through effects such as larger grains disaggregating fine particles through collisions.

The Martian atmospheric dust particles are generally 3 μm in diameter. It is important to note that while the atmosphere of Mars is thinner, Mars also has a lower gravitational acceleration, so the size of particles that will remain in suspension cannot be estimated with atmospheric thickness alone. Electrostatic and van der Waals forces acting among fine particles introduce additional complexities to calculations. Rigorous modeling of all relevant variables suggests that 3 μm diameter particles can remain in suspension indefinitely at most wind speeds, while particles as large as 20 μm diameter can enter suspension from rest at surface wind turbulence as low as 2 ms^{-1} or remain in suspension at 0.8 ms^{-1}.

In July 2018, researchers reported that the largest single source of dust on the planet Mars comes from the Medusae Fossae Formation.

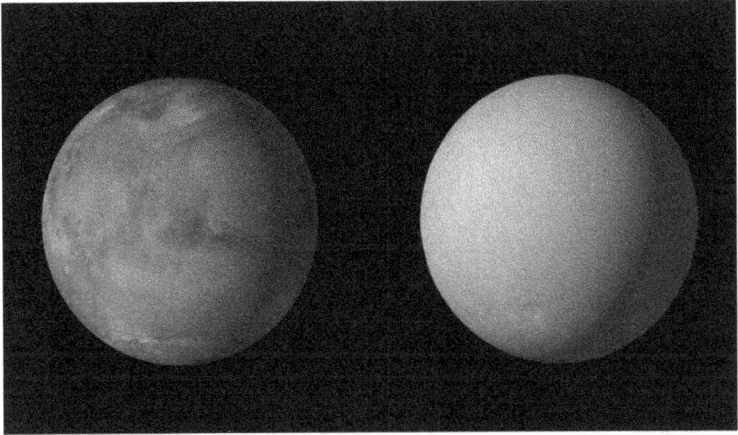

Figure 58:
Mars (before/after) dust storm (July 2018)

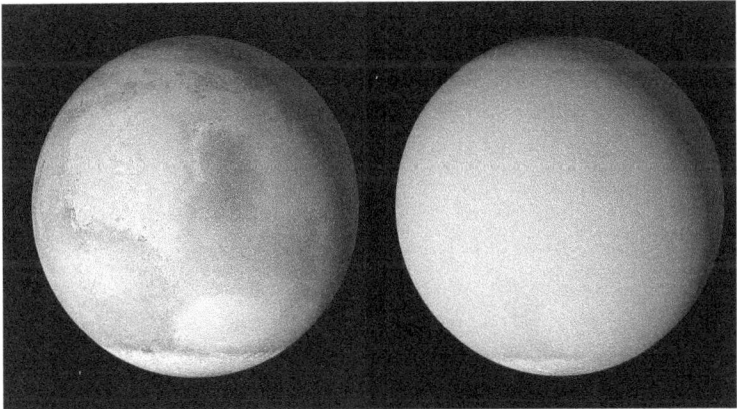

Figure 59:
Mars without a dust storm on June 2001 (on left) and with a global dust storm on July 2001 (on right), as seen by Mars Global Surveyor

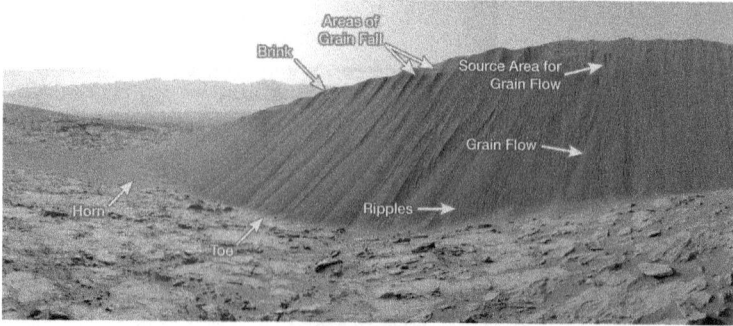

Figure 60:
Namib sand dune (downwind side) on Mars
(Curiosity rover; December 17, 2015).

Gallery

Figure 61: *Martian sand and boulders photographed by NASA's Mars Exploration Rover Spirit (April 13, 2006).*

Figure 62: *"Hottah" rock outcrop (close-up[58]; 3D[59]) (September 12, 2012).*

Figure 63: *"Rocknest" sand on Mars – scoffmark made by the Curiosity rover (MAHLI, October 4, 2012).*

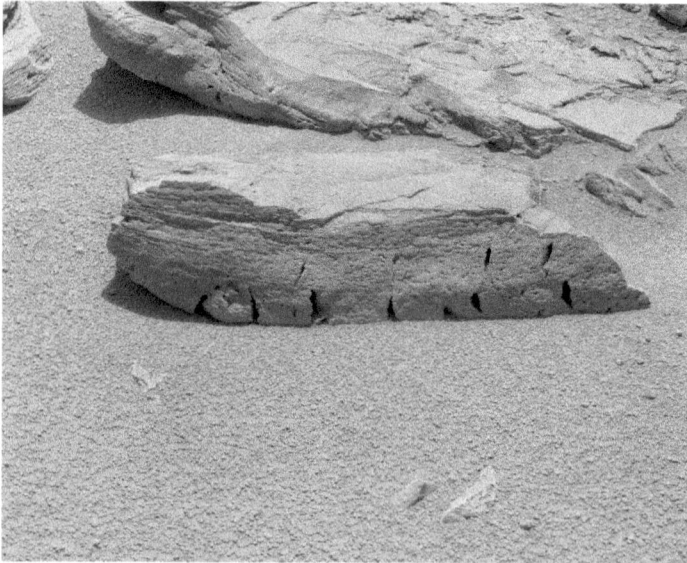

Figure 64: *"Rocknest 3" rock on Mars – as viewed by the MastCam on Curiosity (October 5, 2012).*

Figure 65: *Tracks of the Curiosity rover in the sands of "Hidden Valley" (August 4, 2014).*

Figure 66: *Wheel of the Curiosity rover partially submerged in sand at Hidden Valley (August 6, 2014).*

Figure 67: *Sand moving on Mars – as viewed by Curiosity (January 23, 2017).*

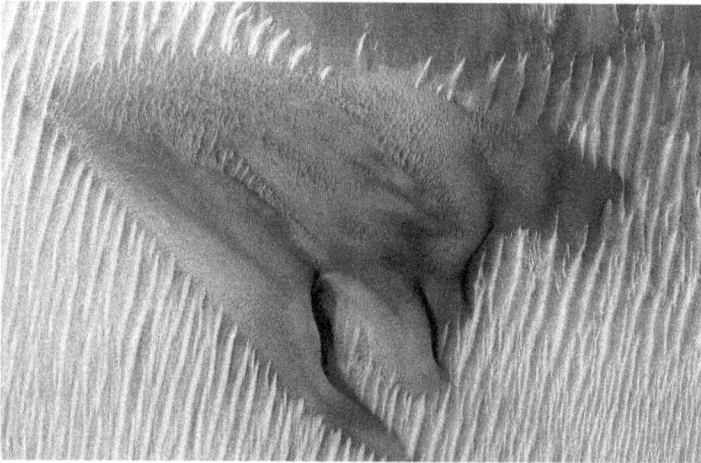

Figure 68: *Blue dune on Mars*
(January 24, 2018).

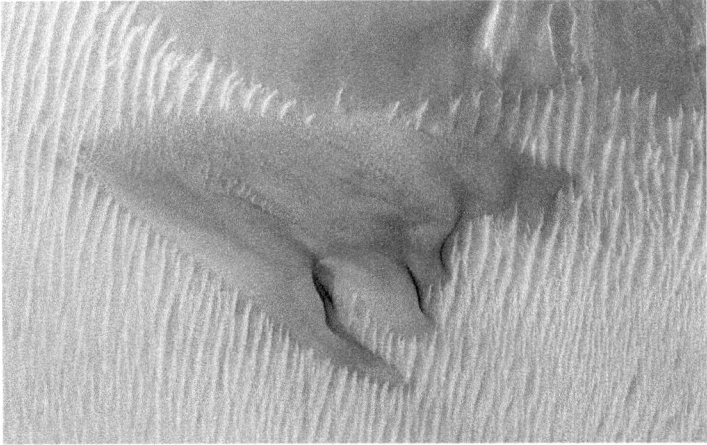

Figure 69: *Blue dune on Mars>*
(Enhanced color; January 24, 2018)

External links

- Video (04:32) - Evidence: Water "Vigorously" Flowed On Mars -
  September, 2012[60]

Water on Mars

Water on Mars

<templatestyles src="Multiple_image/styles.css" />

Mars – Utopia Planitia

Martian terrain

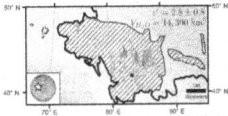

Map of terrain

Scalloped terrain led to the discovery of a large amount of underground ice – enough water to fill Lake Superior (November 22, 2016)

Almost all **water on Mars** today exists as ice, though it also exists in small quantities as vapor in the atmosphere, and occasionally as low-volume liquid brines in shallow Martian soil. The only place where water ice is visible at the surface is at the north polar ice cap. Abundant water ice is also present beneath the permanent carbon dioxide ice cap at the Martian south pole and in the shallow subsurface at more temperate conditions. More than five million cubic kilometers of ice have been identified at or near the surface of modern Mars, enough to cover the whole planet to a depth of 35 meters (115 ft). Even more ice is likely to be locked away in the deep subsurface.[61]

Some liquid water may occur transiently on the Martian surface today, but limited to traces of dissolved moisture from the atmosphere and thin films, which

Figure 70: *An artist's impression of what ancient Mars may have looked like, based on geological data*

are challenging environments for known life. No large standing bodies of liquid water exist on the planet's surface, because the atmospheric pressure there averages just 600 pascals (0.087 psi) – about 0.6% of Earth's mean sea level pressure – leading to either rapid evaporation (sublimation) or rapid freezing. Before about 3.8 billion years ago, Mars may have had a denser atmosphere and higher surface temperatures, allowing vast amounts of liquid water on the surface, possibly including a large ocean that may have covered one-third of the planet. Water has also apparently flowed across the surface for short periods at various intervals more recently in Mars' history.[62] On December 9, 2013, NASA reported that, based on evidence from the *Curiosity* rover studying Aeolis Palus, Gale Crater contained an ancient freshwater lake that could have been a hospitable environment for microbial life.

Many lines of evidence indicate that water ice is abundant on Mars and it has played a significant role in the planet's geologic history. The present-day inventory of water on Mars can be estimated from spacecraft imagery, remote sensing techniques (spectroscopic measurements, radar, etc.), and surface investigations from landers and rovers. Geologic evidence of past water includes enormous outflow channels carved by floods, ancient river valley networks, deltas, and lakebeds; and the detection of rocks and minerals on the surface that could only have formed in liquid water. Numerous geomorphic features suggest the presence of ground ice (permafrost) and the movement of ice in glaciers, both in the recent past and present. Gullies and slope lineae along cliffs and crater walls suggest that flowing water continues to shape the surface of Mars, although to a far lesser degree than in the ancient past.

Although the surface of Mars was periodically wet and could have been hospitable to microbial life billions of years ago, the current environment at the surface is dry and subfreezing, probably presenting an insurmountable obstacle for living organisms. In addition, Mars lacks a thick atmosphere, ozone layer, and magnetic field, allowing solar and cosmic radiation to strike the surface unimpeded. The damaging effects of ionizing radiation on cellular structure is another one of the prime limiting factors on the survival of life on the surface. Therefore, the best potential locations for discovering life on Mars may be in subsurface environments. On November 22, 2016, NASA reported finding a large amount of underground ice on Mars; the volume of water detected is equivalent to the volume of water in Lake Superior. In July 2018, Italian scientists reported the discovery of a subglacial lake on Mars, 1.5 km (0.93 mi) below the southern polar ice cap, and extending sideways about 20 km (12 mi), the first known stable body of water on the planet.

Understanding the extent and situation of water on Mars is vital to assess the planet's potential for harboring life and for providing usable resources for future human exploration. For this reason, "Follow the Water" was the science theme of NASA's Mars Exploration Program (MEP) in the first decade of the 21st century. Discoveries by the 2001 Mars Odyssey, Mars Exploration Rovers (MERs), Mars Reconnaissance Orbiter (MRO), and Mars *Phoenix* lander have been instrumental in answering key questions about water's abundance and distribution on Mars. The ESA's Mars Express orbiter has also provided essential data in this quest.[63] The Mars Odyssey, Mars Express, MER *Opportunity* rover, MRO, and Mars Science Lander *Curiosity* rover are still sending back data from Mars, and discoveries continue to be made.

Historical background

The notion of water on Mars preceded the space age by hundreds of years. Early telescopic observers correctly assumed that the white polar caps and clouds were indications of water's presence. These observations, coupled with the fact that Mars has a 24-hour day, led astronomer William Herschel to declare in 1784 that Mars probably offered its inhabitants "a situation in many respects similar to ours."[64] <templatestyles src="Multiple_image/styles.css" />

Historical map of Mars from Giovanni Schiaparelli.

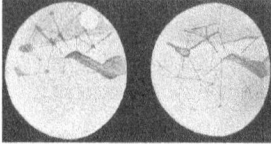

Mars canals illustrated by astronomer Percival Lowell, 1898.

By the start of the 20th century, most astronomers recognized that Mars was far colder and drier than Earth. The presence of oceans was no longer accepted, so the paradigm changed to an image of Mars as a "dying" planet with only a meager amount of water. The dark areas, which could be seen to change seasonally, were then thought to be tracts of vegetation. The man most responsible for popularizing this view of Mars was Percival Lowell (1855–1916), who imagined a race of Martians constructing a network of canals to bring water from the poles to the inhabitants at the equator. Although generating tremendous public enthusiasm, Lowell's ideas were rejected by most astronomers. The majority view of the scientific establishment at the time is probably best summarized by English astronomer Edward Walter Maunder (1851–1928) who compared the climate of Mars to conditions atop a twenty-thousand-foot peak on an arctic island[65] where only lichen might be expected to survive.

In the meantime, many astronomers were refining the tool of planetary spectroscopy in hope of determining the composition of the Martian atmosphere. Between 1925 and 1943, Walter Adams and Theodore Dunham at the Mount Wilson Observatory tried to identify oxygen and water vapor in the Martian atmosphere, with generally negative results. The only component of the Martian atmosphere known for certain was carbon dioxide (CO_2) identified spectroscopically by Gerard Kuiper in 1947.[66] Water vapor was not unequivocally detected on Mars until 1963.

The composition of the polar caps, assumed to be water ice since the time of Cassini (1666), was questioned by a few scientists in the late 1800s who favored CO_2 ice, because of the planet's overall low temperature and apparent lack of appreciable water. This hypothesis was confirmed theoretically by Robert Leighton and Bruce Murray in 1966. Today it is known that the winter caps at both poles are primarily composed of CO_2 ice, but that a permanent (or perennial) cap of water ice remains during the summer at the northern pole.

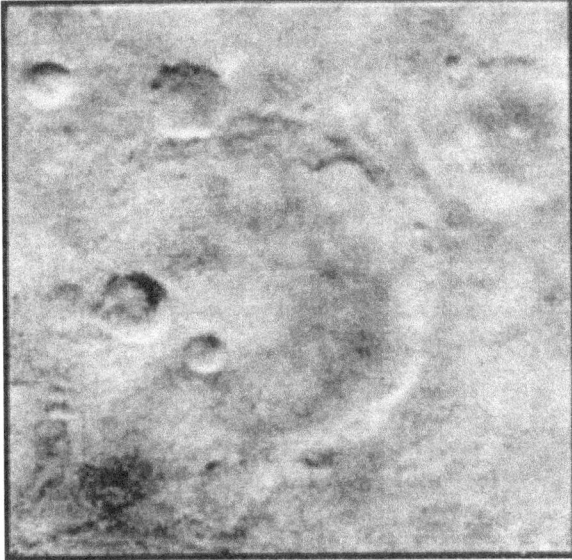

Figure 71: *Mariner 4 acquired this image showing a barren planet (1965)*

At the southern pole, a small cap of CO_2 ice remains during summer, but this cap too is underlain by water ice.

The final piece of the Martian climate puzzle was provided by Mariner 4 in 1965. Grainy television pictures from the spacecraft showed a surface dominated by impact craters, which implied that the surface was very old and had not experienced the level of erosion and tectonic activity seen on Earth. Little erosion meant that liquid water had probably not played a large role in the planet's geomorphology for billions of years. Furthermore, the variations in the radio signal from the spacecraft as it passed behind the planet allowed scientists to calculate the density of the atmosphere. The results showed an atmospheric pressure less than 1% of Earth's at sea level, effectively precluding the existence of liquid water, which would rapidly boil or freeze at such low pressures. Thus, a vision of Mars was born of a world much like the Moon, but with just a wisp of an atmosphere to blow the dust around. This view of Mars would last nearly another decade until Mariner 9 showed a much more dynamic Mars with hints that the planet's past environment was more clement than the present one.

On January 24, 2014, NASA reported that current studies on Mars by the *Curiosity* and *Opportunity* rovers will now be searching for evidence of ancient life, including a biosphere based on autotrophic, chemotrophic and/or chemolithoautotrophic microorganisms, as well as ancient water, including fluvio-lacustrine environments (plains related to ancient rivers or lakes) that may have been habitable.

For many years it was thought that the observed remains of floods were caused by the release of water from a global water table, but research published in 2015 reveals regional deposits of sediment and ice emplaced 450 million years earlier to be the source. "Deposition of sediment from rivers and glacial melt filled giant canyons beneath primordial ocean contained within the planet's northern lowlands. It was the water preserved in these canyon sediments that was later released as great floods, the effects of which can be seen today."

Evidence from rocks and minerals

It is widely accepted that Mars had abundant water very early in its history, but all large areas of liquid water have since disappeared. A fraction of this water is retained on modern Mars as both ice and locked into the structure of abundant water-rich materials, including clay minerals (phyllosilicates) and sulfates. Studies of hydrogen isotopic ratios indicate that asteroids and comets from beyond 2.5 astronomical units (AU) provide the source of Mars' water, that currently totals 6% to 27% of the Earth's present ocean.

Water in weathering products (aqueous minerals)

The primary rock type on the surface of Mars is basalt, a fine-grained igneous rock made up mostly of the mafic silicate minerals olivine, pyroxene, and plagioclase feldspar. When exposed to water and atmospheric gases, these minerals chemically weather into new (secondary) minerals, some of which may incorporate water into their crystalline structures, either as H_2O or as hydroxyl (OH). Examples of hydrated (or hydoxylated) minerals include the iron hydroxide goethite (a common component of terrestrial soils); the evaporate minerals gypsum and kieserite; opalline silica; and phyllosilicates (also called clay minerals), such as kaolinite and montmorillonite. All of these minerals have been detected on Mars.

One direct effect of chemical weathering is to consume water and other reactive chemical species, taking them from mobile reservoirs like the atmosphere and hydrosphere and sequestering them in rocks and minerals. The amount of water in the Martian crust stored in hydrated minerals is currently unknown, but may be quite large. For example, mineralogical models of the rock outcroppings examined by instruments on the *Opportunity* rover at Meridiani Planum suggest that the sulfate deposits there could contain up to 22% water by weight.

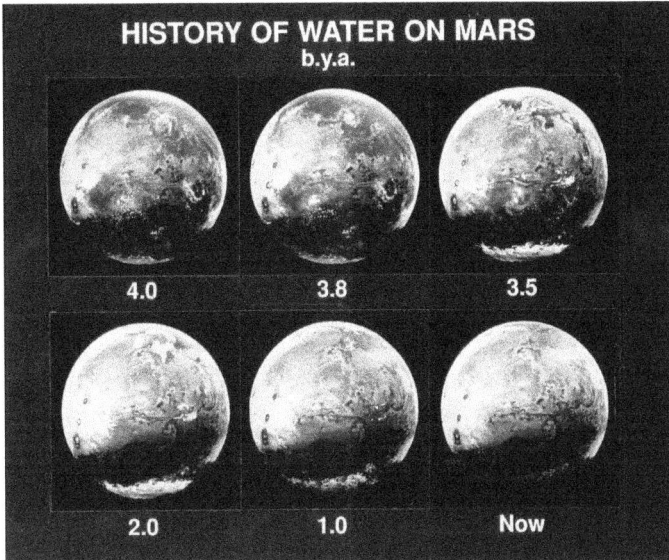

Figure 72: *History of water on Mars. Numbers represent how many billions of years ago*

On Earth, all chemical weathering reactions involve water to some degree. Thus, many secondary minerals do not actually incorporate water, but still require water to form. Some examples of anhydrous secondary minerals include many carbonates, some sulfates (e.g., anhydrite), and metallic oxides such as the iron oxide mineral hematite. On Mars, a few of these weathering products may theoretically form without water or with scant amounts present as ice or in thin molecular-scale films (monolayers). The extent to which such exotic weathering processes operate on Mars is still uncertain. Minerals that incorporate water or form in the presence of water are generally termed "aqueous minerals."

Aqueous minerals are sensitive indicators of the type of environment that existed when the minerals formed. The ease with which aqueous reactions occur (see Gibbs free energy) depends on the pressure, temperature, and on the concentrations of the gaseous and soluble species involved. Two important properties are pH and oxidation-reduction potential (Eh). For example, the sulfate mineral jarosite forms only in low pH (highly acidic) water. Phyllosilicates usually form in water of neutral to high pH (alkaline). Eh is a measure is the oxidation state of an aqueous system. Together Eh and pH indicate the types of minerals that are thermodynamically most likely to form from a given set of aqueous components. Thus, past environmental conditions on Mars, including those conducive to life, can be inferred from the types of minerals present in the rocks.

Hydrothermal alteration

Aqueous minerals can also form in the subsurface by hydrothermal fluids migrating through pores and fissures. The heat source driving a hydrothermal system may be nearby magma bodies or residual heat from large impacts. One important type of hydrothermal alteration in the Earth's oceanic crust is serpentinization, which occurs when seawater migrates through ultramafic and basaltic rocks. The water-rock reactions result in the oxidation of ferrous iron in olivine and pyroxene to produce ferric iron (as the mineral magnetite) yielding molecular hydrogen (H_2) as a byproduct. The process creates a highly alkaline and reducing (low Eh) environment favoring the formation of certain phyllosilicates (serpentine minerals) and various carbonate minerals, which together form a rock called serpentinite. The hydrogen gas produced can be an important energy source for chemosynthtetic organisms or it can react with CO_2 to produce methane gas, a process that has been considered as a non-biological source for the trace amounts of methane reported in the Martian atmosphere. Serpentine minerals can also store a lot of water (as hydroxyl) in their crystal structure. A recent study has argued that hypothetical serpentinites in the ancient highland crust of Mars could hold as much as a 500 metres (1,600 ft)-thick global equivalent layer (GEL) of water. Although some serpentine minerals have been detected on Mars, no widespread outcroppings are evident from remote sensing data. This fact does not preclude the presence of large amounts of sepentinite hidden at depth in the Martian crust.

Weathering rates

The rates at which primary minerals convert to secondary aqueous minerals vary. Primary silicate minerals crystallize from magma under pressures and temperatures vastly higher than conditions at the surface of a planet. When exposed to a surface environment these minerals are out of equilibrium and will tend to interact with available chemical components to form more stable mineral phases. In general, the silicate minerals that crystallize at the highest temperatures (solidify first in a cooling magma) weather the most rapidly.[67] On the Earth and Mars, the most common mineral to meet this criterion is olivine, which readily weathers to clay minerals in the presence of water.

Olivine is widespread on Mars, suggesting that Mars' surface has not been pervasively altered by water; abundant geological evidence suggests otherwise.

Figure 73: *Mars meteorite ALH84001*

Martian meteorites

Over 60 meteorites have been found that came from Mars.[68] Some of them contain evidence that they were exposed to water when on Mars. Some Martian meteorites called basaltic shergottites, appear (from the presence of hydrated carbonates and sulfates) to have been exposed to liquid water prior to ejection into space. It has been shown that another class of meteorites, the nakhlites, were suffused with liquid water around 620 million years ago and that they were ejected from Mars around 10.75 million years ago by an asteroid impact. They fell to Earth within the last 10,000 years. Martian meteorite NWA 7034 has one order of magnitude more water than most other Martian meteorites. It is similar to the basalts studied by rover missions, and it was formed in the early Amazonian epoch.[69]

In 1996, a group of scientists reported the possible presence of microfossils in the Allan Hills 84001, a meteorite from Mars. Many studies disputed the validity of the fossils. It was found that most of the organic matter in the meteorite was of terrestrial origin. In addition, the scientific consensus is that "morphology alone cannot be used unambiguously as a tool for primitive life detection." Interpretation of morphology is notoriously subjective, and its use alone has led to numerous errors of interpretation.

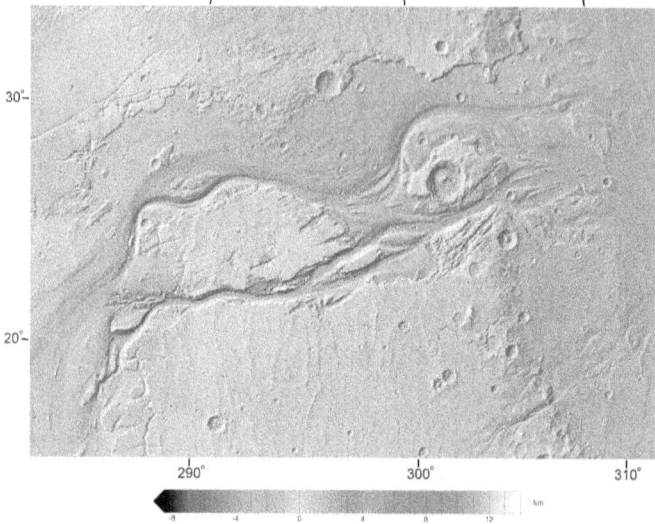

Figure 74: *Kasei Valles—a major outflow channel—seen in MOLA elevation data. Flow was from bottom left to right. Image is approx. 1600 km across. The channel system extends another 1200 km south of this image to Echus Chasma.*

Geomorphic evidence

Lakes and river valleys

The 1971 Mariner 9 spacecraft caused a revolution in our ideas about water on Mars. Huge river valleys were found in many areas. Images showed that floods of water broke through dams, carved deep valleys, eroded grooves into bedrock, and traveled thousands of kilometers. Areas of branched streams, in the southern hemisphere, suggested that rain once fell. The numbers of recognised valleys has increased through time. Research published in June 2010 mapped 40,000 river valleys on Mars, roughly quadrupling the number of river valleys that had previously been identified. Martian water-worn features can be classified into two distinct classes: 1) dendritic (branched), terrestrial-scale, widely distributed, Noachian-age valley networks and 2) exceptionally large, long, single-thread, isolated, Hesperian-age outflow channels. Recent work suggests that there may also be a class of currently enigmatic, smaller, younger (Hesperian to Amazonian) channels in the midlatitudes, perhaps associated with the occasional local melting of ice deposits.

Some parts of Mars show inverted relief. This occurs when sediments are deposited on the floor of a stream and then become resistant to erosion, perhaps

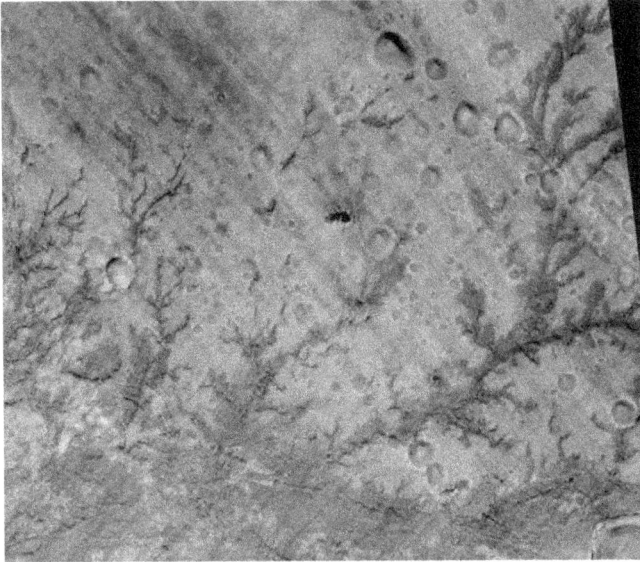

Figure 75: *Inverted stream channels in Anto-niadi Crater. Location is Syrtis Major quadrangle*

by cementation. Later the area may be buried. Eventually, erosion removes the covering layer and the former streams become visible since they are resistant to erosion. Mars Global Surveyor found several examples of this process. Many inverted streams have been discovered in various regions of Mars, especially in the Medusae Fossae Formation, Miyamoto Crater, Saheki Crater, and the Juventae Plateau.

A variety of lake basins have been discovered on Mars. Some are comparable in size to the largest lakes on Earth, such as the Caspian Sea, Black Sea, and Lake Baikal. Lakes that were fed by valley networks are found in the southern highlands. There are places that are closed depressions with river valleys leading into them. These areas are thought to have once contained lakes; one is in Terra Sirenum that had its overflow move through Ma'adim Vallis into Gusev Crater, explored by the Mars Exploration Rover Spirit. Another is near Parana Valles and Loire Vallis. Some lakes are thought to have formed by precipitation, while others were formed from groundwater. Lakes are estimated to have existed in the Argyre basin, the Hellas basin, and maybe in Valles Marineris. It is likely that at times in the Noachian, many craters hosted lakes. These lakes are consistent with a cold, dry (by Earth standards) hydrological environment somewhat like that of the Great Basin of the western USA during the Last Glacial Maximum.[70]

Research from 2010 suggests that Mars also had lakes along parts of the equator. Although earlier research had showed that Mars had a warm and wet early history that has long since dried up, these lakes existed in the Hesperian Epoch, a much later period. Using detailed images from NASA's Mars Reconnaissance Orbiter, the researchers speculate that there may have been increased volcanic activity, meteorite impacts or shifts in Mars' orbit during this period to warm Mars' atmosphere enough to melt the abundant ice present in the ground. Volcanoes would have released gases that thickened the atmosphere for a temporary period, trapping more sunlight and making it warm enough for liquid water to exist. In this study, channels were discovered that connected lake basins near Ares Vallis. When one lake filled up, its waters overflowed the banks and carved the channels to a lower area where another lake would form. These dry lakes would be targets to look for evidence (biosignatures) of past life.

On September 27, 2012, NASA scientists announced that the *Curiosity rover* found direct evidence for an ancient streambed in Gale Crater, suggesting an ancient "vigorous flow" of water on Mars. In particular, analysis of the now dry streambed indicated that the water ran at 3.3 km/h (0.92 m/s), possibly at hip-depth. Proof of running water came in the form of rounded pebbles and gravel fragments that could have only been weathered by strong liquid currents. Their shape and orientation suggests long-distance transport from above the rim of the crater, where a channel named Peace Vallis feeds into the alluvial fan.

Eridania Lake is a theorized ancient lake with a surface area of roughly 1.1 million square kilometers.[71,72,73] Its maximum depth is 2,400 meters and its volume is 562,000 km^2. It was larger than the largest landlocked sea on Earth, the Caspian Sea and contained more water than all the other martian lakes together. The Eridania sea held more than 9 times as much water as all of North America's Great Lakes. The upper surface of the lake was assumed to be at the elevation of valley networks that surround the lake; they all end at the same elevation, suggesting that they emptied into a lake.[74]

Research with CRISM found thick deposits, greater than 400 meters thick, that contained the minerals saponite, talc-saponite, Fe-rich mica (for example, glauconite-nontronite), Fe- and Mg-serpentine, Mg-Fe-Ca-carbonate and probable Fe-sulphide. The Fe-sulphide probably formed in deep water from water heated by volcanoes. Such a process, classified as hydrothermal may have been a place where life on Earth began.

Figure 76: *Delta in Eberswalde crater*

Lake deltas

Researchers have found a number of examples of deltas that formed in Martian lakes. Finding deltas is a major sign that Mars once had a lot of liquid water. Deltas usually require deep water over a long period of time to form. Also, the water level needs to be stable to keep sediment from washing away. Deltas have been found over a wide geographical range, though there is some indication that deltas may be concentrated around the edges of the putative former northern ocean of Mars.

Groundwater

By 1979 it was thought that outflow channels formed in single, catastrophic ruptures of subsurface water reservoirs, possibly sealed by ice, discharging colossal quantities of water across an otherwise arid Mars surface. In addition, evidence in favor of heavy or even catastrophic flooding is found in the giant ripples in the Athabasca Vallis. Many outflow channels begin at Chaos or Chasma features, providing evidence for the rupture that could have breached a subsurface ice seal.

The branching valley networks of Mars are not consistent with formation by sudden catastrophic release of groundwater, both in terms of their dendritic

Groundwater helps to form layers

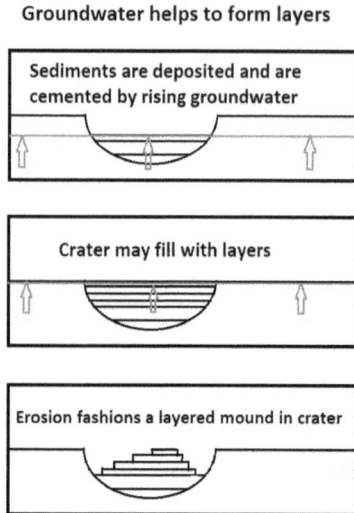

Figure 77: *Layers may be formed by groundwater rising up gradually*

shapes that do not come from a single outflow point, and in terms of the discharges that apparently flowed along them. Instead, some authors have argued that they were formed by slow seepage of groundwater from the subsurface essentially as springs. In support of this interpretation, the upstream ends of many valleys in such networks begin with box canyon or "amphitheater" heads, which on Earth are typically associated with groundwater seepage. There is also little evidence of finer scale channels or valleys at the tips of the channels, which some authors have interpreted as showing the flow appeared suddenly from the subsurface with appreciable discharge, rather than accumulating gradually across the surface. Others have disputed the link between amphitheater heads of valleys and formation by groundwater for terrestrial examples,[75] and have argued that the lack of fine scale heads to valley networks is due to their removal by weathering or impact gardening. Most authors accept that most valley networks were at least partly influenced and shaped by groundwater seep processes.

Groundwater also played a vital role in controlling broad scale sedimentation patterns and processes on Mars. According to this hypothesis, groundwater with dissolved minerals came to the surface, in and around craters, and helped to form layers by adding minerals —especially sulfate— and cementing sediments. In other words, some layers may have been formed by groundwater rising up depositing minerals and cementing existing, loose, aeolian sediments.

Figure 78: *The preservation and cementation of aeolian dune stratigraphy in Burns Cliff in Endurance Crater are thought to have been controlled by flow of shallow groundwater.*

The hardened layers are consequently more protected from erosion. A study published in 2011 using data from the Mars Reconnaissance Orbiter, show that the same kinds of sediments exist in a large area that includes Arabia Terra. It has been argued that areas that are rich in sedimentary rocks are also those areas that most likely experienced groundwater upwelling on a regional scale.

Mars ocean hypothesis

The Mars ocean hypothesis proposes that the Vastitas Borealis basin was the site of an ocean of liquid water at least once, and presents evidence that nearly a third of the surface of Mars was covered by a liquid ocean early in the planet's geologic history. This ocean, dubbed **Oceanus Borealis**, would have filled the Vastitas Borealis basin in the northern hemisphere, a region that lies 4–5 kilometres (2.5–3.1 mi) below the mean planetary elevation. Two major putative shorelines have been suggested: a higher one, dating to a time period of approximately 3.8 billion years ago and concurrent with the formation of the valley networks in the Highlands, and a lower one, perhaps correlated with the younger outflow channels. The higher one, the 'Arabia shoreline', can be traced all around Mars except through the Tharsis volcanic region. The lower, the 'Deuteronilus', follows the Vastitas Borealis formation.

A study in June 2010 concluded that the more ancient ocean would have covered 36% of Mars. Data from the Mars Orbiter Laser Altimeter (MOLA),

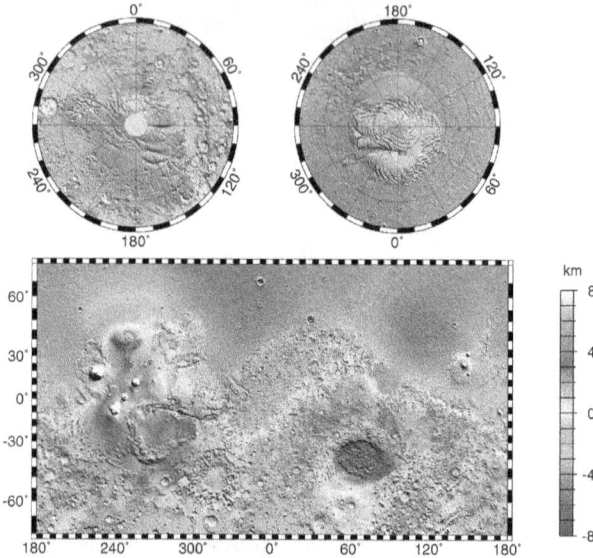

Figure 79: *The blue region of low topography in the Martian northern hemisphere is hypothesized to be the site of a primordial ocean of liquid water.*

which measures the altitude of all terrain on Mars, was used in 1999 to determine that the watershed for such an ocean would have covered about 75% of the planet. Early Mars would have required a warmer climate and denser atmosphere to allow liquid water to exist at the surface. In addition, the large number of valley networks strongly supports the possibility of a hydrological cycle on the planet in the past.

The existence of a primordial Martian ocean remains controversial among scientists, and the interpretations of some features as 'ancient shorelines' has been challenged. One problem with the conjectured 2-billion-year-old (2 Ga) shoreline is that it is not flat—i.e., does not follow a line of constant gravitational potential. This could be due to a change in distribution in Mars' mass, perhaps due to volcanic eruption or meteor impact; the Elysium volcanic province or the massive Utopia basin that is buried beneath the northern plains have been put forward as the most likely causes.

In March 2015, scientists stated that evidence exists for an ancient Martian ocean, likely in the planet's northern hemisphere and about the size of Earth's Arctic Ocean, or approximately 19% of the Martian surface. This finding was derived from the ratio of water and deuterium in the modern Martian atmosphere compared to the ratio found on Earth. Eight times as much deuterium was found at Mars than exists on Earth, suggesting that ancient Mars had significantly higher levels of water. Results from the *Curiosity* rover had previously found a high ratio of deuterium in Gale Crater, though not significantly high enough to suggest the presence of an ocean. Other scientists caution that this new study has not been confirmed, and point out that Martian climate models have not yet shown that the planet was warm enough in the past to support bodies of liquid water.

Additional evidence for a northern ocean was published in May 2016, describing how some of the surface in Ismenius Lacus quadrangle was altered by two tsunamis. The tsunamis were caused by asteroids striking the ocean. Both were thought to have been strong enough to create 30 km diameter craters. The first tsunami picked up and carried boulders the size of cars or small houses. The backwash from the wave formed channels by rearranging the boulders. The second came in when the ocean was 300 m lower. The second carried a great deal of ice which was dropped in valleys. Calculations show that the average height of the waves would have been 50 m, but the heights would vary from 10 m to 120 m. Numerical simulations show that in this particular part of the ocean two impact craters of the size of 30 km in diameter would form every 30 million years. The implication here is that a great northern ocean may have existed for millions of years. One argument against an ocean has been the lack of shoreline features. These features may have been washed away by these tsunami events. The parts of Mars studied in this research are Chryse Planitia and northwestern Arabia Terra. These tsunamis affected some surfaces in the Ismenius Lacus quadrangle and in the Mare Acidalium quadrangle.[76,77]

Present water

<templatestyles src="Multiple_image/styles.css" />

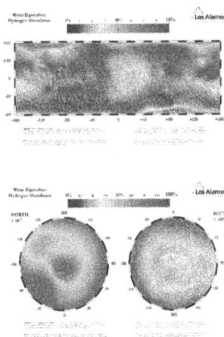

Proportion of water ice present in the upper meter of the Martian surface for lower (top) and higher (bottom) latitudes. The percentages are derived through stoichiometric calculations based on epithermal neutron fluxes. These fluxes were detected by the Neutron Spectrometer aboard the 2001 Mars Odyssey spacecraft.

A significant amount of surface hydrogen has been observed globally by the Mars Odyssey neutron spectrometer and gamma ray spectrometer. This hydrogen is thought to be incorporated into the molecular structure of ice, and through stoichiometric calculations the observed fluxes have been converted into concentrations of water ice in the upper meter of the Martian surface. This process has revealed that ice is both widespread and abundant on the present surface. Below 60 degrees of latitude, ice is concentrated in several regions, particularly around the Elysium volcanoes, Terra Sabaea, and northwest of Terra Sirenum, and exists in concentrations up to 18% ice in the subsurface. Above 60 degrees latitude, ice is highly abundant. Polewards on 70 degrees of latitude, ice concentrations exceed 25% almost everywhere, and approach 100% at the poles. The SHARAD and MARSIS radar sounding instruments have also confirmed that individual surface features are ice rich. Due to the known instability of ice at current Martian surface conditions, it is thought that almost all of this ice is covered by a thin layer of rocky or dusty material.

The Mars Odyssey neutron spectrometer observations indicate that if all the ice in the top meter of the Martian surface were spread evenly, it would give a Water Equivalent Global layer (WEG) of at least ≈ 14 centimetres (5.5 in)—in other words, the globally averaged Martian surface is approximately 14% water. The water ice currently locked in both Martian poles corresponds to a WEG of 30 metres (98 ft), and geomorphic evidence favors significantly larger quantities of surface water over geologic history, with WEG as deep as 500 metres (1,600 ft). It is thought that part of this past water has been lost to the deep subsurface, and part to space, although the detailed mass balance of these processes remains poorly understood. The current atmospheric reservoir of water is important as a conduit allowing gradual migration of ice from one part of the surface to another on both seasonal and longer timescales, but it is insignificant in volume, with a WEG of no more than 10 micrometres (0.00039 in).

Polar ice caps

Both the northern polar cap (Planum Boreum) and the southern polar cap (Planum Australe) have been observed to grow in thickness during the winter and partially sublime during the summer. In 2004, the MARSIS radar sounder on the Mars Express satellite targeted the southern polar cap, and was able to confirm that ice there extends to a depth of 3.7 kilometres (2.3 mi) below the surface. In the same year, the OMEGA instrument on the same orbiter revealed that the cap is divided into three distinct parts, with varying contents

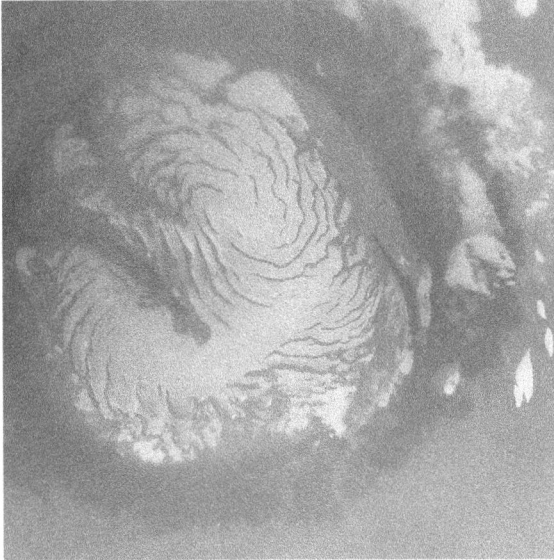

Figure 80: *The Mars Global Surveyor acquired this image of the Martian north polar ice cap in early northern summer.*

of frozen water depending on latitude. The first part is the bright part of the polar cap seen in images, centered on the pole, which is a mixture of 85% CO_2 ice to 15% water ice. The second part comprises steep slopes known as *scarps*, made almost entirely of water ice, that ring and fall away from the polar cap to the surrounding plains. The third part encompasses the vast permafrost fields that stretch for tens of kilometres away from the scarps, and is not obviously part of the cap until the surface composition is analysed. NASA scientists calculate that the volume of water ice in the south polar ice cap, if melted, would be sufficient to cover the entire planetary surface to a depth of 11 meters (36 ft).

An ancient ice sheet that has been proposed for the south polar region may have contained 20 million km^3 of water ice, which is equivalent to a layer 137 m deep over the entire planet.[78,79]

In July 2008, NASA announced that the *Phoenix* lander had confirmed the presence of water ice at its landing site near the northern polar ice cap (at 68.2° latitude). This was the first ever direct observation of ice from the surface. Two years later, the shallow radar on board the Mars Reconnaissance Orbiter took measurements of the north polar ice cap and determined that the total volume of water ice in the cap is 821,000 cubic kilometres (197,000 cu mi).

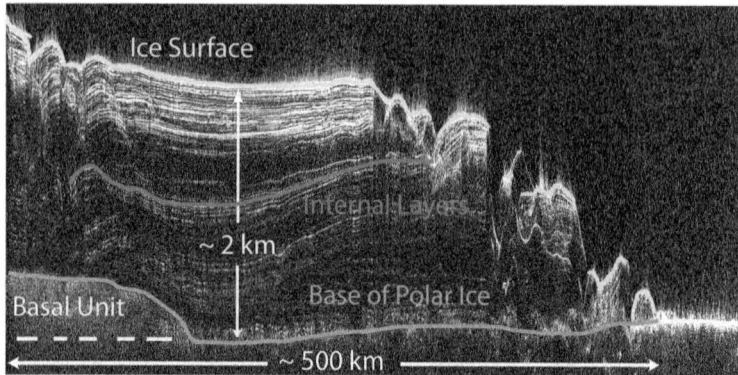

Figure 81: *Cross-section of a portion of the north polar ice cap of Mars, derived from satellite radar sounding.*

That is equal to 30% of the Earth's Greenland ice sheet, or enough to cover the surface of Mars to a depth of 5.6 metres (18 ft). Both polar caps reveal abundant fine internal layers when examined in HiRISE and Mars Global Surveyor imagery. Many researchers have studied this layering to understand the structure, history, and flow properties of the caps, although their interpretation is not straightforward.

Lake Vostok in Antarctica may have implications for liquid water still existing on Mars, because if water existed before the polar ice caps on Mars, it is possible that there is still liquid water below the ice caps.

Subglacial liquid water

In July 2018, scientists from the Italian Space Agency reported the detection of a subglacial lake on Mars, 1.5 kilometres (0.93 mi) below the southern polar ice cap, and spanning 20 kilometres (12 mi) horizontally, the first evidence for a stable body of liquid water on the planet. The evidence for the Mars lake was deduced from a bright spot in the radar echo sounding data, collected between May 2012 and December 2015, using the MARSIS radar on board the European *Mars Express* orbiter.[80] The detected lake is centered at 193°E, 81°S, a flat area that does not exhibit any peculiar topographic characteristics but is surrounded by higher ground, except on its eastern side, where there is a depression. The SHARAD radar on board NASA's *Mars Reconnaissance Orbiter* has seen no sign of the lake, but the team will take another look to try to confirm the finding when its orbital parameters are favorable.[81] It is unlikely that SHARAD will detect the lake, as it has much lower ground-penetrating abilities than MARSIS does.

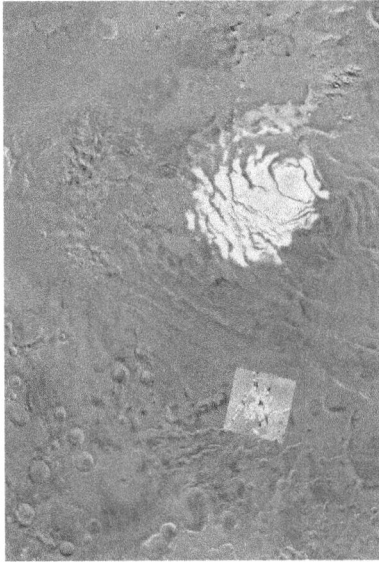

Figure 82: *Site of south polar subglacial water body (reported July 2018)*

Because the temperature at the base of the polar cap is estimated at 205 K (–68 °C; –91 °F), scientists assume that the water may remain liquid by the antifreeze effect of magnesium and calcium perchlorates. The 1.5-kilometre (0.93 mi) ice layer covering the lake is composed of water ice with 10 to 20% admixed dust, and seasonally covered by a 1-metre (3 ft 3 in)-thick layer of CO_2 ice. Since the raw-data coverage of the south polar ice cap is limited, the discoverers stated that "there is no reason to conclude that the presence of subsurface water on Mars is limited to a single location."

The lake may consist of clear water, or may be mixed with soil to form a sludge. The lake's high levels of salt would present difficulties for most lifeforms, but on Earth, organisms called halophiles thrive in salty conditions, though not in dark, cold, concentrated perchlorate solutions.

Ground ice

For many years, various scientists have suggested that some Martian surfaces look like periglacial regions on Earth. By analogy with these terrestrial features, it has been argued for many years that these are regions of permafrost. This would suggest that frozen water lies right beneath the surface. A common feature in the higher latitudes, patterned ground, can occur in a number of shapes, including stripes and polygons. On the Earth, these shapes are caused

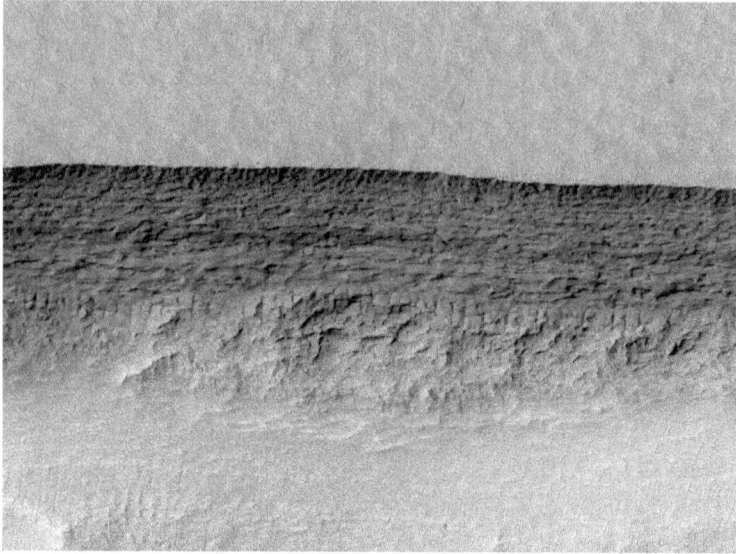

Figure 83: *A cross-section of underground water ice is exposed at the steep slope that appears bright blue in this enhanced-color view from the MRO. The scene is about 500 meters wide. The scarp drops about 128 meters from the level ground. The ice sheets extend from just below the surface to a depth of 100 meters or more.*[82]

by the freezing and thawing of soil. There are other types of evidence for large amounts of frozen water under the surface of Mars, such as terrain softening, which rounds sharp topographical features. Evidence from Mars Odyssey's gamma ray spectrometer and direct measurements with the Phoenix lander have corroborated that many of these features are intimately associated with the presence of ground ice.

Using the HiRISE camera on board the Mars Reconnaissance Orbiter (MRO), researchers found in 2017 at least eight eroding slopes showing exposed water ice sheets as thick as 100 meters, covered by a layer of about 1 or 2 meters thick of soil.[83,84] The sites are at latitudes from about 55 to 58 degrees, suggesting that there is shallow ground ice under roughly a third of the Martian surface. This image confirms what was previously detected with the spectrometer on 2001 Mars Odyssey, the ground-penetrating radars on MRO and on Mars Express, and by the *Phoenix* lander *in situ* excavation. These ice layers hold easily accessible clues about Mars' climate history and make frozen water accessible to future robotic or human explorers. Some researchers suggested these deposits could be the remnants of glaciers that existed millions of years ago when the planet's spin axis and orbit were different. (See section Mars' Ice ages below.)

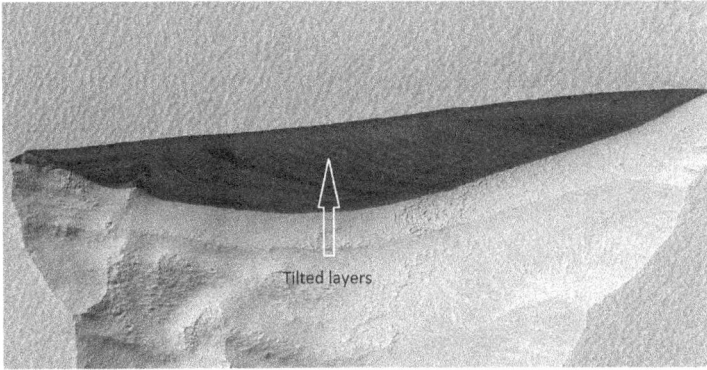

Figure 84: *Close view of wall of triangular depression, as seen by HiRISE layers are visible in the wall. These layers contain ice. The lower layers are tilted, while layers near the surface are more or less horizontal. Such an arrangement of layers is called an "angular unconformity." UNIQ-ref-0-086cfa95efb485b4-QINU*

Figure 85: *Impact crater that may have formed in ice-rich ground, as seen by HiRISE under HiWish program Location is the Ismenius Lacus quadrangle.*

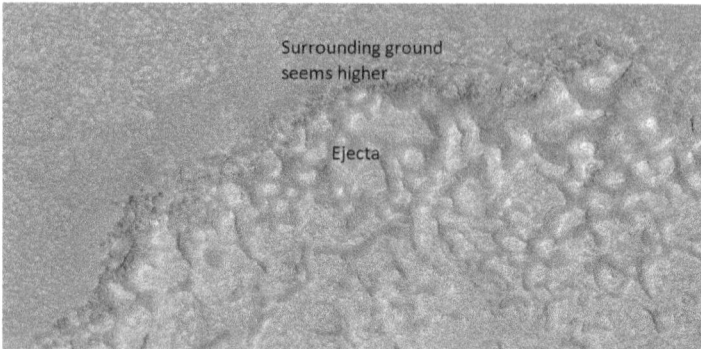

Figure 86: *Close view of impact crater that may have formed in ice-rich ground, as seen by HiRISE under HiWish program. Note that the ejecta seems lower than the surroundings. The hot ejecta may have caused some of the ice to go away; thus lowering the level of the ejecta.*

Scalloped topography

Certain regions of Mars display scalloped-shaped depressions. The depressions are suspected to be the remains of a degrading ice-rich mantle deposit. Scallops are caused by ice sublimating from frozen soil. The landforms of scalloped topography can be formed by the subsurface loss of water ice by sublimation under current Martian climate conditions. A model predicts similar shapes when the ground has large amounts of pure ice, up to many tens of meters in depth.[85] This mantle material was probably deposited from the atmosphere as ice formed on dust when the climate was different due to changes in the tilt of the Mars pole (see "Ice ages", below). The scallops are typically tens of meters deep and from a few hundred to a few thousand meters across. They can be almost circular or elongated. Some appear to have coalesced causing a large heavily pitted terrain to form. The process of forming the terrain may begin with sublimation from a crack. There are often polygonal cracks where scallops form, and the presence of scalloped topography seems to be an indication of frozen ground.

On November 22, 2016, NASA reported finding a large amount of underground ice in the Utopia Planitia region of Mars. The volume of water detected has been estimated to be equivalent to the volume of water in Lake Superior.

The volume of water ice in the region were based on measurements from the ground-penetrating radar instrument on Mars Reconnaissance Orbiter, called SHARAD. From the data obtained from SHARAD, "dielectric permittivity", or the dielectric constant was determined. The dielectric constant value was consistent with a large concentration of water ice.[86,87]

These scalloped features are superficially similar to Swiss cheese features, found around the south polar cap. Swiss cheese features are thought to be due to cavities forming in a surface layer of solid carbon dioxide, rather than water ice—although the floors of these holes are probably H_2O-rich.

Ice patches

On July 28, 2005, the European Space Agency announced the existence of a crater partially filled with frozen water; some then interpreted the discovery as an "ice lake". Images of the crater, taken by the High Resolution Stereo Camera on board the European Space Agency's Mars Express orbiter, clearly show a broad sheet of ice in the bottom of an unnamed crater located on Vastitas Borealis, a broad plain that covers much of Mars' far northern latitudes, at approximately 70.5° North and 103° East. The crater is 35 kilometres (22 mi) wide and about 2 kilometres (1.2 mi) deep. The height difference between the crater floor and the surface of the water ice is about 200 metres (660 ft). ESA scientists have attributed most of this height difference to sand dunes beneath the water ice, which are partially visible. While scientists do not refer to the patch as a "lake", the water ice patch is remarkable for its size and for being present throughout the year. Deposits of water ice and layers of frost have been found in many different locations on the planet.

As more and more of the surface of Mars has been imaged by the modern generation of orbiters, it has become gradually more apparent that there are probably many more patches of ice scattered across the Martian surface. Many of these putative patches of ice are concentrated in the Martian midlatitudes ($\approx$30–60° N/S of the equator). For example, many scientists think that the widespread features in those latitude bands variously described as "latitude dependent mantle" or "pasted-on terrain" consist of dust- or debris-covered ice patches, which are slowly degrading. A cover of debris is required both to explain the dull surfaces seen in the images that do not reflect like ice, and also to allow the patches to exist for an extended period of time without subliming away completely. These patches have been suggested as possible water sources for some of the enigmatic channelized flow features like gullies also seen in those latitudes.

Surface features consistent with existing pack ice have been discovered in the southern Elysium Planitia. What appear to be plates, ranging in size from 30 metres (98 ft) to 30 kilometres (19 mi), are found in channels leading to a large flooded area. The plates show signs of break up and rotation that clearly distinguish them from lava plates elsewhere on the surface of Mars. The source for the flood is thought to be the nearby geological fault Cerberus Fossae that spewed water as well as lava aged some 2 to 10 million years. It was suggested that the water exited the Cerberus Fossae then pooled and froze in the low, level plains and that such frozen lakes may still exist.

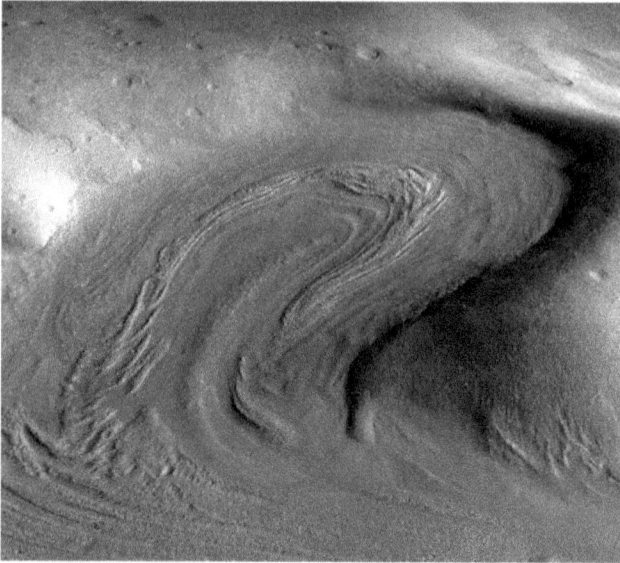

Figure 87: *View of a 5-km-wide, glacial-like lobe deposit sloping up into a box canyon. The surface has moraines, deposits of rocks that show how the glacier advanced.*

Glaciers

Many large areas of Mars either appear to host glaciers, or carry evidence that they used to be present. Much of the areas in high latitudes, especially the Ismenius Lacus quadrangle, are suspected to still contain enormous amounts of water ice. Recent evidence has led many planetary scientists to conclude that water ice still exists as glaciers across much of the Martian mid- and high latitudes, protected from sublimation by thin coverings of insulating rock and/ or dust. An example of this are the glacier-like features called lobate debris aprons in an area called Deuteronilus Mensae, which display widespread evidence of ice lying beneath a few meters of rock debris. Glaciers are associated with fretted terrain, and many volcanoes. Researchers have described glacial deposits on Hecates Tholus, Arsia Mons, Pavonis Mons, and Olympus Mons. Glaciers have also been reported in a number of larger Martian craters in the midlatitudes and above.

Glacier-like features on Mars are known variously as viscous flow features, Martian flow features, lobate debris aprons, or lineated valley fill, depending on the form of the feature, its location, the landforms it is associated with, and the author describing it. Many, but not all, small glaciers seem to be associated

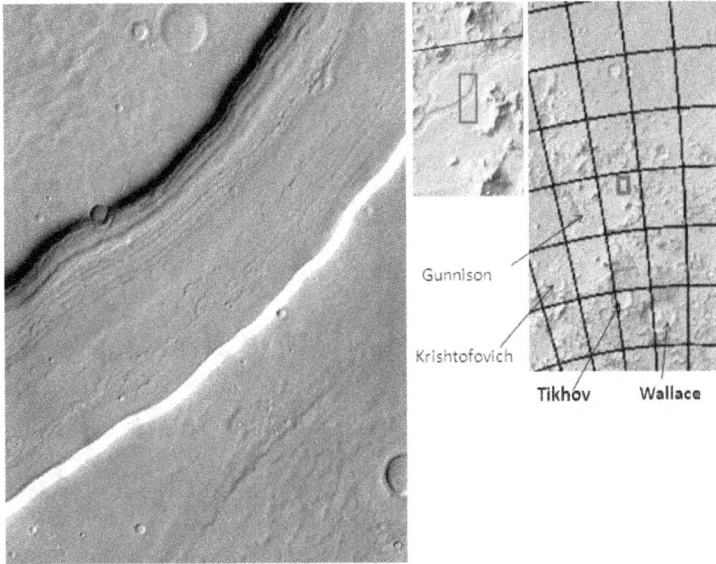

Figure 88: *Reull Vallis with lineated floor deposits. Location is Hellas quadrangle*

with gullies on the walls of craters and mantling material. The lineated deposits known as lineated valley fill are probably rock-covered glaciers that are found on the floors most channels within the fretted terrain found around Arabia Terra in the northern hemisphere. Their surfaces have ridged and grooved materials that deflect around obstacles. Lineated floor deposits may be related to lobate debris aprons, which have been proven to contain large amounts of ice by orbiting radar. For many years, researchers interpreted that features called 'lobate debris aprons' were glacial flows and it was thought that ice existed under a layer of insulating rocks. With new instrument readings, it has been confirmed that lobate debris aprons contain almost pure ice that is covered with a layer of rocks.

Moving ice carries rock material, then drops it as the ice disappears. This typically happens at the snout or edges of the glacier. On Earth, such features would be called moraines, but on Mars they are typically known as *moraine-like ridges*, *concentric ridges*, or *arcuate ridges*. Because ice tends to sublime rather than melt on Mars, and because Mars's low temperatures tend to make glaciers "cold based" (frozen down to their beds, and unable to slide), the remains of these glaciers and the ridges they leave do not appear the exactly same as normal glaciers on Earth. In particular, Martian moraines tend to be deposited without being deflected by the underlying topography, which is thought to reflect the fact that the ice in Martian glaciers is normally frozen

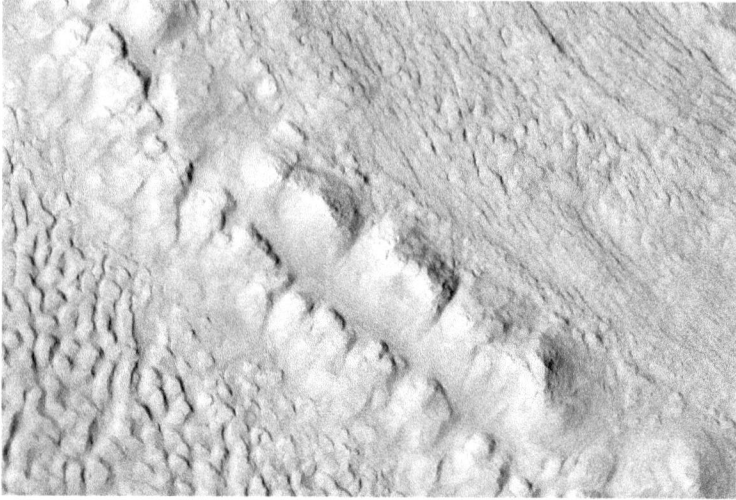

Figure 89: *A ridge interpreted as the terminal moraine of an alpine glacier. Location is Ismenius Lacus quadrangle*

down and cannot slide. Ridges of debris on the surface of the glaciers indicate the direction of ice movement. The surface of some glaciers have rough textures due to sublimation of buried ice. The ice evaporates without melting and leaves behind an empty space. Overlying material then collapses into the void. Sometimes chunks of ice fall from the glacier and get buried in the land surface. When they melt, a more or less round hole remains. Many of these "kettle holes" have been identified on Mars.

Despite strong evidence for glacial flow on Mars, there is little convincing evidence for landforms carved by glacial erosion, e.g., U-shaped valleys, crag and tail hills, arêtes, drumlins. Such features are abundant in glaciated regions on Earth, so their absence on Mars has proven puzzling. The lack of these landforms is thought to be related to the cold-based nature of the ice in most recent glaciers on Mars. Because the solar insolation reaching the planet, the temperature and density of the atmosphere, and the geothermal heat flux are all lower on Mars than they are on Earth, modelling suggests the temperature of the interface between a glacier and its bed stays below freezing and the ice is literally frozen down to the ground. This prevents it from sliding across the bed, which is thought to inhibit the ice's ability to erode the surface.

Figure 90: *Dry channels near Warrego Valles*

Development of Mars' water inventory

The variation in Mars's surface water content is strongly coupled to the evolution of its atmosphere and may have been marked by several key stages.

Early Noachian era (4.6 Ga to 4.1 Ga)

The early Noachian era was characterized by atmospheric loss to space from heavy meteoritic bombardment and hydrodynamic escape. Ejection by meteorites may have removed $\sim$60% of the early atmosphere. Significant quantities of phyllosilicates may have formed during this period requiring a sufficiently dense atmosphere to sustain surface water, as the spectrally dominant phyllosilicate group, smectite, suggests moderate water-to-rock ratios. However, the pH-pCO_2 between smectite and carbonate show that the precipitation of smectite would constrain pCO_2 to a value not more than 1×10^{-2} atm (1.0 kPa). As a result, the dominant component of a dense atmosphere on early Mars becomes uncertain, if the clays formed in contact with the Martian atmosphere, particularly given the lack of evidence for carbonate deposits. An additional complication is that the $\sim$25% lower brightness of the young Sun would have required an ancient atmosphere with a significant greenhouse effect to raise surface temperatures to sustain liquid water. Higher CO_2 content alone would have been insufficient, as CO_2 precipitates at partial pressures exceeding 1.5 atm (1,500 hPa), reducing its effectiveness as a greenhouse gas.

Middle to late Noachian era (4.1 Ga to 3.8 Ga)

During the middle to late Noachian era, Mars underwent potential formation of a secondary atmosphere by outgassing dominated by the Tharsis volcanoes, including significant quantities of H_2O, CO_2, and SO_2. Martian valley networks date to this period, indicating globally widespread and temporally sustained surface water as opposed to catastrophic floods. The end of this period coincides with the termination of the internal magnetic field and a spike in meteoritic bombardment. The cessation of the internal magnetic field and subsequent weakening of any local magnetic fields allowed unimpeded atmospheric stripping by the solar wind. For example, when compared with their terrestrial counterparts, $^{38}Ar/^{36}Ar$, $^{15}N/^{14}N$, and $^{13}C/^{12}C$ ratios of the Martian atmosphere are consistent with $\sim$60% loss of Ar, N_2, and CO_2 by solar wind stripping of an upper atmosphere enriched in the lighter isotopes via Rayleigh fractionation. Supplementing the solar wind activity, impacts would have ejected atmospheric components in bulk without isotopic fractionation. Nevertheless, cometary impacts in particular may have contributed volatiles to the planet.

Hesperian to Amazonian era (present) ($\sim$3.8 Ga to present)

Atmospheric enhancement by sporadic outgassing events were countered by solar wind stripping of the atmosphere, albeit less intensely than by the young Sun. Catastrophic floods date to this period, favoring sudden subterranean release of volatiles, as opposed to sustained surface flows. While the earlier portion of this era may have been marked by aqueous acidic environments and Tharsis-centric groundwater discharge dating to the late Noachian, much of the surface alteration processes during the latter portion is marked by oxidative processes including the formation of Fe^{3+} oxides that impart a reddish hue to the Martian surface. Such oxidation of primary mineral phases can be achieved by low-pH (and possibly high temperature) processes related to the formation of palagonitic tephra, by the action of H_2O_2 that forms photochemically in the Martian atmosphere, and by the action of water, none of which require free O_2. The action of H_2O_2 may have dominated temporally given the drastic reduction in aqueous and igneous activity in this recent era, making the observed Fe^{3+} oxides volumetrically small, though pervasive and spectrally dominant. Nevertheless, aquifers may have driven sustained, but highly localized surface water in recent geologic history, as evident in the geomorphology of craters such as Mojave. Furthermore, the Lafayette Martian meteorite shows evidence of aqueous alteration as recently as 650 Ma.

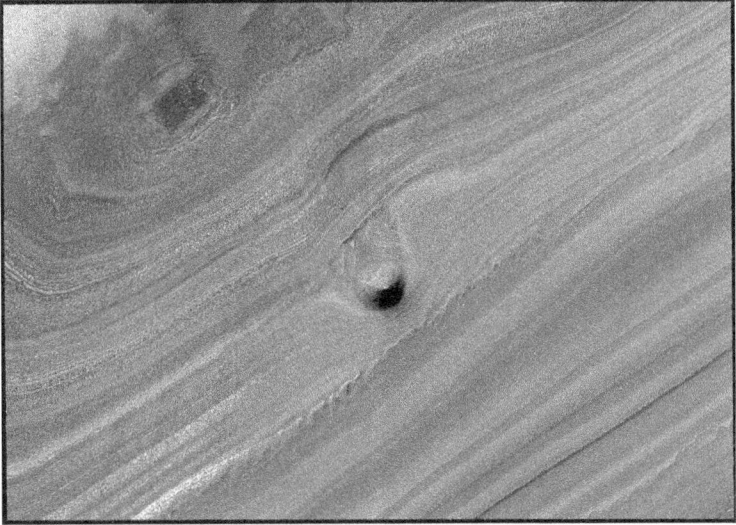

Figure 91: *North polar layered deposits of ice and dust*

Ice ages

Mars has experienced about 40 large scale changes in the amount and distribution of ice on its surface over the past five million years, with the most recent happening about 2.1 to 0.4 Myr ago, during the Late Amazonian glaciation at the dichotomy boundary. These changes are known as ice ages. Ice ages on Mars are very different from the ones that the Earth experiences. Ice ages are driven by changes in Mars's orbit and tilt —also known as obliquity. Orbital calculations show that Mars wobbles on its axis far more than Earth does. The Earth is stabilized by its proportionally large moon, so it only wobbles a few degrees. Mars may change its tilt by many tens of degrees. When this obliquity is high, its poles get much more direct sunlight and heat; this causes the ice caps to warm and become smaller as ice sublimes. Adding to the variability of the climate, the eccentricity of the orbit of Mars changes twice as much as Earth's eccentricity. As the poles sublime, the ice is redeposited closer to the equator, which receive somewhat less solar insolation at these high obliquities. Computer simulations have shown that a 45° tilt of the Martian axis would result in ice accumulation in areas that display glacial landforms.

The moisture from the ice caps travels to lower latitudes in the form of deposits of frost or snow mixed with dust. The atmosphere of Mars contains a great deal of fine dust particles, the water vapor condenses on these particles that then fall down to the ground due to the additional weight of the water coating.

Figure 92: *Warm-season flows on slope in Newton Crater*

When ice at the top of the mantling layer returns to the atmosphere, it leaves behind dust that serves to insulate the remaining ice. The total volume of water removed is a few percent of the ice caps, or enough to cover the entire surface of the planet under one meter of water. Much of this moisture from the ice caps results in a thick smooth mantle with a mixture of ice and dust. This ice-rich mantle, that can be 100 meters thick at mid-latitudes, smoothes the land at lower latitudes, but in places it displays a bumpy texture or patterns that give away the presence of water ice underneath.

Evidence for recent flows

Pure liquid water cannot exist in a stable form on the surface of Mars with its present low atmospheric pressure and low temperature, except at the lowest elevations for a few hours. So, a geological mystery commenced in 2006 when observations from NASA's *Mars Reconnaissance Orbiter* revealed gully deposits that were not there ten years prior, possibly caused by flowing liquid brine during the warmest months on Mars. The images were of two craters called Terra Sirenum and Centauri Montes that appear to show the presence of flows (wet or dry) on Mars at some point between 1999 and 2001.

There is disagreement in the scientific community as to whether or not gullies are formed by liquid water. It is also possible that the flows that carve gullies

Figure 93: *Branched gullies*

Figure 94: *Group of deep gullies*

are dry grains, or perhaps lubricated by carbon dioxide. Some studies attest that gullies forming in the southern highlands could not be formed by water due to improper conditions. The low pressure, non-geothermal, colder regions would not give way to liquid water at any point in the year but would be ideal for solid carbon dioxide. The carbon dioxide melting in the warmer summer would yield liquid carbon dioxide which would then form the gullies. Even if gullies are carved by flowing water at the surface, the exact source of the water and the mechanisms behind its motion are not understood.

The dry gullies are deep grooves etched into the slopes that are there year-round. There are many other features on Mars, and some of them change seasonally.

In August 2011, NASA announced the discovery by undergraduate student Lujendra Ojha of current seasonal changes on steep slopes below rocky outcrops near crater rims in the Southern hemisphere. These dark streaks, now called recurrent slope lineae (RSL), were seen to grow downslope during the warmest part of the Martian Summer, then to gradually fade through the rest of the year, recurring cyclically between years. The researchers suggested these marks were consistent with salty water (brines) flowing downslope and then evaporating, possibly leaving some sort of residue. The CRISM spectroscopic instrument has since made direct observations of hydrous salts appearing at the same time that these recurrent slope lineae form, confirming in 2015 that these lineae are produced by the flow of liquid brines through shallow soils. The lineae contain hydrated chlorate and perchlorate salts (ClO_4^-), which contain liquid water molecules. The lineae flow downhill in Martian summer, when the temperature is above $-23\ °C$ ($-9\ °F$; $250\ K$). However, the source of the water remains unknown. However, neutron spectrometer data by the *Mars Odyssey* orbiter obtained over one decade, was published in December 2017, and shows no evidence of water (hydrogenated regolith) at the active sites, so its authors also support the hypotheses of either short-lived atmospheric water vapour deliquesence, or dry granular flows. They conclude that liquid water on today's Mars may be limited to traces of dissolved moisture from the atmosphere and thin films, which are challenging environments for life as we know it.[88]

Habitability assessment

Liquid water is a necessary but not sufficient condition for life as we know it, as habitability is a function of a multitude of environmental parameters.[89] Present day life on Mars could occur kilometers below the surface in a hypothetical hydrosphere, or in subsurface geothermal hot spots, or it could occur near the surface. The permafrost layer on Mars is only a couple of centimeters below the surface. Salty brines can be liquid a few centimeters below that but not far down. Most of the proposed surface habitats are within centimeters of the surface. Any life deeper than that is likely to be dormant. Water is close to its boiling point even at the deepest points in the Hellas basin, and so cannot remain liquid for long on the surface of Mars in its present state, except when covered in ice or after a sudden release of water. Liquid water on today's Mars may be limited to traces of dissolved moisture from the atmosphere and thin films, which are challenging environments for life as we know it.

So far, NASA has pursued a "follow the water" strategy on Mars and has not searched for biosignatures for life there directly since the *Viking* landers in July 1976. The observations by *Phoenix* lander in 2008 of potential drops of liquid brines forming on its legs led to a renewed interest in the potential habitability of the surface of Mars. Since then, experiments have led to many suggestions for potential habitats on the surface of Mars. However, though liquid water is now confirmed to occur there in brine layers, it is not yet known whether any of it is habitable. This depends on factors such as the exact mix of salts, temperature, energy sources, and the radiation environment on Mars.

Proposed surface habitats for Earth microorganisms

For purposes of planetary protection, scientists are trying to identify potential habitats where stowaway bacteria from Earth on spacecraft could contaminate Mars.

- Ice and salt — Ice and salt are both common in the higher latitudes of Mars, but these transient brine films may not offer the needed water activity, thermal, or UV protection to microbial life brought on contaminated landers.
- Warm seasonal flows — Most of these are thought to be due to dry ice (frozen CO_2), dry granular flows, liquid brines and wind effects. They form on Sun-facing slopes in the summer when the local temperatures rise above 0 °C. There is strong evidence now that they are associated with liquid brines, or hydrated salts. They are currently classified as "Uncertain regions to be treated as Special Regions" for purposes of planetary protection.
- Atmospheric water — Perchlorates are found over most of the surface of Mars,[90] and could, potentially, take up water from the atmosphere more readily (deliquescence). Thin layers of salty perchlorate rich brines could form a short way below the surface at night and in the early morning. Perchlorates are toxic to most lifeforms. However, some Haloarchaea are able to tolerate them, and some can use them as a source of energy as well. The brines detected by *Curiosity*, are thought to be too salty for

life, and when the water activity is high enough, they are too cold. Other deliquescing salts on Mars may be more habitable for Earth life. A series of experiments by DLR (German Aerospace Center) in Mars simulation chambers and on the ISS show that some Earth life (lichens and strains of chrooccocidiopsis, a green algae) could survive some Mars surface conditions and photosynthesize and metabolize, slowly, in absence of any water at all. They could make use of the humidity of the Mars atmosphere.[91,92,93,94] The lichens studied in these experiments have protection from UV light due to special pigments only found in lichens, such as parietin and antioxidants such as b-carotene in epilithic lichens.

- Sun warmed dust grains embedded in ice: Möhlmann originally suggested this process in 2011 as a possible way for liquid water to form on Mars, based on a mechanism that produces liquid water in similar conditions in Antarctica. As the sunlight hits the ice, it would warm up any heat-absorbing dust grains trapped inside. These grains would then trap heat and melt the surrounding ice, forming a transient water coat that may protect the dust grain from the vacuum conditions of the atmosphere.[95] They developed this model as a hypothesis to explain presence of extensive deposits of gypsum in the Northern polar ice cap and the dune fields around it, a process that has been observed in Antarctica.[96]

- Shallow interfacial layers a few molecules thick: These interfacial layers occur on boundaries between ice and rock due to intermolecular forces that depress the freezing point of the water. The water flows and acts as a solvent. They may be used by microbes in Arctic permafrost, which have been found to metabolize at temperatures as low as −20 °C. Liquid water may be possible in layers as thin as three monolayers, and the model by Stephen Jepsen et al. obtained 10^9 cells/g at −20 °C, though the microbes would spend most of their time in survival mode. Models show that interfacial water should form in some regions of Mars, for instance in Richardson crater.

- Advancing sand dunes bioreactor: This hypothesis suggests that the constantly moving sand dunes of Mars may be able to create a potential environment for life by replenishing nutrients, and the chemical disequilibrium needed for life maintained through churning of the sand by the winds. In this model, the water source is deliquescing salts.[97]

Proposed subsurface habitats

- Geological hot spots: There is some evidence that Mars may still have some deep geological activity. This includes, small scale volcanic features associated with some of the volcanoes on Mars which may have formed about two million years ago.[98] There is also isotopic evidence from the *Phoenix* lander of release of CO

$_2$ some millions of years ago.[99] It may be possible that there are magma plumes deep underground, associated with the occasional surface volcanism on the geological timescale of millions of years. And given that there has been activity on Olympus Mons as recently as four million years ago, it seems unlikely that all activity has stopped permanently. So far no currently active volcanism has been observed, nor have any present day warm areas have ever been found on the surface, in extensive searches.[100] Another way to search for volcanic activity is through searches of trace gases produced in volcanic eruptions. One of the instruments on the 2016 ExoMars Trace Gas Orbiter is NOMAD (Nadir and Occultation for Mars Discovery), which is searching for trace gases indicating current volcanic activity, as well as searching directly for methane plumes.[101]

If these hot spots exist, they could keep water liquid through geothermal heating. The water could be trapped under overlying deposits and kept at a pressure high enough to stay liquid. They could also be a source for intermittent surface or near surface water (for instance one of the hypotheses for the RSL is that they may be occur over geological hot spots deep below the surface that indirectly supply them with water). Another possibility is a volcanic ice tower – a column of ice that can form around volcanic vents, as shown on Mount Erebus, Ross Island, Antarctica.[102] Since they are underground, and would be only a few degrees higher in temperature than the surrounding ground, they would be easy to miss in thermal images from orbit.[103,104]

- Potential for cave habitats on Mars: Caves, as well as lava tubes could offer protection from some of the harsh surface conditions. Some possibilities for cave formation are:[105] "(1) diversion of channel courses in underground conduits; (2) fractures of surface drainage patterns; chaotic terrain and collapsed areas in general; (4) seepage face in valley walls and/ or headwaters; (5) inactive hydrothermal vents and lava tubes." There's also the possibility of liquid CO2 (which forms under pressure, at depth, e.g. in a cliff wall) forming caves. The lava tubes on Mars are far larger than the ones on the Earth. Also Mars could have sublimational caves caused by dry ice and ordinary ice subliming directly into the atmosphere. Some cave habitats on Earth, such as in the toxic sulfur cave Cueva de Villa Luz, the microbial snottites flourish on hydrogen sulfide gas.
- Hypothetical hydrosphere: It has been speculated that a hydrosphere may exist below the icy layer (cryosphere) of permafrost. In higher latitudes, the permafrost starts a few cm below the surface, and may continue down for several kilometers. If the Mars hydrosphere exists, it lies below the cryosphere, and would be a layer where the water is kept liquid by geothermal heating, and prevented from evaporating by the overlying

layers of ice. There may be evidence of a deep subsurface cryosphere in the form of hydrogen/deuterium isotope ratios in Martian meteorites, which give indirect evidence that Mars might have a subsurface frozen water reservoir. If it exists, a modeling from 2013 put its depth at about 5 kilometers below the surface. Whether this layer exists or not depends on the presence or otherwise of perchlorates, and clathrates, and it also depends on the total inventory of water on Mars, so there are many unknowns in the models.[106] If this hydrosphere exists, then it may be more habitable than similar depth zones on Earth because of the lower gravity, leading to larger pore size. Possible metabolisms at this depth could use hydrogen, carbon dioxide, and possibly abiotic hydrocarbons. The carbon for biomass could come from magmatic carbon in basalts which has been detected in Martian meteorites. It could also support methanogens feeding off methane released from serpentinization, and the alteration of basalt could also be a basis for iron respiration.[107] If the habitat exists it could replenish surface areas of Mars with microbial life – a process that is known to happen beneath arctic permafrost layers on Earth.[108] One prime place to visit to search for evidence of the deep hydrosphere is McLaughlin Crater, which contained an ancient lake in the past and seems to display alteration minerals rich in Fe and Mg, and the detection of carbonates there suggests that the fluids were alkaline, and are consistent with the expected composition of fluids that emerged from a deep subsurface hydrosphere.

Dormant subsurface life

Curiosity measured ionizing radiation levels of 76 mGy a year. This level of ionizing radiation is sterilizing for life on the surface of Mars. However, the radiation varies considerably depending on Mars' orbital eccentricity and the tilt of its axis. If any surface life has been reanimated as recently as 450,000 years ago, then a rover on Mars could potentially find dormant but still viable life at a depth of one meter below the surface, according to an estimate.

Galactic cosmic rays

The level of 76 mGy per year of galactic cosmic rays measured by *Curiosity* is similar to levels inside the ISS. In the 2014 Findings of the Second MEPAG Special Regions Science Analysis Group, their conclusion was:

"From MSL RAD measurements, ionizing radiation from galactic cosmic rays (GCRs) at Mars is so low as to be negligible. Intermittent solar proton events (SPEs) can increase the atmospheric ionization down to ground level and increase the total dose, but these events are sporadic and last at most a few (2–5) days. These facts are not used to distinguish Special Regions on Mars." (A "Special Region" is a region where Earth life could potentially survive.)

UV radiation

On UV radiation, the report finds

"The martian UV radiation environment is rapidly lethal to unshielded microbes but can be attenuated by global dust storms and shielded completely by < 1 mm of regolith or by other organisms."

Perchlorates

The discovery in 2008 of perchlorates on the Martian surface prompted speculation of what their influence would be on habitability. Research published in 2017 showed that perchlorate adds a tenfold toxicity (bacteriocidal effects) because of the UV irradiation, in addition of the combined affects of low pressure, low temperatures, oxidants, and iron oxides.[109,110] These studies strongly suggest that bacteria that hitch a ride on landers sent to Mars will be swiftly destroyed on the surface, alleviating most concerns about contaminating the Martian surface.

Recurrent Slope Lineae (RSL)

These features form on sun-facing slopes at times of the year when the local temperatures reach above the melting point for ice. The streaks grow in spring, widen in late summer and then fade away in autumn. This is hard to model in any other way except as involving liquid water in some form, though the streaks themselves are thought to be a secondary effect and not a direct indication of dampness of the regolith. Although these features are now confirmed to involve liquid water in some form, the water could be either too cold or too salty for life. At present they are treated as potentially habitable, as "Uncertain Regions, to be treated as Special Regions".

The "Special Regions" assessment says of them:

- "Although no single model currently proposed for the origin of RSL adequately explains all observations, they are currently best interpreted as being due to the seepage of water at > 250 K, with a_w [water activity] unknown and perhaps variable. As such they meet the criteria for Uncertain Regions, to be treated as Special Regions. There are other features on Mars with characteristics similar to RSL, but their relationship to possible liquid water is much less likely"

They were first reported in the paper by McEwan in Science, August 5, 2011. They were already suspected as involving flowing brines back then, as all the other models available involved liquid water in some form. Finally proven pretty much conclusively to involve liquid water in some form, possibly habitable if temperatures and salinity are right – after detection of hydrated salts that change their hydration state rapidly, reported in a paper published on September 28, 2015 along with a press conference. MRO is in a slowly precessing Sun-synchronous orbit inclined at 93 degrees (orbital period 1 hr 52 minutes). Each time it crosses the Mars equator on the sunny side, South to North, the time is 3:00 pm, in the local solar time on the surface, all year round. This is the worst time of day to spot brines from orbit.

An analysis of data from the Mars Odyssey Neutron Spectrometer revealed that the RSL sites do not contain any more water than found at anywhere else at similar latitudes. The authors concluded that RSL are not supplied by large, near-surface briny aquifers. It is still possible with this data that water vapor from deeply buried ice, from the atmosphere, or from small deeply buried aquifers supply humidity.

Findings by probes

Mariner 9

The images acquired by the Mariner 9 Mars orbiter, launched in 1971, revealed the first direct evidence of past water in the form of dry river beds, canyons (including the Valles Marineris, a system of canyons over about 4,020 kilometres (2,500 mi) long), evidence of water erosion and deposition, weather fronts, fogs, and more. The findings from the Mariner 9 missions underpinned the later Viking program. The enormous Valles Marineris canyon system is named after Mariner 9 in honor of its achievements.

Viking program

By discovering many geological forms that are typically formed from large amounts of water, the two Viking orbiters and the two landers caused a revolution in our knowledge about water on Mars. Huge outflow channels were found in many areas. They showed that floods of water broke through dams, carved deep valleys, eroded grooves into bedrock, and traveled thousands of kilometers. Large areas in the southern hemisphere contained branched valley networks, suggesting that rain once fell. Many craters look as if the impactor fell into mud. When they were formed, ice in the soil may have melted, turned the ground into mud, then the mud flowed across the surface. Regions, called "Chaotic Terrain," seemed to have quickly lost great volumes of water that

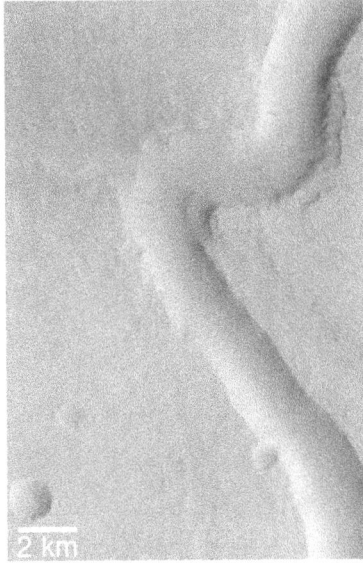

Figure 95: *Meander in Scamander Vallis, as seen by Mars Global Surveyor. Such images implied that large amounts of water once flowed on the surface of Mars.*

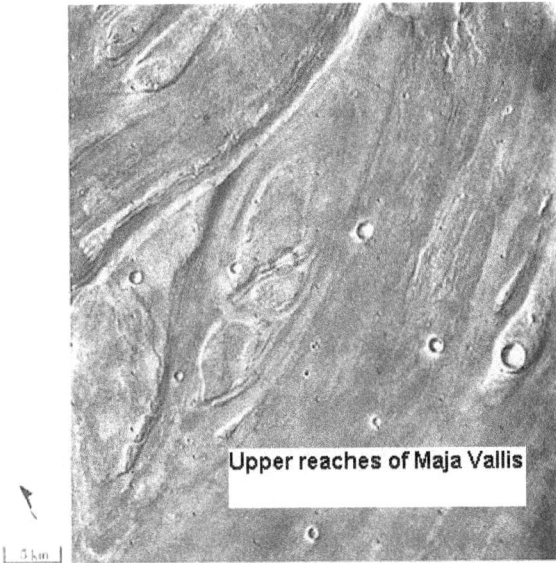

Upper reaches of Maja Vallis

Figure 96: *Streamlined islands in Maja Valles suggest that large floods occurred on Mars*

TES Hematite Abundance

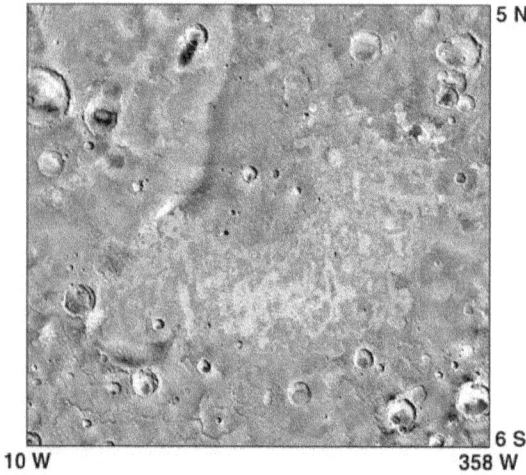

5 N

10 W

6 S

358 W

Figure 97: *Map showing the distribution of hematite in Si-nus Meridiani. This data was used to target the landing of the Opportunity rover that found definite evidence of past water.*

caused large channels to form downstream. Estimates for some channel flows run to ten thousand times the flow of the Mississippi River. Underground volcanism may have melted frozen ice; the water then flowed away and the ground collapsed to leave chaotic terrain. Also, general chemical analysis by the two Viking landers suggested the surface has been either exposed to or submerged in water in the past.

Mars Global Surveyor

The Mars Global Surveyor's Thermal Emission Spectrometer (TES) is an instrument able to determine the mineral composition on the surface of Mars. Mineral composition gives information on the presence or absence of water in ancient times. TES identified a large (30,000 square kilometres (12,000 sq mi)) area in the Nili Fossae formation that contains the mineral olivine. It is thought that the ancient asteroid impact that created the Isidis basin resulted in faults that exposed the olivine. The discovery of olivine is strong evidence that parts of Mars have been extremely dry for a long time. Olivine was also discovered in many other small outcrops within 60 degrees north and south of the equator. The probe has imaged several channels that

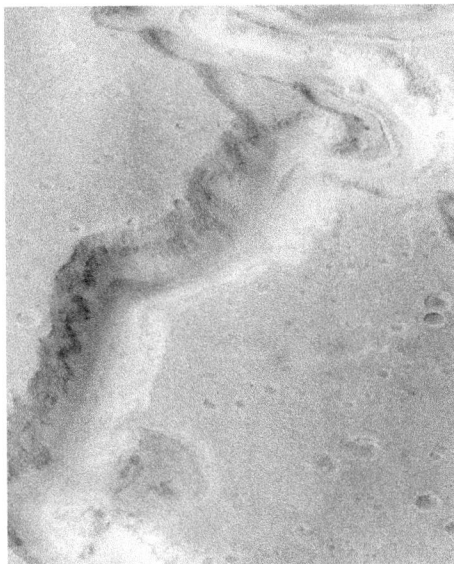

Figure 98: *Inner channel (near top of the image) on floor of Nanedi Valles that suggests that water flowed for a fairly long period. Image from Lunae Palus quadrangle.*

suggest past sustained liquid flows, two of them are found in Nanedi Valles and in Nirgal Vallis.

Mars Pathfinder

The Pathfinder lander recorded the variation of diurnal temperature cycle. It was coldest just before sunrise, about –78 °C (–108 °F; 195 K), and warmest just after Mars noon, about –8 °C (18 °F; 265 K). At this location, the highest temperature never reached the freezing point of water (0 °C (32 °F; 273 K)), too cold for pure liquid water to exist on the surface.

The atmospheric pressure measured by the Pathfinder on Mars is very low —about 0.6% of Earth's, and it would not permit pure liquid water to exist on the surface.

Other observations were consistent with water being present in the past. Some of the rocks at the Mars Pathfinder site leaned against each other in a manner geologists term imbricated. It is suspected that strong flood waters in the past pushed the rocks around until they faced away from the flow. Some pebbles were rounded, perhaps from being tumbled in a stream. Parts of the ground

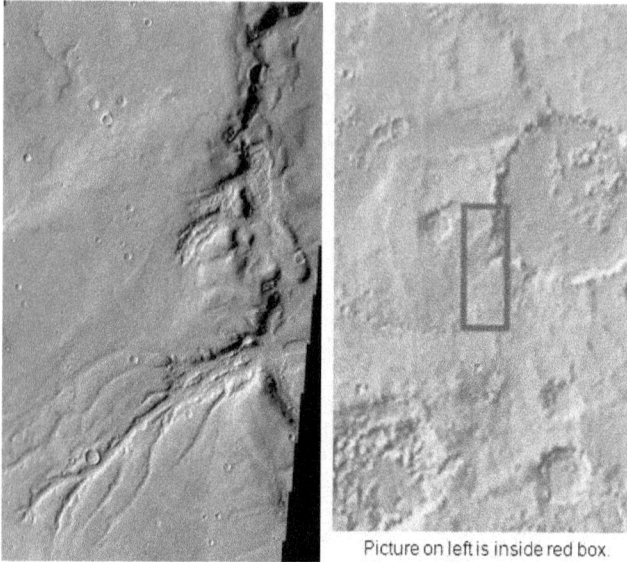

Picture on left is inside red box.

Figure 99: *Complex drainage system in Semeykin Crater. Location is Ismenius Lacus quadrangle*

are crusty, maybe due to cementing by a fluid containing minerals. There was evidence of clouds and maybe fog.

Mars Odyssey

The 2001 Mars Odyssey found much evidence for water on Mars in the form of images, and with its neutron spectrometer, it proved that much of the ground is loaded with water ice. Mars has enough ice just beneath the surface to fill Lake Michigan twice. In both hemispheres, from 55° latitude to the poles, Mars has a high density of ice just under the surface; one kilogram of soil contains about 500 grams (18 oz) of water ice. But close to the equator, there is only 2% to 10% of water in the soil. Scientists think that much of this water is also locked up in the chemical structure of minerals, such as clay and sulfates. Although the upper surface contains a few percent of chemically-bound water, ice lies just a few meters deeper, as it has been shown in Arabia Terra, Amazonis quadrangle, and Elysium quadrangle that contain large amounts of water ice. The orbiter also discovered vast deposits of bulk water ice near the surface of equatorial regions. Evidence for equatorial hydration is both morphological and compositional and is seen at both the Medusae Fossae formation and the Tharsis Montes. Analysis of the data suggests that the southern hemisphere

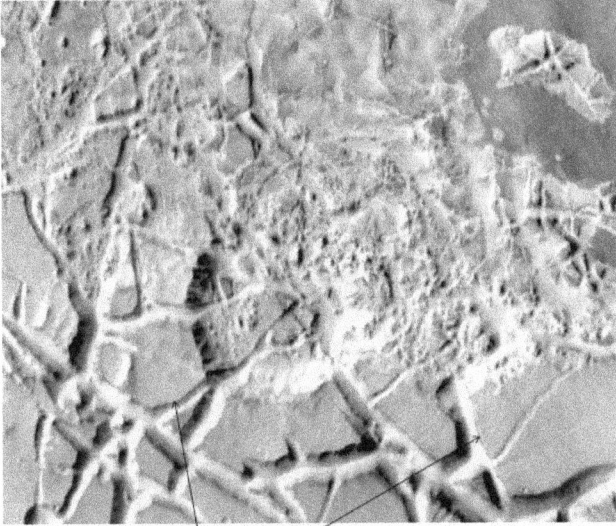

Ground ice may still be beneath the blocks and mesas.

Figure 100: *Blocks in Aram showing a possible ancient source of water. Location is Oxia Palus quadrangle*

may have a layered structure, suggestive of stratified deposits beneath a now extinct large water mass.

The instruments aboard the *Mars Odyssey* are able to study the top meter of soil. In 2002, available data were used to calculate that if all soil surfaces were covered by an even layer of water, this would correspond to a global layer of water (GLW) 0.5–1.5 kilometres (0.31–0.93 mi).

Thousands of images returned from *Odyssey* orbiter also support the idea that Mars once had great amounts of water flowing across its surface. Some images show patterns of branching valleys; others show layers that may have been formed under lakes; even river and lake deltas have been identified. For many years researchers suspected that glaciers exist under a layer of insulating rocks. Lineated valley fill is one example of these rock-covered glaciers. They are found on the floors of some channels. Their surfaces have ridged and grooved materials that deflect around obstacles. Lineated floor deposits may be related to lobate debris aprons, which have been shown by orbiting radar to contain large amounts of ice.

Figure 101: *Permafrost polygons imaged by the Phoenix lander*

Phoenix

The *Phoenix* lander also confirmed the existence of large amounts of water ice in the northern region of Mars. This finding was predicted by previous orbital data and theory, and was measured from orbit by the Mars Odyssey instruments. On June 19, 2008, NASA announced that dice-sized clumps of bright material in the "Dodo-Goldilocks" trench, dug by the robotic arm, had vaporized over the course of four days, strongly indicating that the bright clumps were composed of water ice that sublimes following exposure. Even though CO_2 (dry ice) also sublimes under the conditions present, it would do so at a rate much faster than observed. On July 31, 2008, NASA announced that *Phoenix* further confirmed the presence of water ice at its landing site. During the initial heating cycle of a sample, the mass spectrometer detected water vapor when the sample temperature reached 0 °C (32 °F; 273 K). Liquid water cannot exist on the surface of Mars with its present low atmospheric pressure and temperature, except at the lowest elevations for short periods.[111]

Perchlorate (ClO_4), a strong oxidizer, was confirmed to be in the soil. The chemical, when mixed with water, can lower the water freezing point in a manner similar to how salt is applied to roads to melt ice.

When *Phoenix* landed, the retrorockets splashed soil and melted ice onto the vehicle. Photographs showed the landing had left blobs of material stuck to

Figure 102: *View underneath Phoenix lander show-ing water ice exposed by the landing retrorockets*

the landing struts. The blobs expanded at a rate consistent with deliquescence, darkened before disappearing (consistent with liquefaction followed by dripping), and appeared to merge. These observations, combined with thermodynamic evidence, indicated that the blobs were likely liquid brine droplets. Other researchers suggested the blobs could be "clumps of frost." In 2015 it was confirmed that perchlorate plays a role in forming recurring slope lineae on steep gullies.

For about as far as the camera can see, the landing site is flat, but shaped into polygons between 2–3 metres (6 ft 7 in–9 ft 10 in) in diameter which are bounded by troughs that are 20–50 centimetres (7.9–19.7 in) deep. These shapes are due to ice in the soil expanding and contracting due to major temperature changes. The microscope showed that the soil on top of the polygons is composed of rounded particles and flat particles, probably a type of clay. Ice is present a few inches below the surface in the middle of the polygons, and along its edges, the ice is at least 8 inches (200 mm) deep.

Snow was observed to fall from cirrus clouds. The clouds formed at a level in the atmosphere that was around –65 °C (–85 °F; 208 K), so the clouds would have to be composed of water-ice, rather than carbon dioxide-ice (CO_2 or dry ice), because the temperature for forming carbon dioxide ice is much lower than –120 °C (–184 °F; 153 K). As a result of mission observations, it is now suspected that water ice (snow) would have accumulated later in the year at this location. The highest temperature measured during the mission, which took place during the Martian summer, was –19.6 °C (–3.3 °F; 253.6 K), while the coldest was –97.7 °C (–143.9 °F; 175.5 K). So, in this region the temperature remained far below the freezing point (0 °C (32 °F; 273 K)) of water.

Figure 103: *Close-up of a rock outcrop*

Mars Exploration Rovers

The Mars Exploration Rovers, *Spirit* and *Opportunity* found a great deal of evidence for past water on Mars. The Spirit rover landed in what was thought to be a large lake bed. The lake bed had been covered over with lava flows, so evidence of past water was initially hard to detect. On March 5, 2004, NASA announced that *Spirit* had found hints of water history on Mars in a rock dubbed "Humphrey".

As *Spirit* traveled in reverse in December 2007, pulling a seized wheel behind, the wheel scraped off the upper layer of soil, uncovering a patch of white ground rich in silica. Scientists think that it must have been produced in one of two ways. One: hot spring deposits produced when water dissolved silica at one location and then carried it to another (i.e. a geyser). Two: acidic steam rising through cracks in rocks stripped them of their mineral components, leaving silica behind. The *Spirit* rover also found evidence for water in the Columbia Hills of Gusev crater. In the Clovis group of rocks the Mössbauer spectrometer (MB) detected goethite, that forms only in the presence of water, iron in the oxidized form Fe^{3+}, carbonate-rich rocks, which means that regions of the planet once harbored water.

The *Opportunity* rover was directed to a site that had displayed large amounts of hematite from orbit. Hematite often forms from water. The rover indeed

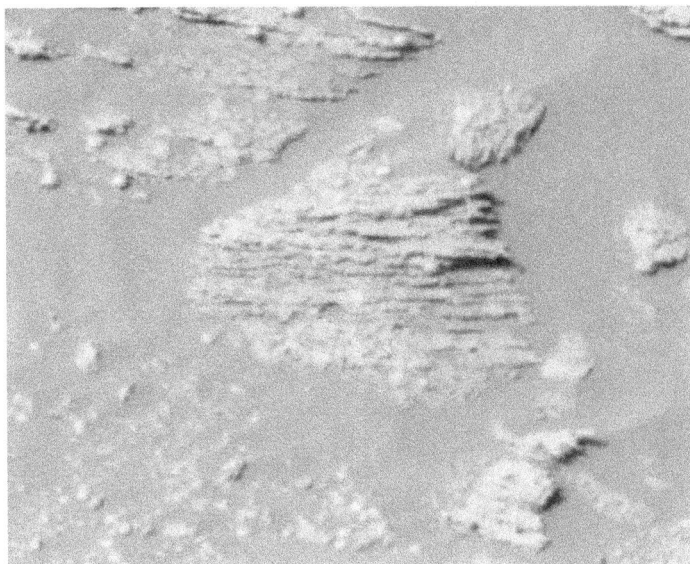

Figure 104: *Thin rock layers, not all parallel to each other*

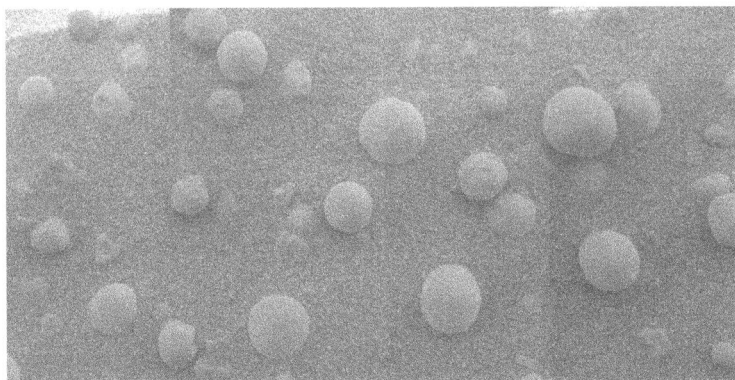

Figure 105: *Hematite spherules*

Figure 106: *Partly embedded spherules*

found layered rocks and marble- or blueberry-like hematite concretions. Else-where on its traverse, *Opportunity* investigated aeolian dune stratigraphy in Burns Cliff in Endurance Crater. Its operators concluded that the preserva-tion and cementation of these outcrops had been controlled by flow of shallow groundwater. In its years of continuous operation, *Opportunity* is still sending back evidence that this area on Mars was soaked in liquid water in the past.

The MER rovers had been finding evidence for ancient wet environments that were very acidic. In fact, what *Opportunity* has mostly discovered, or found evidence for, was sulphuric acid, a harsh chemical for life. But on May 17, 2013, NASA announced that *Opportunity* found clay deposits that typically form in wet environments that are near neutral acidity. This find provides additional evidence about a wet ancient environment possibly favorable for life.

Mars Reconnaissance Orbiter

The Mars Reconnaissance Orbiter's HiRISE instrument has taken many images that strongly suggest that Mars has had a rich history of water-related processes. A major discovery was finding evidence of ancient hot springs. If they have hosted microbial life, they may contain biosignatures. Research published in January 2010, described strong evidence for sustained precipitation in the area

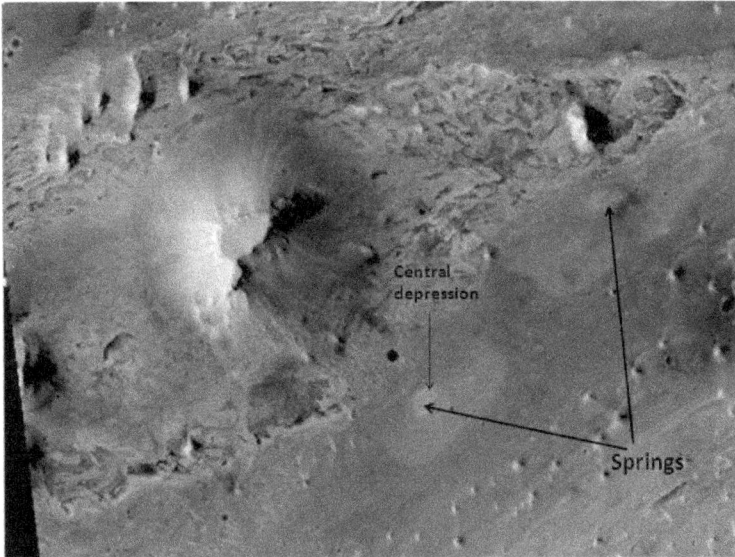

Figure 107: *Springs in Vernal Crater, as seen by HIRISE. These springs may be good places to look for evidence of past life, because hot springs can preserve evidence of life forms for a long time. Location is Oxia Palus quadrangle.*

around Valles Marineris. The types of minerals there are associated with water. Also, the high density of small branching channels indicates a great deal of precipitation.

Rocks on Mars have been found to frequently occur as layers, called strata, in many different places. Layers form by various ways, including volcanoes, wind, or water. Light-toned rocks on Mars have been associated with hydrated minerals like sulfates and clay.

The orbiter helped scientists determine that much of the surface of Mars is covered by a thick smooth mantle that is thought to be a mixture of ice and dust.

The ice mantle under the shallow subsurface is thought to result from frequent, major climate changes. Changes in Mars' orbit and tilt cause significant changes in the distribution of water ice from polar regions down to latitudes equivalent to Texas. During certain climate periods water vapor leaves polar ice and enters the atmosphere. The water returns to the ground at lower latitudes as deposits of frost or snow mixed generously with dust. The atmosphere of Mars contains a great deal of fine dust particles. Water vapor condenses on the particles, then they fall down to the ground due to the additional weight of

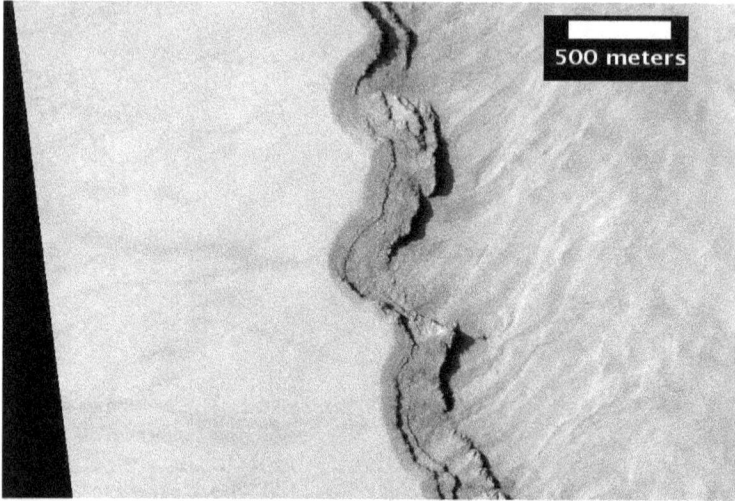

Figure 108: *Layers on the west slope of Asimov Crater. Location is Noachis quadrangle.*

the water coating. When ice at the top of the mantling layer goes back into the atmosphere, it leaves behind dust, which insulates the remaining ice.

In 2008, research with the Shallow Radar on the Mars Reconnaissance Orbiter provided strong evidence that the lobate debris aprons (LDA) in Hellas Planitia and in mid northern latitudes are glaciers that are covered with a thin layer of rocks. Its radar also detected a strong reflection from the top and base of LDAs, meaning that pure water ice made up the bulk of the formation. The discovery of water ice in LDAs demonstrates that water is found at even lower latitudes.

Research published in September 2009, demonstrated that some new craters on Mars show exposed, pure water ice. After a time, the ice disappears, evaporating into the atmosphere. The ice is only a few feet deep. The ice was confirmed with the Compact Imaging Spectrometer (CRISM) on board the Mars Reconnaissance Orbiter.

Curiosity rover

Very early in its ongoing mission, NASA's *Curiosity* rover discovered unambiguous fluvial sediments on Mars. The properties of the pebbles in these outcrops suggested former vigorous flow on a streambed, with flow between ankle- and waist-deep. These rocks were found at the foot of an alluvial fan system descending from the crater wall, which had previously been identified from orbit.

Figure 109: *"Hottah" rock outcrop – an ancient streambed discovered by the Curiosity rover team (September 14, 2012) (close-up[112]) (3-D version[113]).*

Figure 110: *Rock outcrop on Mars – compared with a terrestrial fluvial conglomerate – suggesting water "vigorously" flowing in a stream.*

In October 2012, the first X-ray diffraction analysis of a Martian soil was performed by *Curiosity*. The results revealed the presence of several minerals, including feldspar, pyroxenes and olivine, and suggested that the Martian soil in the sample was similar to the weathered basaltic soils of Hawaiian volcanoes. The sample used is composed of dust distributed from global dust storms and local fine sand. So far, the materials *Curiosity* has analyzed are consistent with the initial ideas of deposits in Gale Crater recording a transition through time from a wet to dry environment.

In December 2012, NASA reported that *Curiosity* performed its first extensive soil analysis, revealing the presence of water molecules, sulfur and chlorine in the Martian soil. And in March 2013, NASA reported evidence of mineral hydration, likely hydrated calcium sulfate, in several rock samples including the broken fragments of "Tintina" rock and "Sutton Inlier" rock as well as in veins and nodules in other rocks like "Knorr" rock and "Wernicke" rock. Analysis using the rover's DAN instrument provided evidence of subsurface water, amounting to as much as 4% water content, down to a depth of 60 cm (2.0 ft), in the rover's traverse from the *Bradbury Landing* site to the *Yellowknife Bay* area in the *Glenelg* terrain.

On September 26, 2013, NASA scientists reported the Mars *Curiosity* rover detected abundant chemically-bound water (1.5 to 3 weight percent) in soil samples at the Rocknest region of Aeolis Palus in Gale Crater. In addition, NASA reported the rover found two principal soil types: a fine-grained mafic type and a locally derived, coarse-grained felsic type. The mafic type, similar to other martian soils and martian dust, was associated with hydration of the amorphous phases of the soil. Also, perchlorates, the presence of which may make detection of life-related organic molecules difficult, were found at the *Curiosity* rover landing site (and earlier at the more polar site of the Phoenix lander) suggesting a "global distribution of these salts". NASA also reported that Jake M rock, a rock encountered by *Curiosity* on the way to Glenelg, was a mugearite and very similar to terrestrial mugearite rocks.

On December 9, 2013, NASA reported that Mars once had a large freshwater lake inside Gale Crater, that could have been a hospitable environment for microbial life.

On December 16, 2014, NASA reported detecting an unusual increase, then decrease, in the amounts of methane in the atmosphere of the planet Mars; in addition, organic chemicals were detected in powder drilled from a rock by the *Curiosity* rover. Also, based on deuterium to hydrogen ratio studies, much of the water at Gale Crater on Mars was found to have been lost during ancient times, before the lakebed in the crater was formed; afterwards, large amounts of water continued to be lost.

On April 13, 2015, *Nature* published an analysis of humidity and ground temperature data collected by *Curiosity*, showing evidence that films of liquid brine water form in the upper 5 cm of Mars's subsurface at night. The water activity and temperature remain below the requirements for reproduction and metabolism of known terrestrial microorganisms.

On October 8, 2015, NASA confirmed that lakes and streams existed in Gale crater 3.3 – 3.8 billion years ago delivering sediments to build up the lower layers of Mount Sharp.

Mars Express

The *Mars Express Orbiter*, launched by the European Space Agency, has been mapping the surface of Mars and using radar equipment to look for evidence of sub-surface water. Between 2012 and 2015, the *Orbiter* scanned the area beneath the ice caps on the Planum Australe. Scientists determined by 2018 that the readings indicated a sub-surface lake bearing water about 20 kilometres (12 mi) wide. The top of the lake is located 1.5 kilometres (0.93 mi) under the planet's surface; how much deeper the liquid water extends remains unknown.

Bibliography

* Boyce, Joseph, M. (2008). *The Smithsonian Book of Mars;* Konecky & Konecky: Old Saybrook, CT, ISBN 978-1-58834-074-0
* Carr, Michael, H. (1996). *Water on Mars;* Oxford University Press: New York, ISBN 0-19-509938-9.
* Carr, Michael, H. (2006). *The Surface of Mars;* Cambridge University Press: Cambridge, UK, ISBN 978-0-521-87201-0.
* Hartmann, William, K. (2003). *A Traveler's Guide to Mars: The Mysterious Landscapes of the Red Planet;* Workman: New York, ISBN 0-7611-2606-6.
* Hanlon, Michael (2004). *The Real Mars: Spirit, Opportunity, Mars Express and the Quest to Explore the Red Planet;* Constable: London, ISBN 1-84119-637-1.
* Kargel, Jeffrey, S. (2004). *Mars: A Warmer Wetter Planet;* Springer-Praxis: London, ISBN 1-85233-568-8.
* Morton, Oliver (2003). *Mapping Mars: Science, Imagination, and the Birth of a World;* Picador: New York, ISBN 0-312-42261-X.
* Sheehan, William (1996). *The Planet Mars: A History of Observation and Discovery;* University of Arizona Press: Tucson, AZ, ISBN 0-8165-1640-5.
* Viking Orbiter Imaging Team (1980). *Viking Orbiter Views of Mars,* C.R. Spitzer, Ed.; NASA SP-441: Washington DC.

External links

Wikimedia Commons has media related to *Water on Mars*.

Wikinews has related news: *NASA announces water on Mars*

- NASA – *Curiosity* Rover Finds Evidence For An Ancient Streambed – September, 2012[114]
- Images – Signs Of Water On Mars[115] (HiRISE)
- Video (02:01) – Liquid Flowing Water Discovered on Mars – August, 2011[116]
- Video (04:32) – Evidence: Water "Vigorously" Flowed On Mars – September, 2012[117]
- Video (03:56) – Measuring Mars' Ancient Ocean – March, 2015[118]

Polar caps

Martian polar ice caps

<templatestyles src="Multiple_image/styles.css" />

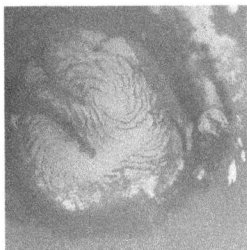

North polar cap in 1999

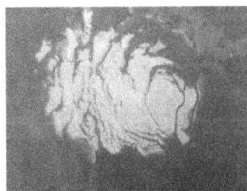

South polar cap in 2000

The planet Mars has two permanent polar ice caps. During a pole's winter, it lies in continuous darkness, chilling the surface and causing the deposition of 25–30% of the atmosphere into slabs of CO_2 ice (dry ice). When the poles are again exposed to sunlight, the frozen CO_2 sublimes. These seasonal actions transport large amounts of dust and water vapor, giving rise to Earth-like frost and large cirrus clouds. Clouds of water-ice were photographed by the *Opportunity* rover in 2004.

The caps at both poles consist primarily of water ice. Frozen carbon dioxide accumulates as a comparatively thin layer about one metre thick on the north cap in the northern winter, while the south cap has a permanent dry ice cover about 8 m thick. The northern polar cap has a diameter of about 1000 km during the northern Mars summer, and contains about 1.6 million cubic km of ice, which if spread evenly on the cap would be 2 km thick. (This compares to a volume of 2.85 million cubic km (km^3) for the Greenland ice sheet.) The southern polar cap has a diameter of 350 km and a thickness of 3 km. The total volume of ice in the south polar cap plus the adjacent layered deposits has also been estimated at 1.6 million cubic km. Both polar caps show spiral troughs, which recent analysis of SHARAD ice penetrating radar has shown are a result of roughly perpendicular katabatic winds that spiral due to the Coriolis Effect.

The seasonal frosting of some areas near the southern ice cap results in the formation of transparent 1 m thick slabs of dry ice above the ground. With the arrival of spring, sunlight warms the subsurface and pressure from subliming CO_2 builds up under a slab, elevating and ultimately rupturing it. This leads to geyser-like eruptions of CO_2 gas mixed with dark basaltic sand or dust. This process is rapid, observed happening in the space of a few days, weeks or months, a rate of change rather unusual in geology—especially for Mars. The gas rushing underneath a slab to the site of a geyser carves a spider-like pattern of radial channels under the ice.

In July 2018, Italian scientists reported the discovery of a subglacial lake on Mars, 1.5 km (0.93 mi) below the surface of the southern polar layered deposits (not under the visible permanent ice cap), and about 20 km (12 mi) across, the first known stable body of water on the planet.

Freezing of atmosphere

Research based on slight changes in the orbits of spacecraft around Mars over 16 years found that when one hemisphere experiences winter, approximately 3 trillion to 4 trillion tons of carbon dioxide freezes out of the atmosphere onto the northern and southern polar caps. This represents 12 to 16 percent of the mass of the entire Martian atmosphere. These observations support predictions from the Mars Global Reference Atmospheric Model—2010.

Layers

Both polar caps show layered features, called polar-layered deposits, that result from seasonal melting and deposition of ice together with dust from Martian dust storms. Information about the past climate of Mars may be eventually revealed in these layers, just as tree ring patterns and ice core data do on Earth. Both polar caps also display grooved features, probably caused by wind flow patterns. The grooves are also influenced by the amount of dust.[119] The more dust, the darker the surface. The darker the surface, the more melting. Dark surfaces absorb more light energy. There are other theories that attempt to explain the large grooves.

Layers in northern ice cap, as seen by HiRISE under HiWish program

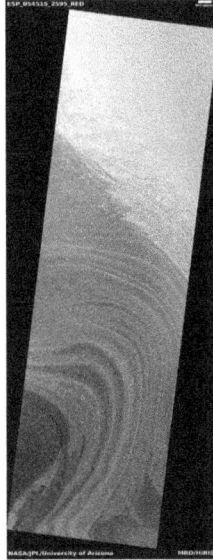

Figure 111: *Layers exposed in northern ice cap, as seen by HiRISE under HiWish program*

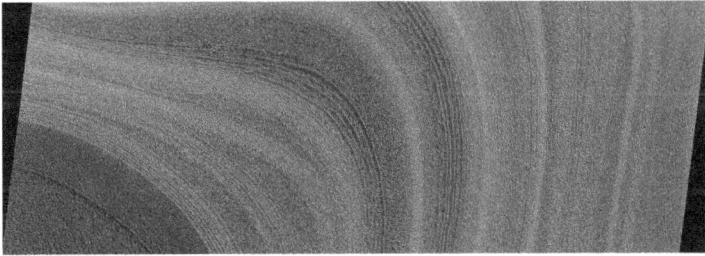

Figure 112: *Close view of layers exposed in northern ice cap, as seen by HiRISE under HiWish program*

North polar cap

The bulk of the northern ice cap consists of water ice; it also has a thin seasonal veneer of dry ice, solid carbon dioxide. Each winter the ice cap grows by adding 1.5 to 2 m of dry ice. In summer, the dry ice sublimates (goes directly from a solid to a gas) into the atmosphere. Mars has seasons that are similar to Earth's, because its rotational axis has a tilt close to our own Earth's (25.19° for Mars, 23.44° for Earth).

During each year on Mars as much as a third of Mars' thin carbon dioxide (CO_2) atmosphere "freezes out" during the winter in the northern and southern hemispheres. Scientists have even measured tiny changes in the gravity field of Mars due to the movement of carbon dioxide.

The ice cap in the north is of a lower altitude (base at -5000 m, top at -2000 m) than the one in the south (base at 1000 m, top at 3500 m). It is also warmer, so all the frozen carbon dioxide disappears each summer. The part of the cap that survives the summer is called the north residual cap and is made of water ice. This water ice is believed to be as much as three kilometers thick. The much thinner seasonal cap starts to form in the late summer to early fall when a variety of clouds form. Called the polar hood, the clouds drop precipitation which thickens the cap. The north polar cap is symmetrical around the pole and covers the surface down to about 60 degrees latitude. High resolution images taken with NASA's Mars Global Surveyor show that the northern polar cap is covered mainly by pits, cracks, small bumps and knobs that give it a cottage cheese look. The pits are spaced close together relative to the very different depressions in the south polar cap.

Both polar caps show layered features that result from seasonal melting and deposition of ice together with dust from Martian dust storms. These polar layered deposits lie under the permanent polar caps. Information about the past climate of Mars may be eventually revealed in these layers, just as tree ring patterns and ice core data do on Earth. Both polar caps also display grooved features, probably caused by wind flow patterns and sun angles, although there are several theories that have been advanced. The grooves are also influenced by the amount of dust. The more dust, the darker the surface. The darker the surface, the more melting. Dark surfaces absorb more light energy. One large valley, Chasma Boreale runs halfway across the cap. It is about 100 km wide and up to 2 km deep—that's deeper than Earth's Grand Canyon.

When the tilt or obliquity changes the size of the polar caps change. When the tilt is at its highest, the poles receive far more sunlight and for more hours each day. The extra sunlight causes the ice to melt, so much so that it could cover parts of the surface in 10 m of ice. Much evidence has been found for glaciers that probably formed when this tilt-induced climate change occurred.

Research reported in 2009 shows that the ice rich layers of the ice cap match models for Martian climate swings. NASA's Mars Reconnaissance Orbiter's radar instrument can measure the contrast in electrical properties between layers. The pattern of reflectivity reveals the pattern of material variations within the layers. Radar produced a cross-sectional view of the north-polar layered deposits of Mars. High-reflectivity zones, with multiple contrasting layers, alternate with zones of lower reflectivity. Patterns of how these two types of zones alternate can be correlated to models of changes in the tilt of Mars. Since the top zone of the north-polar layered deposits—the most recently deposited portion—is strongly radar-reflective, the researchers propose that such sections of high-contrast layering correspond to periods of relatively small swings in the planet's tilt because the Martian axis has not varied much recently. Dustier layers appear to be deposited during periods when the atmosphere is dustier.

Research, published in January 2010 using HiRISE images, says that understanding the layers is more complicated than was formerly believed. The brightness of the layers does not just depend on the amount of dust. The angle of the sun together with the angle of the spacecraft greatly affect the brightness seen by the camera. This angle depends on factors such as the shape of the trough wall and its orientation. Furthermore, the roughness of the surface can greatly change the albedo (amount of reflected light). In addition, many times what one is seeing is not a real layer, but a fresh covering of frost. All of these factors are influenced by the wind which can erode surfaces. The HiRISE camera did not reveal layers that were thinner than those seen by the Mars Global Surveyor. However, it did see more detail within layers.

Radar measurements of the north polar ice cap found the volume of water ice in the layered deposits of the cap was 821,000 cubic kilometers (197,000 cubic miles). That's equal to 30% of the Earth's Greenland ice sheet. (The layered deposits overlie an additional basal deposit of ice.) The radar is on board the Mars Reconnaissance Orbiter.

SHARAD radar data when combined to form a 3D model reveal buried craters. These may be used to date certain layers.

In February 2017, ESA released a new view of Mars's North Pole. It was a mosaic made from 32 individual orbits of the Mars Express.[120,121]

South polar cap

The south polar permanent cap is much smaller than the one in the north. It is 400 km in diameter, as compared to the 1100 km diameter of the northern cap. Each southern winter, the ice cap covers the surface to a latitude of 50°. Part of the ice cap consists of dry ice, solid carbon dioxide. Each winter the ice cap grows by adding 1.5 to 2 meters of dry ice from precipitation from a polar-hood of clouds. In summer, the dry ice sublimates (goes directly from a solid to a gas) into the atmosphere. During each year on Mars as much as a third of Mars' thin carbon dioxide (CO_2) atmosphere "freezes out" during the winter in the northern and southern hemispheres. Scientists have even measured tiny changes in the gravity field of Mars due to the movement of carbon dioxide. In other words, the winter buildup of ice changes the gravity of the planet. Mars has seasons that are similar to Earth's because its rotational axis has a tilt close to our own Earth's (25.19° for Mars, 23.45° for Earth). The south polar cap is higher in altitude and colder than the one in the north.

The residual southern ice cap is displaced; that is, it is not centered on the south pole. However, the south seasonal cap is centered near the geographic pole. Studies have shown that the off center cap is caused by much more snow falling on one side than the other. On the western hemisphere side of the south pole a low pressure system forms because the winds are changed by the Hellas Basin. This system produces more snow. On the other side, there is less snow and more frost. Snow tends to reflect more sunlight in the summer, so not much melts or sublimates (Mars climate causes snow to go directly from a solid to a gas). Frost, on the other hand has a rougher surface and tends to trap more sunlight, resulting in more sublimation. In other words, areas with more of the rougher frost are warmer.[122]

Research, published in April 2011, described a large deposit of frozen carbon dioxide near the south pole. Most of this deposit probably enters Mars' atmosphere when the planet's tilt increases. When this occurs, the atmosphere thickens, winds get stronger, and larger areas on the surface can support liquid water.[123] Analysis of data showed that if these deposits were all changed into gas, the atmospheric pressure on Mars would double.[124] There are three layers of these deposits; each is capped with a 30-meter layer of water ice that prevents the CO_2 from sublimating into the atmosphere. In sublimation a solid material goes directly into a gas phase. These three layers are linked to periods when the atmosphere collapsed when the climate changed.[125]

A large field of eskers exist around the south pole, called the Dorsa Argentea Formation, it is believed to be the remains of a giant ice sheet.[126] This large polar ice sheet is believed to have covered about 1.5 million square kilometers. That area is twice the area of the state of Texas.[127]WP:NOTRS[128]

In July 2018 ESA discovered indications of liquid salt water buried under layers of ice and dust by analyzing the reflection of radar pulses generated by Mars Express.

Swiss cheese appearance

While the north polar cap of Mars has a flat, pitted surface resembling cottage cheese, the south polar cap has larger pits, troughs and flat mesas that give it a Swiss cheese appearance.[129,130,131,132] The upper layer of the Martian south polar residual cap has been eroded into flat-topped mesas with circular depressions.[133] Observations made by Mars Orbiter Camera in 2001 have shown that the scarps and pit walls of the south polar cap had retreated at an average rate of about 3 meters (10 feet) since 1999. In other words, they were retreating 3 meters per Mars year. In some places on the cap, the scarps retreat less than 3 meters a Mars year, and in others it can retreat as much as 8 meters (26 feet) per Martian year. Over time, south polar pits merge to become plains, mesas turn into buttes, and buttes vanish forever. The round shape is probably aided in its formation by the angle of the sun. In the summer, the sun moves around the sky, sometimes for 24 hours each day, just above the horizon. As a result, the walls of a round depression will receive more intense sunlight than the floor; the wall will melt far more than the floor. The walls melt and recede, while the floor remains the same.[134,135]

Later research with the powerful HiRISE showed that the pits are in a 1-10 meter thick layer of dry ice that is sitting on a much larger water ice cap. Pits have been observed to begin with small areas along faint fractures. The circular pits have steep walls that work to focus sunlight, thereby increasing erosion. For a pit to develop a steep wall of about 10 cm and a length of over 5 meters in necessary.[136]

The pictures below show why it is said the surface resembles Swiss cheese; one can also observe the differences over a two-year period.

Starburst channels or spiders

Starburst channels are patterns of channels that radiate out into feathery extensions. They are caused by gas which escapes along with dust. The gas builds up beneath translucent ice as the temperature warms in the spring.[137] Typically 500 meters wide and 1 meter deep, the spiders may undergo observable changes in just a few days.[138] One model for understanding the formation of the spiders says that sunlight heats dust grains in the ice. The warm dust grains settle by melting through the ice while the holes are annealed behind them. As a result, the ice becomes fairly clear. Sunlight then reaches the dark bottom of the slab of ice and changes the solid carbon dioxide ice into a gas which flows toward higher regions that open to the surface. The gas rushes out carrying dark dust with it. Winds at the surface will blow the escaping gas and dust into dark fans that we observe with orbiting spacecraft. The physics of this model is similar to ideas put forth to explain dark plumes erupting from the surface of Triton.

Research, published in January 2010 using HiRISE images, found that some of the channels in spiders grow larger as they go uphill since gas is doing the erosion. The researchers also found that the gas flows to a crack that has occurred at a weak point in the ice. As soon as the sun rises above the horizon, gas from the spiders blows out dust which is blown by wind to form a dark fan shape. Some of the dust gets trapped in the channels. Eventually frost covers all the fans and channels until the next spring when the cycle repeats.

Layers

Chasma Australe, a major valley, cuts across the layered deposits in the South Polar cap. On the 90 E side, the deposits rest on a major basin, called Prometheus.

Some of the layers in the south pole also show polygonal fracturing in the form of rectangles. It is thought that the fractures were caused by the expansion and contraction of water ice below the surface.[139]

Polar ice cap deuterium enrichment

Evidence that Mars once had enough water to create a global ocean at least 137 m deep has been obtained from measurement of the HDO to H_2O ratio over the north polar cap. In March 2015, a team of scientists published results showing that the polar cap ice is about eight times as enriched with deuterium, heavy hydrogen, as water in Earth's oceans. This means that Mars has lost a volume of water 6.5 times as large as that stored in today's polar caps. The water for a time may have formed an ocean in the low-lying Vastitas Borealis and adjacent lowlands (Acidalia, Arcadia and Utopia planitiae). Had the water ever all been liquid and on the surface, it would have covered 20% of the planet and in places would have been almost a mile deep.

This international team used ESO's Very Large Telescope, along with instruments at the W. M. Keck Observatory and the NASA Infrared Telescope Facility, to map out different isotopic forms of water in Mars's atmosphere over a six-year period.

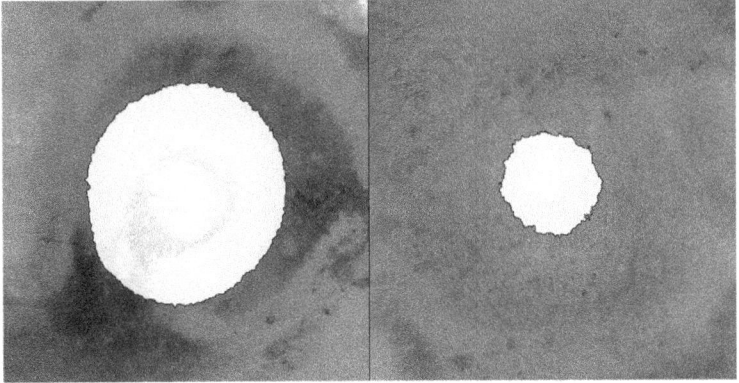

Figure 113:
Extents of north (left) and south (right) polar CO_2 ice during a martian year

Gallery

Ice cap images

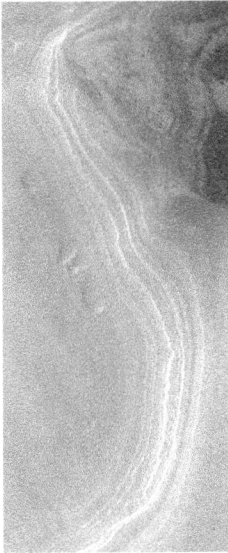

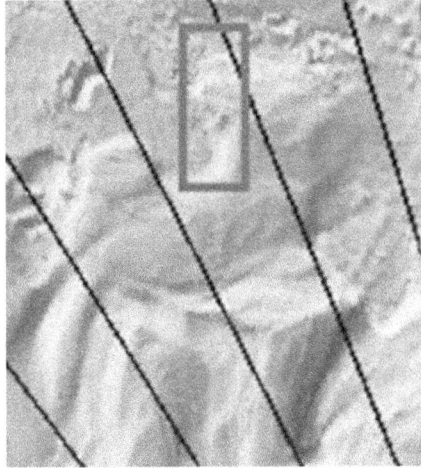

Picture to the left is in red box.

Figure 114: *South polar layers, as seen by THEMIS.*

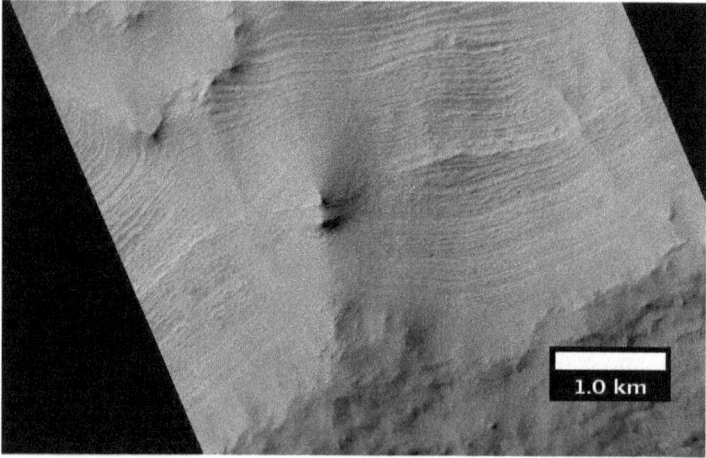

Figure 115: *Close-up of layers in wall of McMurdo crater, as seen by HiRISE.*

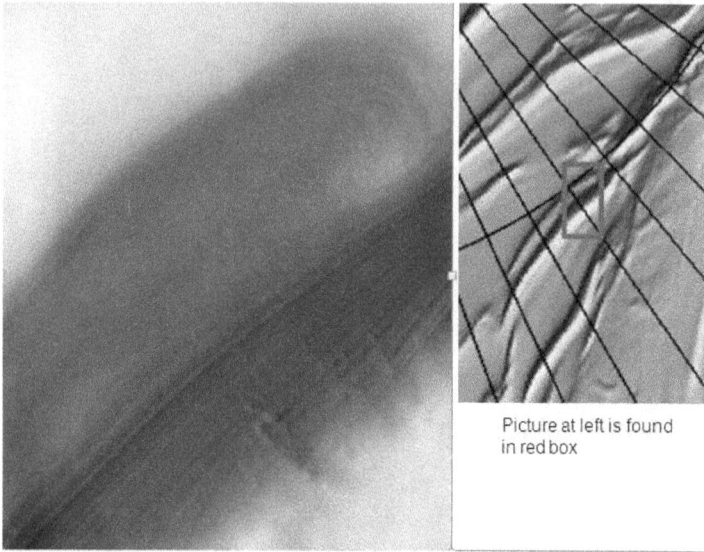

Picture at left is found
in red box

Figure 116: *Layers exposed in a valley on the north polar
ice cap as observed by Mars Odyssey. Click on image to en-
large to see clouds of dust caused by winds coming off the cap.*

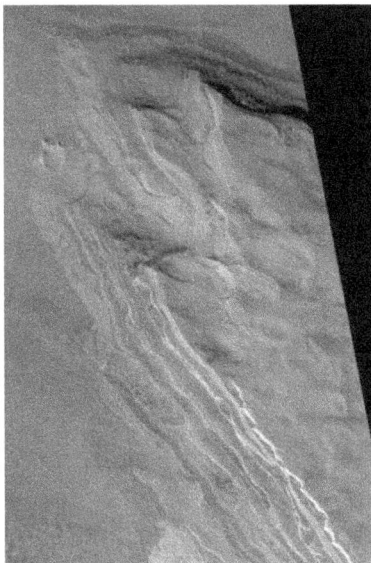

Figure 117: *Chasma Boreale streamlined feature, as seen by HiRISE.*

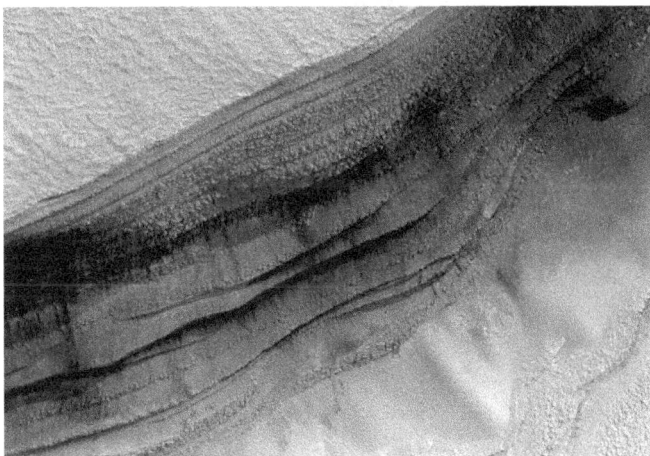

Figure 118: *Chasma Boreale, as seen by HiRISE.*

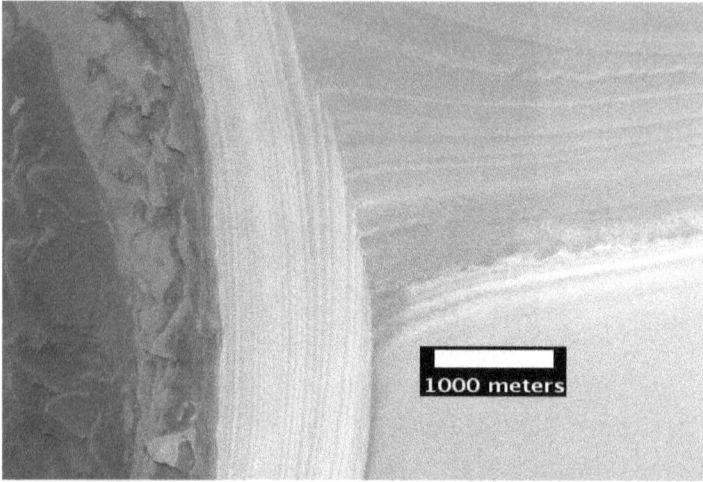

Figure 119: *North polar layers on the side of a valley, as seen by HiRISE. Layers erode differently, depending on what direction they face. On one side they are straight, as if cut by a knife.*

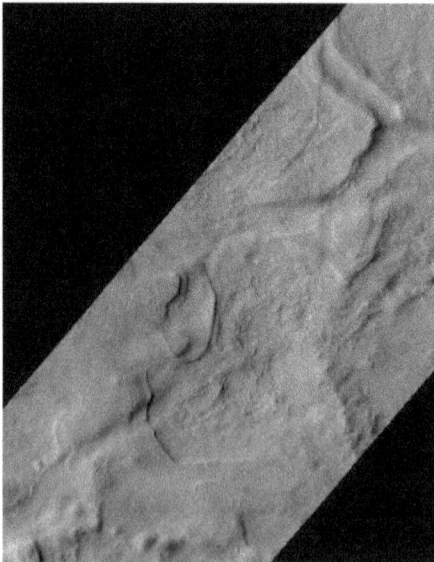

Figure 120: *Chasma Boreale channels, as seen by HiRISE.*

External links

- http://seg.org/podcast/Post/4604/Episode-10-Remote-sensing-on-Mars Podcast describing the use of SHARAD radar data to explore the ice caps
- https://sharad.psi.edu/3D/movies/SHARAD_PB3D_depth_20161223. mp4 High-resolution movie showing the interior of the north polar ice cap in 3D, as determined with SHARAD radar data
- https://sharad.psi.edu/3D/movies/SHARAD_PA3D_depth_20170105. mp4 High-resolution movie showing the interior of the south polar ice cap in 3D, as determined with SHARAD radar data

Geography of Mars

Geography of Mars

The **geography of Mars**, also known as **areography**, entails the delineation and characterization of regions on Mars. Martian geography is mainly focused on what is called physical geography on Earth; that is the distribution of physical features across Mars and their cartographic representations.

History

The first observations of Mars were from ground-based telescopes. The history of these observations are marked by the oppositions of Mars, when the planet is closest to Earth and hence is most easily visible, which occur every couple of years. Even more notable are the perihelic oppositions of Mars which occur approximately every 16 years, and are distinguished because Mars is close to perihelion making it even closer to Earth.

In September 1877, (a perihelic opposition of Mars occurred on September 5), Italian astronomer Giovanni Schiaparelli published the first detailed map of Mars. These maps notably contained features he called *canali* ("channels"), that were later shown to be an optical illusion. These *canali* were supposedly long straight lines on the surface of Mars to which he gave names of famous rivers on Earth. His term was popularly mistranslated as *canals*, and so started the Martian canal controversy.

Following these observations, it was a long-held belief that Mars contained vast seas and vegetation. It was not until spacecraft visited the planet during NASA's Mariner missions in the 1960s that these myths were dispelled. Some maps of Mars were made using the data from these missions, but it wasn't until the Mars Global Surveyor mission, launched in 1996 and ending in late 2006, that complete, extremely detailed maps were obtained. These maps are now available online at http://www.google.com/mars/

Figure 121: *High-resolution colorized map of Mars based on Viking orbiter images. Surface frost and water ice fog brighten the impact basin Hellas to the right of lower center; Syrtis Major just above it is darkened by winds that sweep dust off its basaltic surface. Residual north and south polar ice caps are shown at upper and lower right as they appear in early summer and at minimum size, respectively.*

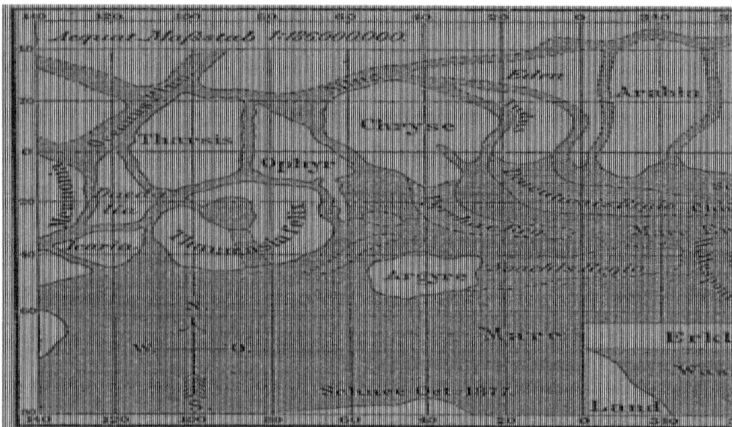

Figure 122: *Map of Mars by Giovanni Schiaparelli. North is at the top of this map; however, in most maps of Mars drawn before space exploration the convention among astronomers was to put south at the top because the telescopic image of a planet is inverted.*

Figure 123: *High resolution topographic map of Mars based on the Mars Global Surveyor laser altimeter research led by Maria Zuber and David Smith. North is at the top. Notable features include the Tharsis volcanoes in the west (including Olympus Mons), Valles Marineris to the east of Tharsis, and Hellas basin in the southern hemisphere.*

Cartography

The United States Geological Survey defines thirty cartographic quadrangles for the surface of Mars. These can be seen below.

File:MGS_MOC_Wide_Angle_Map_of_Mars_PIA03467.jpg

The thirty cartographic quadrangles of Mars, defined by the United States Geological Survey. The quadrangles are numbered with the prefix "MC" for "Mars Chart." Click on a quadrangle name link and you will be taken to the corresponding article. North is at the top; 0°N 180°W[140] is at the far left on the equator. The map images were taken by the Mars Global Surveyor.

[

- view
- talk

]

Topography

Given that it is a planet, the geography of Mars varies considerably. However, the dichotomy of **Martian topography** is striking: northern plains flattened by lava flows contrast with the southern highlands, pitted and cratered by ancient impacts. The surface of Mars as seen from Earth is consequently divided into two kinds of areas, with differing albedo. The paler plains covered with

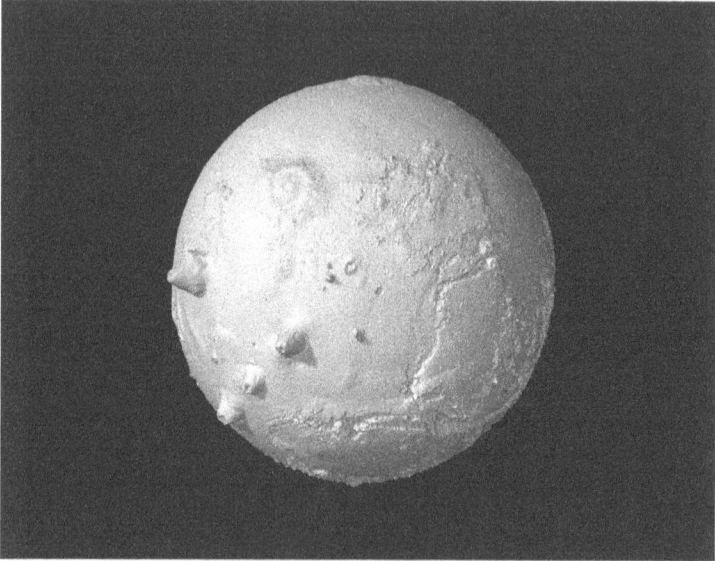

Figure 124: *STL 3D model of Mars with 20× elevation exaggeration using data from the Mars Global Surveyor Mars Orbiter Laser Altimeter.*

Figure 125: *Mars, 2001, with the southern polar ice cap visible on the bottom.*

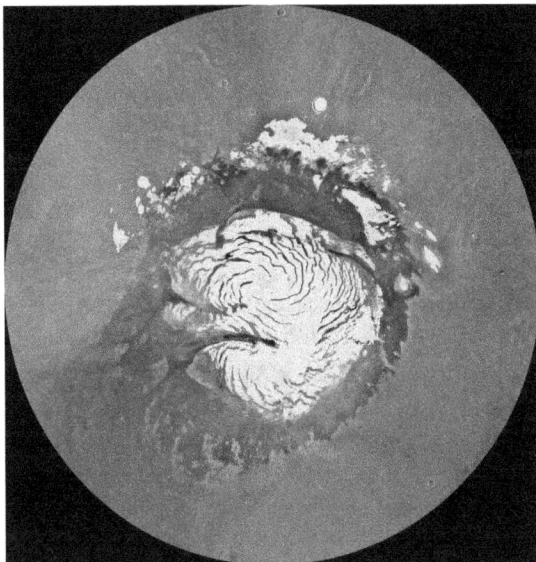

Figure 126: *North Polar region with icecap.*

dust and sand rich in reddish iron oxides were once thought of as Martian 'continents' and given names like Arabia Terra (*land of Arabia*) or Amazonis Planitia (*Amazonian plain*). The dark features were thought to be seas, hence their names Mare Erythraeum, Mare Sirenum and Aurorae Sinus. The largest dark feature seen from Earth is Syrtis Major Planum.

The shield volcano, Olympus Mons (*Mount Olympus*), rises 22 km above the surrounding volcanic plains, and is the highest known mountain on any planet in the solar system.[141] It is in a vast upland region called Tharsis, which contains several large volcanos. See list of mountains on Mars. The Tharsis region of Mars also has the solar system's largest canyon system, Valles Marineris or the *Mariner Valley*, which is 4,000 km long and 7 km deep. Mars is also scarred by countless impact craters. The largest of these is the Hellas impact basin. See list of craters on Mars.

Mars has two permanent polar ice caps, the northern one located at Planum Boreum and the southern one at Planum Australe.

The difference between Mars' highest and lowest points is nearly 30 km (from the top of Olympus Mons at an altitude of 21.2 km to the bottom of the Hellas impact basin at an altitude of 8.2 km below the datum). In comparison, the difference between Earth's highest and lowest points (Mount Everest and the

Mariana Trench) is only 19.7 km. Combined with the planets' different radii, this means Mars is nearly three times "rougher" than Earth.

The International Astronomical Union's Working Group for Planetary System Nomenclature is responsible for naming Martian surface features.

Zero elevation

On Earth, the zero elevation datum is based on sea level. Since Mars has no oceans and hence no 'sea level', it is convenient to define an arbitrary zero-elevation level or "datum" for mapping the surface. The datum for Mars is arbitrarily defined in terms of a constant atmospheric pressure.

From the Mariner 9 mission up until 2001, this was chosen as 610.5 Pa (6.105 mbar), on the basis that below this pressure liquid water can never be stable (i.e., the triple point of water is at this pressure). This value is only 0.6% of the pressure at sea level on Earth. Note that the choice of this value does not mean that liquid water does exist below this elevation, just that it could were the temperature to exceed 273.16 K (0.01 degrees C, 32.018 degrees F).

In 2001, Mars Orbiter Laser Altimeter data led to a new convention of zero elevation defined as the equipotential surface (gravitational plus rotational) whose average value at the equator is equal to the mean radius of the planet.

Zero meridian

Mars' equator is defined by its rotation, but the location of its prime meridian was specified, as was Earth's, by choice of an arbitrary point which was accepted by later observers. The German astronomers Wilhelm Beer and Johann Heinrich Mädler selected a small circular feature in the Sinus Meridiani ('Middle Bay' or 'Meridian Bay') as a reference point when they produced the first systematic chart of Mars features in 1830–32. In 1877, their choice was adopted as the prime meridian by the Italian astronomer Giovanni Schiaparelli when he began work on his notable maps of Mars. In 1909 the ephemeris makers decided that it was more important to maintain continuity of the ephemerides as a guide to observations and this definition was "virtually abandoned."

After the Mariner spacecraft provided extensive imagery of Mars, in 1972 the Mariner 9 Geodesy/Cartography Group proposed that the prime meridian passed through the center of a small 500 m diameter crater (named Airy-0), located in Sinus Meridiani along the meridian line of Beer and Mädler, thus defining 0.0° longitude with a precision of 0.001°. This model used the planetographic control point network developed by Merton Davies of the RAND Corporation.

As radiometric techniques increased the precision with which objects could be located on the surface of Mars, the center of a 500 m circular crater was considered to be insufficiently precise for exact measurements. The IAU Working Group on Cartographic Coordinates and Rotational Elements therefore recommended setting the longitude of the Viking 1 lander, for which there were extensive radiometric tracking data, as marking the standard longitude of 47.95137° west. This definition maintains the position of the center of Airy-0 at 0° longitude, within the tolerance of current cartographic uncertainties.

Martian dichotomy

Observers of Martian topography will notice a dichotomy between the northern and southern hemispheres. Most of the northern hemisphere is flat, with few impact craters, and lies below the conventional 'zero elevation' level. In contrast, the southern hemisphere is mountains and highlands, mostly well above zero elevation. The two hemispheres differ in elevation by 1 to 3 km. The border separating the two areas is very interesting to geologists.

One distinctive feature is the fretted terrain.[142] It contains mesas, knobs, and flat-floored valleys having walls about a mile high. Around many of the mesas and knobs are lobate debris aprons that have been shown to be rock-covered glaciers.[143]

Other interesting features are the large river valleys and outflow channels that cut through the dichotomy.[144,145,146]

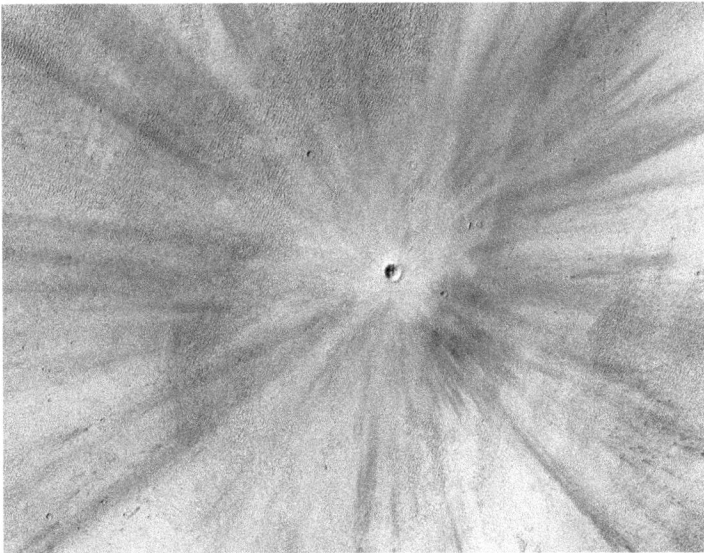

Figure 127: *Fresh Impact crater on Mars (November 19, 2013).*

Fretted terrain north of Arabia

Fretted terrain contains a mix of cliffs, mesas, buttes, straight-walled and
sinuous canyons. The cliffs are sometimes more than a mile high.

Image credit: Malin Space ScienceSystems/NASA

Figure 128: *Fretted terrain of Ismenius Lacus showing flat floored valleys and
cliffs. Photo taken with Mars Orbiter Camera (MOC) on the Mars Global Surveyor.*

Cliff from bottom of S02-
00191. This cliff is over a half
mile high. Some cliffs in this
region are over a mile high.
The wrinkles may be evidence
of movement. Such cliffs or
scarps are common in fretted
terrain on Mars.

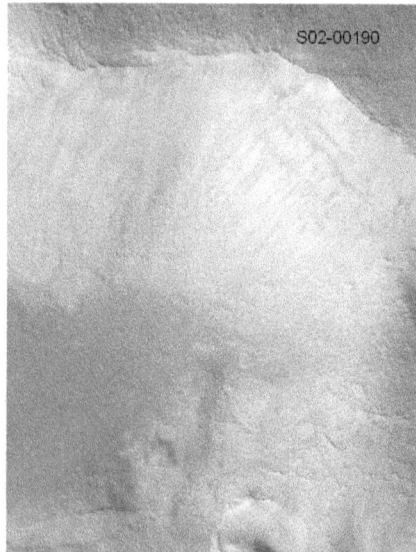

Photo credit: Malin
Space Science
Systems/NASA

Figure 129: *Enlargement of the photo on the left showing cliff. Photo
taken with high resolution camera of Mars Global Surveyor (MGS).*

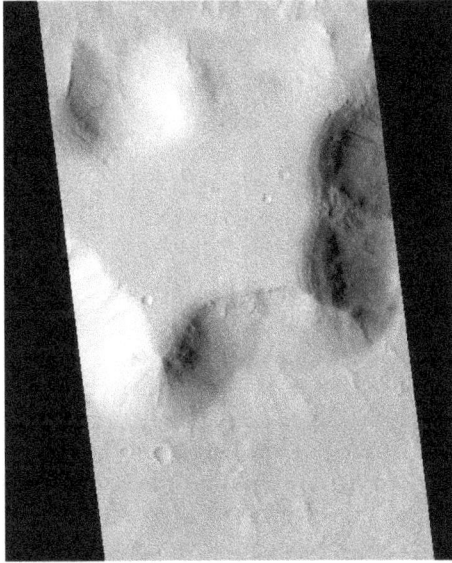

Figure 130: *View of lobate debris apron along a slope. Image located in Arcadia quadrangle.*

500 meters

Figure 131: *Place where a lobate debris apron begins. Note stripes which indicate movement. Image located in Ismenius Lacus quadrangle.*

The northern lowlands comprise about one-third of the surface of Mars and are relatively flat, with occasional impact craters. The other two-thirds of the Martian surface are the southern highlands. The difference in elevation between the hemispheres is dramatic. Because of the density of impact craters, scientists believe the southern hemisphere to be far older than the northern plains.[147] Much of heavily cratered southern highlands date back to the period of heavy bombardment, the Noachian.

Multiple hypotheses have been proposed to explain the differences. The three most commonly accepted are a single mega-impact, multiple impacts, and endogenic processes such as mantle convection.[148] Both impact-related hypotheses involve processes that could have occurred before the end of the primordial bombardment, implying that the crustal dichotomy has its origins early in the history of Mars.

The giant impact hypothesis, originally proposed in the early 1980s, was met with skepticism due to the impact area's non-radial (elliptical) shape, where a circular pattern would be stronger support for impact by larger object(s). But a 2008 study[149] provided additional research that supports a single giant impact. Using geologic data, researchers found support for the single impact of a large object hitting Mars at approximately a 45-degree angle. Additional evidence analyzing Martian rock chemistry for post-impact upwelling of mantle material would further support the giant impact theory.

Nomenclature

Early nomenclature

Although better remembered for mapping the Moon starting in 1830, Johann Heinrich Mädler and Wilhelm Beer were the first "areographers". They started off by establishing once and for all that most of the surface features were permanent, and pinned down Mars' rotation period. In 1840, Mädler combined ten years of observations and drew the first map of Mars ever made. Rather than giving names to the various markings they mapped, Beer and Mädler simply designated them with letters; Meridian Bay (Sinus Meridiani) was thus feature "a".

Over the next twenty years or so, as instruments improved and the number of observers also increased, various Martian features acquired a hodge-podge of names. To give a couple of examples, Solis Lacus was known as the "Oculus" (the Eye), and Syrtis Major was usually known as the "Hourglass Sea" or the "Scorpion". In 1858, it was also dubbed the "Atlantic Canale" by the Jesuit astronomer Angelo Secchi. Secchi commented that it "seems to play the role of the Atlantic which, on Earth, separates the Old Continent from the New" —this was the first time the fateful *canale*, which in Italian can mean either "channel" or "canal", had been applied to Mars.

In 1867, Richard Anthony Proctor drew up a map of Mars-based, somewhat crudely, on the Rev. William Rutter Dawes' earlier drawings of 1865, then the best ones available. Proctor explained his system of nomenclature by saying, "I have applied to the different features the names of those observers who have studied the physical peculiarities presented by Mars." Here are some of his names, paired with those later used by Schiaparelli in his Martian map created between 1877 and 1886.[150] Schiaparelli's names were generally adopted and are the names actually used today:

Proctor nomenclature	Schiaparelli nomenclature
Kaiser Sea	Syrtis Major
Lockyer Land	Hellas Planitia
Main Sea	Lacus Moeris
Herschel II Strait	Sinus Sabaeus
Dawes Continent	Aeria and Arabia
De La Rue Ocean	Mare Erythraeum
Lockyer Sea	Solis Lacus
Dawes Sea	Tithonius Lacus
Madler Continent	Chryse Planitia, Ophir, Tharsis
Maraldi Sea	Maria Sirenum and Cimmerium
Secchi Continent	Memnonia
Hooke Sea	Mare Tyrrhenum
Cassini Land	Ausonia
Herschel I Continent	Zephyria, Aeolis, Aethiopis
Hind Land	Libya

Proctor's nomenclature has often been criticized, mainly because so many of his names honored English astronomers, but also because he used many names more than once. In particular, Dawes appeared no fewer than *six* times (Dawes Ocean, Dawes Continent, Dawes Sea, Dawes Strait, Dawes Isle, and Dawes Forked Bay). Even so, Proctor's names are not without charm, and for all their shortcomings they were a foundation on which later astronomers would improve.

Modern nomenclature

Today, names of Martian features derive from a number of sources, but the names of the large features are derived primarily from the maps of Mars made in 1886 by the Italian astronomer Giovanni Schiaparelli. Schiaparelli named the larger features of Mars primarily using names from Greek mythology and to a lesser extent the Bible. Mars Large albedo features retain many of the older names, but are often updated to reflect new knowledge of the nature of the features. For example, 'Nix Olympica' (the snows of Olympus) has become Olympus Mons (Mount Olympus).

Large Martian craters are named after important scientists and science fiction writers; smaller ones are named after towns and villages on Earth.

Various landforms studied by the Mars Exploration Rovers are given temporary names or nicknames to identify them during exploration and investigation. However, it is hopedWikipedia:Attribution needed that the International Astronomical Union will make permanent the names of certain major features, such as the Columbia Hills, which were named after the seven astronauts who died in the Space Shuttle *Columbia* disaster.

Further reading

- Sheehan, William, "The Planet Mars: A History of Observation and Discovery"[151] (Full text online) The University of Arizona Press, Tucson. 1996.

External links

- Google Mars[152] – Google Maps for Mars, with various surface features and interesting places pointed out
- Mars Maps[153] – Maps of Mars
- MEC-1 Prototype[154]
- Historical Globes of the Red Planet[155]
- 3D Map of Mars[156] – 3D Map of Mars
- Presents distances and altitudes of features/NASA[157]
- The Origin of Mars Crater Names[158]

Volcanology of Mars

Volcanology of Mars

Volcanic activity, or volcanism, has played a significant role in the geologic evolution of Mars.[159] Scientists have known since the Mariner 9 mission in 1972 that volcanic features cover large portions of the Martian surface. These features include extensive lava flows, vast lava plains, and the largest known volcanoes in the Solar System. Martian volcanic features range in age from Noachian (>3.7 billion years) to late Amazonian (< 500 million years), indicating that the planet has been volcanically active throughout its history, and some speculate it probably still is so today. Both Earth and Mars are large, differentiated planets built from similar chondritic materials.[160] Many of the same magmatic processes that occur on Earth also occurred on Mars, and both planets are similar enough compositionally that the same names can be applied to their igneous rocks and minerals.

Volcanism is a process in which magma from a planet's interior rises through the crust and erupts on the surface. The erupted materials consist of molten rock (lava), hot fragmental debris (tephra or ash), and gases. Volcanism is a principal way that planets release their internal heat. Volcanic eruptions produce distinctive landforms, rock types, and terrains that provide a window on the chemical composition, thermal state, and history of a planet's interior.[161]

Magma is a complex, high-temperature mixture of molten silicates, suspended crystals, and dissolved gases. Magma on Mars likely ascends in a similar manner to that on Earth. It rises through the lower crust in diapiric bodies that are less dense than the surrounding material. As the magma rises, it eventually reaches regions of lower density. When the magma density matches that of the host rock, buoyancy is neutralized and the magma body stalls. At this point, it may form a magma chamber and spread out laterally into a network of dikes and sills. Subsequently, the magma may cool and solidify to form intrusive igneous bodies (plutons). Geologists estimate that about 80% of the magma generated on Earth stalls in the crust and never reaches the surface.[162]

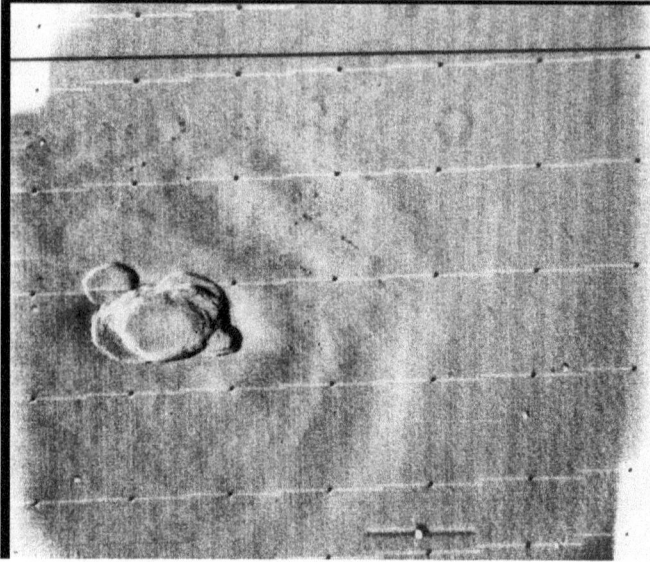

Figure 132: *Mariner 9 image of Ascraeus Mons. This is one of the first images to show that Mars has large volcanoes.*

Figure 133: *THEMIS image of lava flows. Note the lobate shape of the edges.*

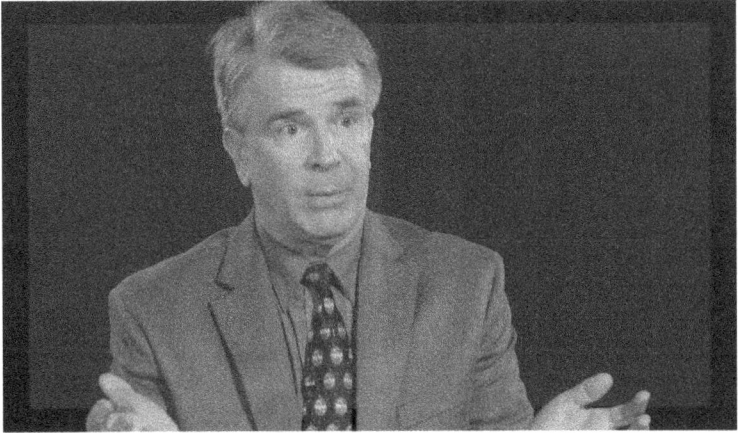

Figure 134: *Using Earth to understand how water may have affected volcanoes on Mars.*

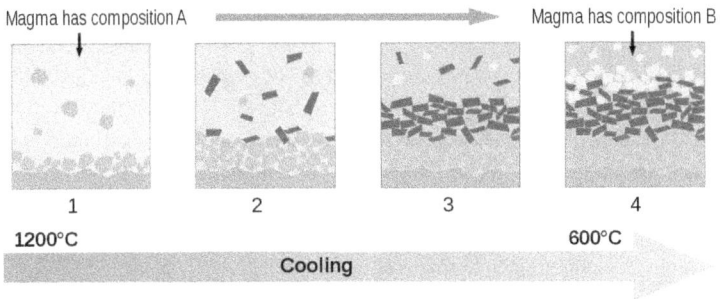

Figure 135: *Schematic diagrams showing the principles behind fractional crystallisation in a magma. While cooling, the magma evolves in composition because different minerals crystallize from the melt. 1: olivine crystallizes; 2: olivine and pyroxene crystallize; 3: pyroxene and plagioclase crystallize; 4: plagioclase crystallizes. At the bottom of the magma reservoir, a cumulate rock forms.*

As magma rises and cools, it undergoes many complex and dynamic compositional changes. Heavier minerals may crystallize and settle to the bottom of the magma chamber. The magma may also assimilate portions of host rock or mix with other batches of magma. These processes alter the composition of the remaining melt, so that any magma reaching the surface may be chemically quite different from its parent melt. Magmas that have been so altered are said to be "evolved" to distinguish them from "primitive" magmas that more closely resemble the composition of their mantle source. (See igneous differentiation and fractional crystallization.) More highly evolved magmas are usually felsic, that is enriched in silica, volatiles, and other light elements compared to iron- and magnesium-rich (mafic) primitive magmas. The degree and extent to which magmas evolve over time is an indication of a planet's level of internal heat and tectonic activity. The Earth's continental crust is made up of evolved granitic rocks that developed through many episodes of magmatic reprocessing. Evolved igneous rocks are much less common on cold, dead bodies such as the Moon. Mars, being intermediate in size between the Earth and the Moon, is thought to be intermediate in its level of magmatic activity.

At shallower depths in the crust, the lithostatic pressure on the magma body decreases. The reduced pressure can cause gases (volatiles), such as carbon dioxide and water vapor, to exsolve from the melt into a froth of gas bubbles. The nucleation of bubbles causes a rapid expansion and cooling of the surrounding melt, producing glassy shards that may erupt explosively as tephra (also called pyroclastics). Fine-grained tephra is commonly referred to as volcanic ash. Whether a volcano erupts explosively or effusively as fluid lava depends on the composition of the melt. Felsic magmas of andesitic and rhyolitic composition tend to erupt explosively. They are very viscous (thick and sticky) and rich in dissolved gases. Mafic magmas, on the other hand, are low in volatiles and commonly erupt effusively as basaltic lava flows. However, these are only generalizations. For example, magma that comes into sudden contact with groundwater or surface water may erupt violently in steam explosions called hydromagmatic (phreatomagmatic or phreatic) eruptions. Erupting magmas may also behave differently on planets with different interior compositions, atmospheres, and gravitational fields.

Differences in volcanic styles between Earth and Mars

The most common form of volcanism on the Earth is basaltic. Basalts are extrusive igneous rocks derived from the partial melting of the upper mantle. They are rich in iron and magnesium (mafic) minerals and commonly

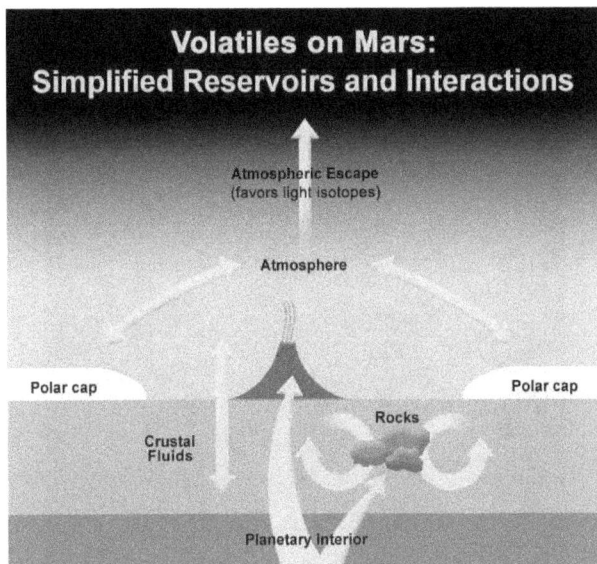

Figure 136: *Planet Mars – volatile gases (Curiosity rover, October 2012)*

dark gray in color. The principal type of volcanism on Mars is almost certainly basaltic too. On Earth, basaltic magmas commonly erupt as highly fluid flows, which either emerge directly from vents or form by the coalescence of molten clots at the base of fire fountains (Hawaiian eruption). These styles are also common on Mars, but the lower gravity and atmospheric pressure on Mars allow nucleation of gas bubbles (see above) to occur more readily and at greater depths than on Earth. As a consequence, Martian basaltic volcanoes are also capable of erupting large quantities of ash in Plinian-style eruptions. In a Plinian eruption, hot ash is incorporated into the atmosphere, forming a huge convective column (cloud). If insufficient atmosphere is incorporated, the column may collapse to form pyroclastic flows. Plinian eruptions are rare in basaltic volcanoes on Earth where such eruptions are most commonly associated with silica-rich andesitic or rhyolitic magmas (e.g., Mount St. Helens).

Because the lower gravity of Mars generates less buoyancy forces on magma rising through the crust, the magma chambers that feed volcanoes on Mars are thought to be deeper and much larger than those on Earth. If a magma body on Mars is to reach close enough to the surface to erupt before solidifying, it must be big. Consequently, eruptions on Mars are less frequent than on Earth, but are of enormous scale and eruptive rate when they do occur. Somewhat paradoxically, the lower gravity of Mars also allows for longer and more widespread lava flows. Lava eruptions on Mars may be unimaginably huge. A

Figure 137: *First X-ray diffraction view of Martian soil – CheMin analysis reveals minerals (including feldspar, pyroxenes and olivine) suggestive of "weathered basaltic soils" of Hawaiian volcanoes (Curiosity rover at "Rocknest", October 17, 2012). Each ring is a diffraction peak that corresponds to a specific atom-atom distance, which are unique enough to identify minerals. Smaller rings corresponds to larger features and vice versa.*

vast lava flow the size of the state of Oregon has recently been described in western Elysium Planitia. The flow is believed to have been emplaced turbulently over the span of several weeks and thought to be one of the youngest lava flows on Mars.

The tectonic settings of volcanoes on Earth and Mars are very different. Most active volcanoes on Earth occur in long, linear chains along plate boundaries, either in zones where the lithosphere is spreading apart (divergent boundaries) or being subducted back into the mantle (convergent boundaries). Because Mars currently lacks plate tectonics, volcanoes there do not show the same global pattern as on Earth. Martian volcanoes are more analogous to terrestrial mid-plate volcanoes, such as those in the Hawaiian Islands, which are thought to have formed over a stationary mantle plume.[163] (See hot spot.) The paragonetic tephra from a Hawaiian cinder cone has been mined to create Martian regolith simulant for researchers to use since 1998.

The largest and most conspicuous volcanoes on Mars occur in Tharsis and Elysium regions. These volcanoes are strikingly similar to shield volcanoes on Earth. Both have shallow-sloping flanks and summit calderas. The main difference between Martian shield volcanoes and those on Earth is in size: Martian shield volcanoes are truly colossal. For example, the tallest volcano on Mars, Olympus Mons, is 550 km across and 21 km high. It is nearly 100 times greater in volume than Mauna Loa in Hawaii, the largest shield volcano on Earth. Geologists think one of the reasons that volcanoes on Mars are able to grow so large is because Mars lacks plate tectonics. The Martian lithosphere does not slide over the upper mantle (asthenosphere) as on Earth, so lava from a stationary hot spot is able to accumulate at one location on the surface for a billion years or longer.

On 17 October 2012, the Curiosity rover on the planet Mars at "Rocknest" performed the first X-ray diffraction analysis of Martian soil. The results from the rover's CheMin analyzer revealed the presence of several minerals, including feldspar, pyroxenes and olivine, and suggested that the Martian soil in the sample was similar to the "weathered basaltic soils" of Hawaiian volcanoes. In July 2015, the same rover identified tridymite in a rock sample from Gale Crater, leading scientists to believe that silicic volcanism might have played a much more prevalent role in the planet's volcanic history than previously thought.

Tharsis volcanic province

The western hemisphere of Mars is dominated by a massive volcano-tectonic complex known as the Tharsis region or the Tharsis bulge. This immense, elevated structure is thousands of kilometers in diameter and covers up to 25% of the planet's surface. Averaging 7–10 km above datum (Martian "sea" level), Tharsis contains the highest elevations on the planet. Three enormous volcanoes, Ascraeus Mons, Pavonis Mons, and Arsia Mons (collectively known as the Tharsis Montes), sit aligned northeast–southwest along the crest of the bulge. The vast Alba Mons (formerly Alba Patera) occupies the northern part of the region. The huge shield volcano Olympus Mons lies off the main bulge, at the western edge of the province.

Built up by countless generations of lava flows and ash, the Tharsis bulge contains some of the youngest lava flows on Mars, but the bulge itself is believed to be very ancient. Geologic evidence indicates that most of the mass of Tharsis was in place by the end of the Noachian Period, about 3.7 billion years ago (Gya). Tharsis is so massive that it has placed tremendous stresses on the planet's lithosphere, generating immense extensional fractures (grabens and rift valleys) that extend halfway around the planet.[164] The mass of Tharsis could have even altered the orientation of Mars' rotational axis, causing climate changes.

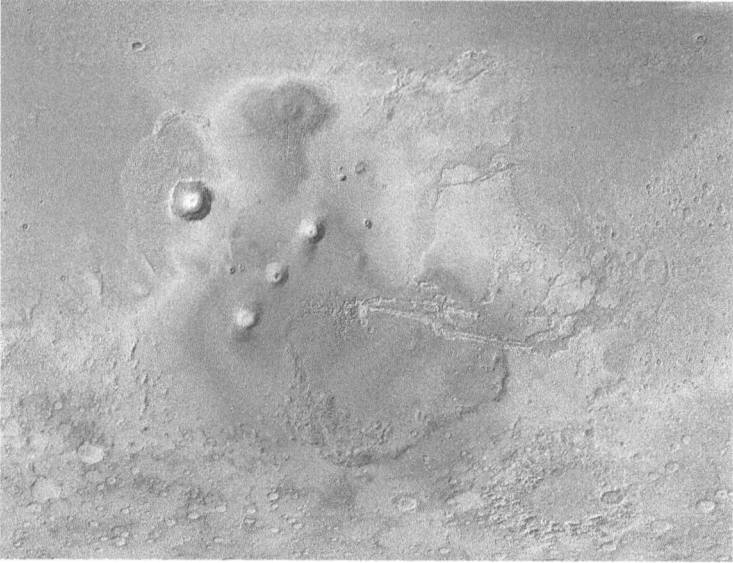

Figure 138: *MOLA colorized shaded-relief map of western hemisphere of Mars showing Tharsis bulge (shades of red and brown). Tall volcanoes appear white.*

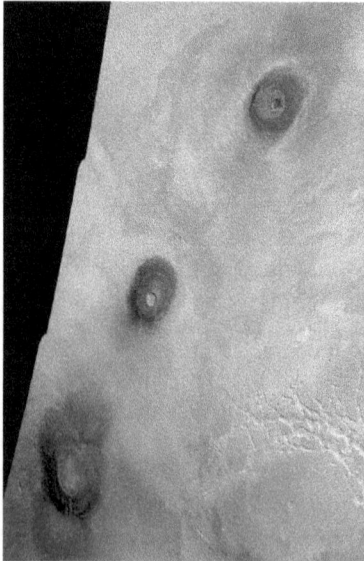

Figure 139: *Viking orbiter image of the three Tharsis Montes: Arsia Mons (bottom), Pavonis Mons (center), and Ascraeus Mons (top)*

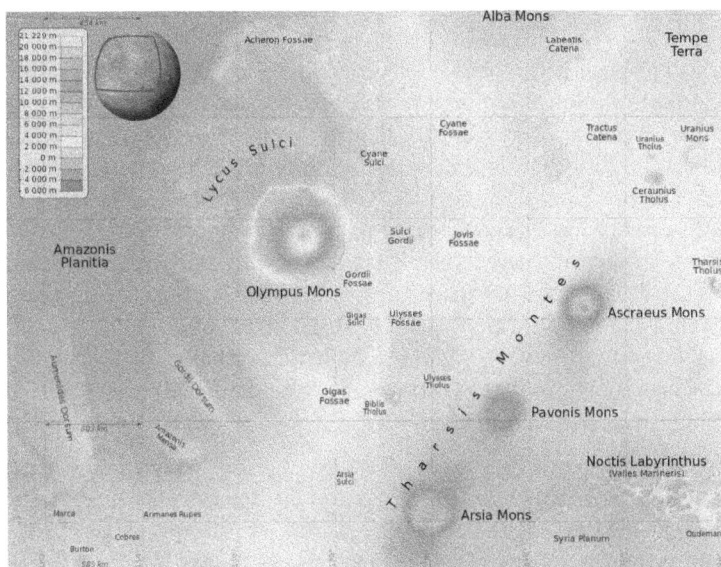

Figure 140: *Topographic map centered on Olympus and Tharsis*

Tharsis Montes

The three Tharsis Montes are shield volcanoes centered near the equator at longitude 247°E. All are several hundred kilometers in diameter and range in height from 14 to 18 km. Arsia Mons, the southernmost of the group, has a large summit caldera that is 130 kilometres (81 mi) across and 1.3 kilometres (0.81 mi) deep. Pavonis Mons, the middle volcano, has two nested calderas with the smaller one being almost 5 kilometres (3.1 mi) deep. Ascraeus Mons in the north, has a complex set of internested calderas and a long history of eruption that is believed to span most of Mars' history.

The three Tharsis Montes are about 700 kilometres (430 mi) apart. They show a distinctive northeast–southwest alignment that has been the source of some interest. Ceraunius Tholus and Uranius Mons follow the same trend to the northeast, and aprons of young lava flows on the flanks of all three Tharsis Montes are aligned in the same northeast–southwest orientation. This line clearly marks a major structural feature in the Martian crust, but its origin is uncertain.

Tholi and paterae

In addition to the large shield volcanoes, Tharsis contains a number of smaller volcanoes called tholi and paterae. The tholi are dome-shaped edifices with flanks that are much steeper than the larger Tharsis shields. Their central calderas are also quite large in proportion to their base diameters. The density of impact craters on many of the tholi indicate they are older than the large shields, having formed between late Noachian and early Hesperian times. Ceraunius Tholus and Uranius Tholus have densely channeled flanks, suggesting that the flank surfaces are made up of easily erodible material, such as ash. The age and morphology of the tholi provide strong evidence that the tholi represent the summits of old shield volcanoes that have been largely buried by great thicknesses of younger lava flows. By one estimate the Tharsis tholi may be buried by up to 4 km of lava.

Patera (pl. paterae) is Latin for a shallow drinking bowl. The term was applied to certain ill-defined, scalloped-edged craters that appeared in early spacecraft images to be large volcanic calderas. The smaller paterae in Tharsis appear to be morphologically similar to the tholi, except for having larger calderas. Like the tholi, the Tharsis paterae probably represent the tops of larger, now buried shield volcanoes. Historically, the term patera has been used to describe the entire edifice of certain volcanoes on Mars (e.g., Alba Patera). In 2007, the International Astronomical Union (IAU) redefined the terms Alba Patera, Uranius Patera, and Ulysses Patera to refer only to the central calderas of these volcanoes.

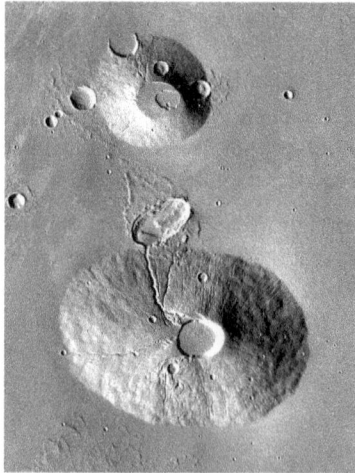

Figure 141: *2001 Mars Odyssey THEMIS mosaic of Uranius Tholus (upper volcano) and Ceraunius Tholus (lower volcano). The latter is about as high as Earth's Mount Everest.*

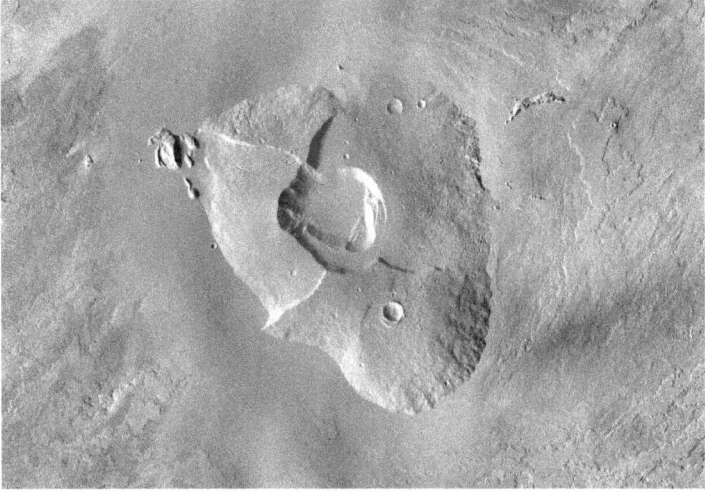

Figure 142: *2001 Mars Odyssey THEMIS mosaic of Tharsis Tholus.*

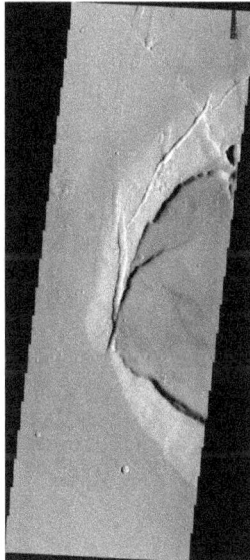

Figure 143: *Western part of Jovis Tholus, as seen by THEMIS.*

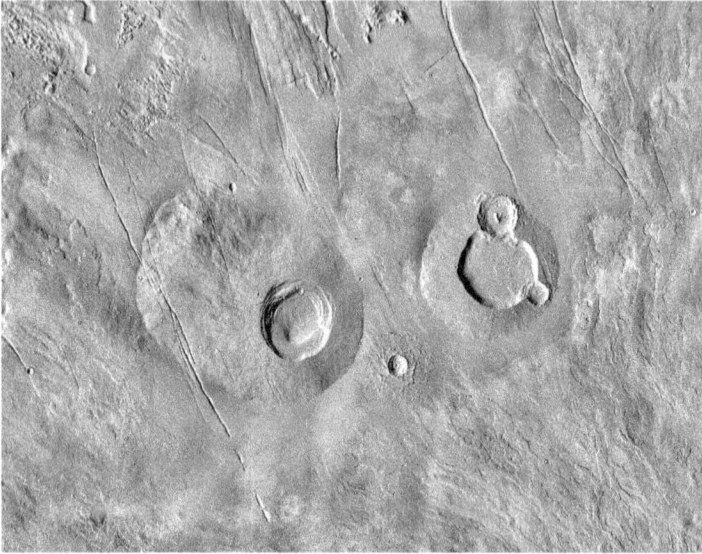

Figure 144: *Neighboring Biblis and Ulysses tholi (THEMIS daytime IR mosaic).*

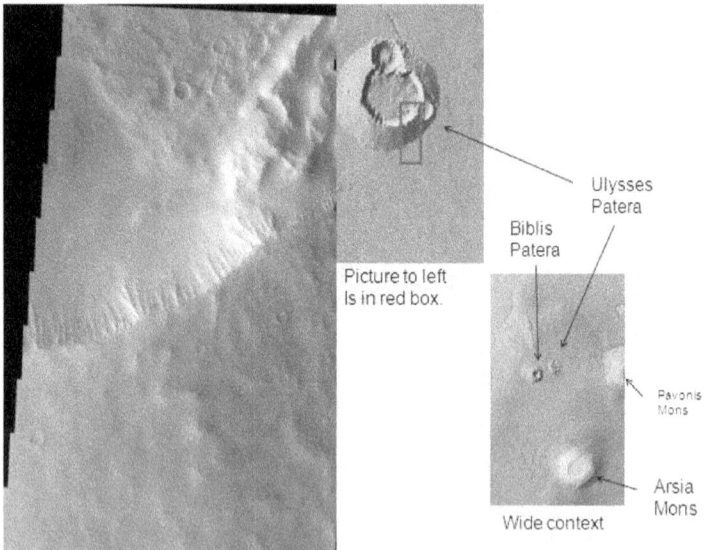

Figure 145: *Ulysses Tholus, with its location in relation to other volcanoes shown (photo by THEMIS).*

Olympus Mons

Olympus Mons is the youngest and tallest large volcano on Mars. It is located 1200 km northwest of the Tharsis Montes, just off the western edge of the Tharsis bulge. Its summit is 21 km above datum (Mars "sea" level) and has a central caldera complex consisting of six nested calderas that together form a depression 72 x 91 km wide and 3.2 km deep. As a shield volcano, it has an extremely low profile with shallow slopes averaging between 4–5 degrees. The volcano was built up by many thousands of individual flows of highly fluid lava. An irregular escarpment, in places up to 8 km tall, lies at the base of the volcano, forming a kind of pedestal on which the volcano sits. At various locations around the volcano, immense lava flows can be seen extending into the adjacent plains, burying the escarpment. In medium resolution images (100 m/pixel), the surface of the volcano has a fine radial texture due to the innumerable flows and leveed lava channels that line its flanks.

Figure 146: *Wide view of lava flowing over cliff around Olympus Mons, as seen by CTX*

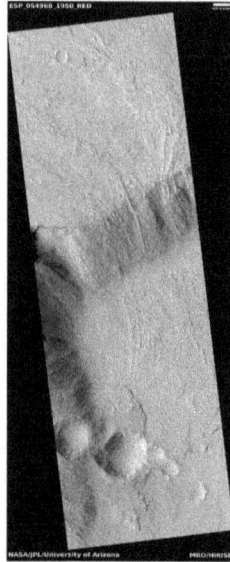

Figure 147: *Close view of lava moving over cliff around Olympus Mons, as seen by HiRISE under HiWish program*

Alba Mons (Alba Patera)

Alba Mons, located in the northern Tharsis region, is a unique volcanic structure, with no counterpart on Earth or elsewhere on Mars. The flanks of the volcano have extremely low slopes characterized by extensive lava flows and channels. The average flank slope on Alba Mons is only about 0.5°, over five times lower than the slopes on the other Tharsis volcanoes. The volcano has a central edifice 350 km wide and 1.5 km high with a double caldera complex at the summit. Surrounding the central edifice is an incomplete ring of fractures. Flows related to the volcano can be traced as far north as 61°N and as far south as 26°N. If one counts these widespread flow fields, the volcano stretches an immense 2000 km north–south and 3000 km east–west, making it one of the most areally extensive volcanic features in the Solar System. Most geological models suggest that Alba Mons is composed of highly fluid basaltic lava flows, but some researchers have identified possible pyroclastic deposits on the volcano's flanks.[165] Because Alba Mons lies antipodal to the Hellas impact basin, some researchers have conjectured that the volcano's formation may have been related to crustal weakening from the Hellas impact, which produced strong seismic waves that focused on the opposite side of the planet.

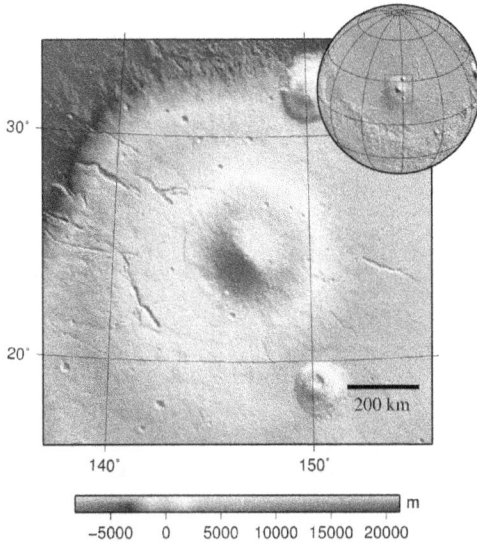

Figure 148: *MOLA view of Elysium province. Elysium Mons is in the center. Albor Tholus and Hecates Tholus are at bottom and top, respectively.*

Elysium volcanic province

A smaller volcanic center lies several thousand kilometers west of Tharsis in Elysium. The Elysium volcanic complex is about 2,000 kilometers in diameter and consists of three main volcanoes, Elysium Mons, Hecates Tholus, and Albor Tholus. The northwestern edge of the province is characterized by large channels (Granicus and Tinjar Valles) that emerge from several grabens on the flanks of Elysium Mons. The grabens may have formed from subsurface dikes. The dikes may have fractured the cryosphere, releasing large volumes of ground water to form the channels. Associated with the channels are widespread sedimentary deposits that may have formed from mudflows or lahars. The Elysium group of volcanoes is thought to be somewhat different from the Tharsis Montes, in that development of the former involved both lavas and pyroclastics.

Elysium Mons is the largest volcanic edifice in the province. It is 375 km across (depending on how one defines the base) and 14 km high. It has single, simple caldera at its summit that measures 14 km wide and 100 m deep. The volcano is distinctly conical in profile, leading some to call it a stratocone; however, given the predominantly low slopes, it is probably a shield. Elysium Mons is only about one-fifth the volume of Arsia Mons.

Hecates Tholus is 180 km across and 4.8 km high. The slopes of the volcano are heavily dissected with channels, suggesting that the volcano is composed of easily erodible material such as volcanic ash. The origin of the channels is unknown; they have been atrributed to lava, ash flows, or even water from snow or rainfall. Albor Tholus, the southernmost of the Elysium volcanoes, is 150 km in diameter and 4.1 km high. Its slopes are smoother and less heavily cratered than the slopes of the other Elysium volcanoes.

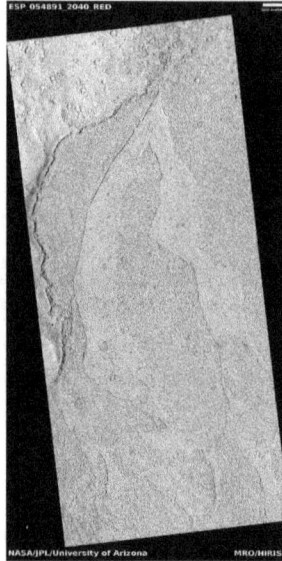

Figure 149: *Lava rafts, as seen by HiRISE under Hi-Wish program Location is the Elysium quadrangle.*

Syrtis Major

Syrtis Major Planum is a vast Hesperian-aged shield volcano located within the albedo feature bearing the same name. The volcano is 1200 km in diameter but only 2 km high. It has two calderas, Meroe Patera and Nili Patera. Studies involving the regional gravity field suggest a solidified magma chamber at least 5 km thick lies under the surface. Syrtis Major is of interest to geologists because dacite and granite have been detected there from orbiting spacecraft. Dacites and granites are silica-rich rocks that crystallize from a magma that is more chemically evolved and differentiated than basalt. They may form at the top of a magma chamber after the heavy minerals, such as olivine and pyroxene (those containing iron and magnesium), have settled to the bottom. Dacites and granites are very common on Earth but rare on Mars.

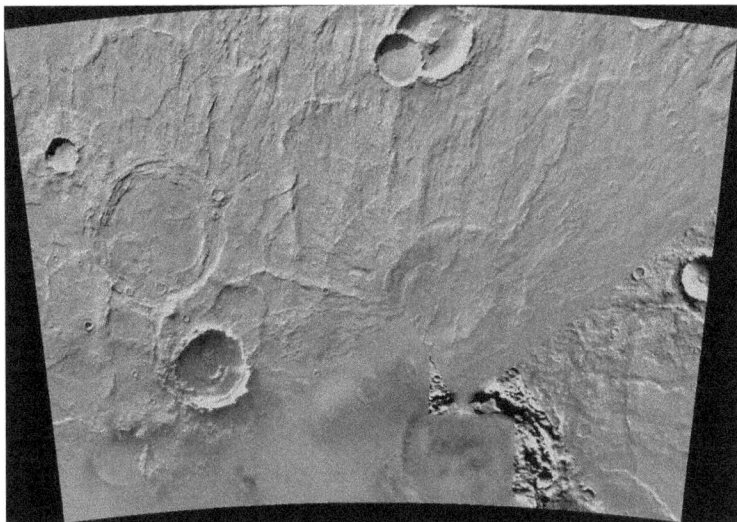

Figure 150: *Viking orbiter view of Peneus Patera (left) and Amphitrites Patera (right). Both are ancient volcanic edifices southwest of Hellas.*

Arabia Terra

Arabia Terra is a large upland region in the north of Mars that lies mostly in the Arabia quadrangle. Several irregularly shaped craters found within the region represent a type of highland volcanic construct which, all together, represent a martian igneous province. Low-relief paterae within the region possess a range of geomorphic features, including structural collapse, effusive volcanism and explosive eruptions, that are similar to terrestrial supervolcanoes. The enigmatic highland ridged plains in the region may have been formed, in part, by the related flow of lavas.

Highland paterae

In the southern hemisphere, particularly around the Hellas impact basin, are several flat-lying volcanic structures called highland paterae These volcanoes are some of the oldest identifiable volcanic edifices on Mars.[166] They are characterized by having extremely low profiles with highly eroded ridges and channels that radiate outward from a degraded, central caldera complex. They include Hadriaca Patera, Amphitrites Patera, Tyrrhena Patera, Peneus Patera, and Pityusa Patera. Geomorphologic evidence suggests that the highland patera were produced through a combination of lava flows and pyroclastics from

the interaction of magma with water. Some researchers speculate that the location of the highland paterae around Hellas is due to deep-seated fractures
caused by the impact that provided conduits for magma to rise to the surface.
Although they are not very high, some paterae cover large areas—Amphritrites
Patera, for example, covers a larger area than Olympus Mons.

Volcanic plains

Volcanic plains are widespread on Mars. Two types of plains are commonly
recognized: those where lava flow features are common, and those where flow
features are generally absent but a volcanic origin is inferred by other characteristics. Plains with abundant lava flow features occur in and around the large
volcanic provinces of Tharsis and Elysium. Flow features include both sheet
flow and tube- and channel-fed flow morphologies. Sheet flows show complex, overlapping flow lobes and may extend for many hundreds of kilometers
from their source areas. Lava flows can form a lava tube when the exposed
upper layers of lava cool and solidify to form a roof while the lava underneath
continues flowing. Often, when all the remaining lava leaves the tube, the roof
collapses to make a channel or line of pit craters (catena).

An unusual type of flow feature occurs in the Cerberus plains south of Elysium and in Amazonis. These flows have a broken platey texture, consisting
of dark, kilometer-scale slabs embedded in a light-toned matrix. They have
been attributed to rafted slabs of solidified lava floating on a still-molten subsurface. Others have claimed the broken slabs represent pack ice that froze
over a sea that pooled in the area after massive releases of groundwater from
the Cerberus Fossae area.

The second type of volcanic plains (ridged plains) are characterized by abundant wrinkle ridges. Volcanic flow features are rare or absent. The ridged
plains are believed to be regions of extensive flood basalts, by analogy with
the lunar maria. Ridged plains make up about 30% of the Martian surface and
are most prominent in Lunae, Hesperia, and Malea Plana, as well as throughout much of the northern lowlands. Ridged plains are all Hesperian in age and
represent a style of volcanism globally predominant during that time period.
The Hesperian Period is named after the ridged plains in Hesperia Planum.

Potential current volcanism

Scientists have never recorded an active volcano eruption on the surface of
Mars; moreover, searches for thermal signatures and surface changes within
the last decade have not yielded any positive evidence for active volcanism.

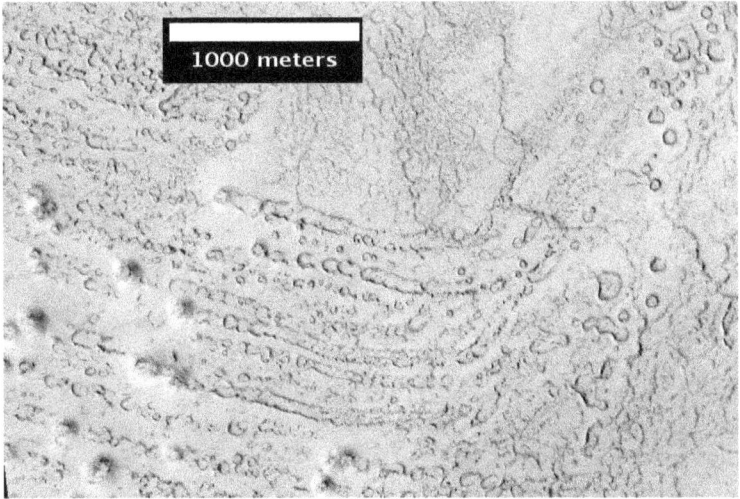

Figure 151: *HiRISE image of possible rootless cones east of Elysium region. The chains of rings are interpreted to be caused by steam explosions when lava moved over ground that was rich in water ice.*

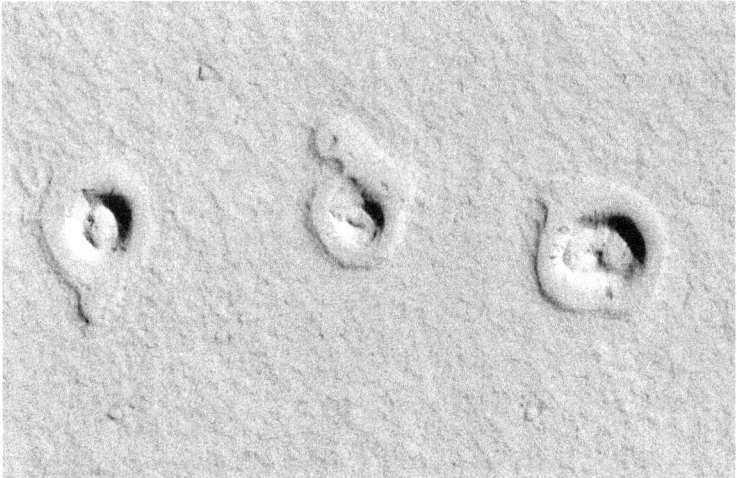

Figure 152: *"Rootless Cones" on Mars – due to lava flows interacting with water (MRO, January 4, 2013) (21.965°N 197.807°E[167])*

However, the European Space Agency's Mars Express orbiter photographed lava flows interpreted in 2004 to have occurred within the past two million years, suggesting a relatively recent geologic activity. An updated study in 2011 estimated that the youngest lava flows occurred in the last few tens of millions of years. The authors consider this age makes it possible that Mars is not yet volcanically extinct.

The upcoming *InSight* lander mission will determine if there is any seismic activity, measure the amount of heat flow from the interior, estimate the size of Mars' core and whether the core is liquid or solid.

Volcanoes and ice

Large amounts of water ice are believed to be present in the Martian subsurface. The interaction of ice with molten rock can produce distinct landforms. On Earth, when hot volcanic material comes into contact with surface ice, large amounts of liquid water and mud may form that flow catastrophically down slope as massive debris flows (lahars). Some channels in Martian volcanic areas, such as Hrad Vallis near Elysium Mons, may have been similarly carved or modified by lahars.[168] Lava flowing over water-saturated ground can cause the water to erupt violently in an explosion of steam (see phreatic eruption), producing small volcano-like landforms called pseudocraters, or rootless cones. Features that resemble terrestrial rootless cones occur in Elysium, Amazonis, and Isidis and Chryse Planitiae.[169] Also, phreatomagmatism produce tuff rings or tuff cones on Earth and existence of similar landforms on Mars is expected too. Their existence was suggested from Nepenthes/Amenthes region. Finally, when a volcano erupts under an ice sheet, it can form a distinct, mesa-like landform called a tuya or table mountain. Some researchers[170] cite geomorphic evidence that many of the layered interior deposits in Valles Marineris may be the Martian equivalent of tuyas.

Tectonic boundaries

Tectonic boundaries have been discovered on Mars. Valles Marineris is a horizontally sliding tectonic boundary that divides two major partial or complete plates of Mars. The recent finding suggests that Mars is geologically active with occurrences in the millions of years, and there is additional speculation. There has been previous evidence of Mars' geologic activity. The Mars Global Surveyor (MGS) discovered magnetic stripes in the crust of Mars, especially in the Phaethontis and Eridania quadrangles. The magnetometer on MGS discovered 100 km wide stripes of magnetized crust running roughly parallel for up to 2000 km. These stripes alternate in polarity with the north magnetic pole of one pointing up from the surface and the north magnetic pole of the next

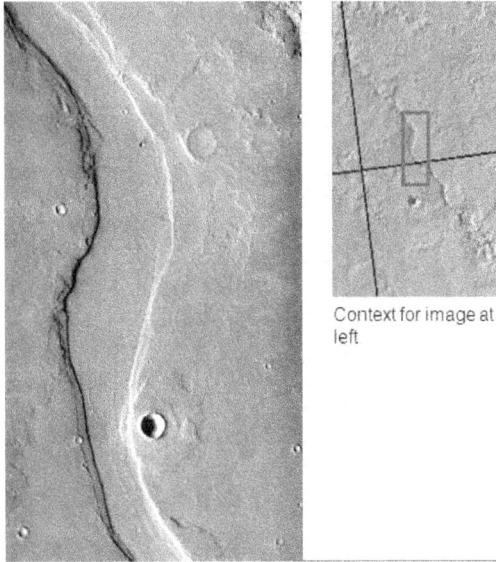

Context for image at left

Figure 153: *THEMIS image of Hrad Vallis. This valley may have formed when eruptions in the Elysium Mons volcanic complex melted ground or surface ice.*

pointing down. When similar stripes were discovered on Earth in the 1960s, they were taken as evidence of plate tectonics. However, there are some differences, between the magnetic stripes on Earth and those on Mars. The Martian stripes are wider, much more strongly magnetized, and do not appear to spread out from a middle crustal spreading zone. Because the area with the magnetic stripes is about 4 billion years old, it is believed that the global magnetic field probably lasted for only the first few hundred million years of Mars' life. At that time the temperature of the molten iron in the planet's core might have been high enough to mix it into a magnetic dynamo. Younger rock does not show any stripes. When molten rock containing magnetic material, such as hematite (Fe_2O_3), cools and solidifies in the presence of a magnetic field, it becomes magnetized and takes on the polarity of the background field. This magnetism is lost only if the rock is subsequently heated above the Curie temperature, which is 770 °C for pure iron, but lower for oxides such as hematite (approximately 650 °C) or magnetite (approximately 580 °C). The magnetism left in rocks is a record of the magnetic field when the rock solidified.

Mars' volcanic features can be likened to Earth's geologic hotspots. Pavonis Mons is the middle of three volcanoes (collectively known as Tharsis Montes) on the Tharsis bulge near the equator of the planet Mars. The other Tharsis

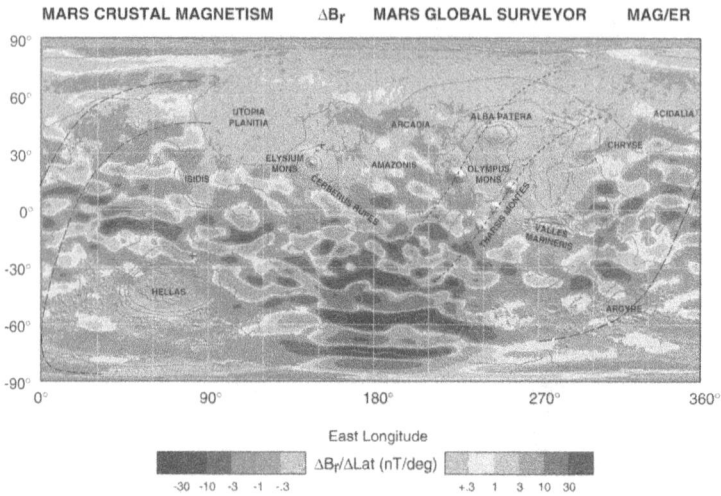

Figure 154: *Mars crustal magnetism*

volcanoes are Ascraeus Mons and Arsia Mons. The three Tharsis Montes, to-gether with some smaller volcanoes to the north, form a straight line. This arrangement suggests that they were formed by a crustal plate moving over a hot spot. Such an arrangement exists in the Earth's Pacific Ocean as the Hawaiian Islands. The Hawaiian Islands are in a straight line, with the youngest in the south and the oldest in the north. So geologists believe the plate is moving while a stationary plume of hot magma rises and punches through the crust to produce volcanic mountains. However, the largest volcano on the planet, Olympus Mons, is thought to have formed when the plates were not moving. Olympus Mons may have formed just after the plate motion stopped. The mare-like plains on Mars are roughly 3 to 3.5 billion years old. The giant shield volcanoes are younger, formed between 1 and 2 billion years ago. Olympus Mons may be "as young as 200 million years."

Norman H. Sleep, professor of geophysics at Stanford University, described how the three volcanoes that form a line along the Tharsis Ridge may be extinct island arc volcanoes like the Japanese Island chain.[171]

Bibliography

- Carr, Michael H. (2006). *The Surface of Mars*. New York: Cambridge University Press. ISBN 978-0-521-87201-0.
- Boyce, J.M. (2008). *The Smithsonian Book of Mars*. Old Saybrook, CT: Konecky & Konecky. ISBN 978-1-58834-074-0.

External links

- The Volcanoes of Mars[172]
- Geology of Mars – Volcanism[173]

Atmosphere of Mars

Atmosphere of Mars

Atmosphere of Mars

Image of Mars with sandstorm visible, taken by
the *Hubble Space Telescope* on 28 October 2005

General information	
Chemical species	Mole fraction
Composition	
Carbon dioxide	95.97%
Argon	1.93%
Nitrogen	1.89%
Oxygen	0.146%
Carbon monoxide	0.0557%

The atmosphere of the planet Mars is composed mostly of carbon dioxide. The atmospheric pressure on the Martian surface averages 600 pascals (0.087 psi; 6.0 mbar), about 0.6% of Earth's mean sea level pressure of 101.3 kilopascals (14.69 psi; 1.013 bar). It ranges from a low of 30 pascals (0.0044 psi; 0.30 mbar) on Olympus Mons's peak to over 1,155 pascals (0.1675 psi; 11.55 mbar) in the depths of Hellas Planitia. This pressure is

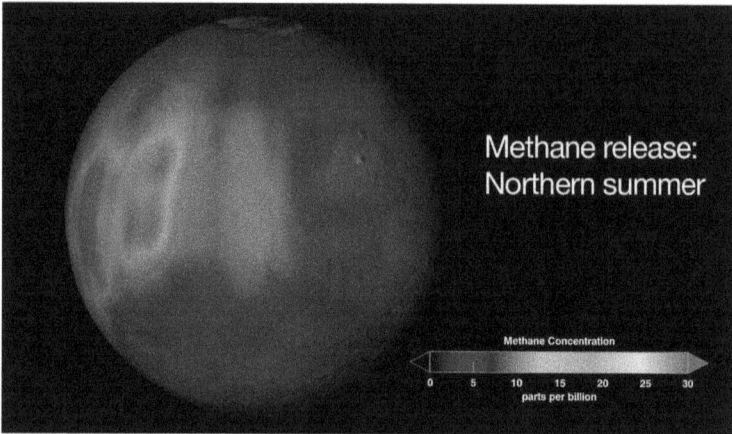

Figure 155: *One mystery is the source of Mars methane, detection shown here*

well below the Armstrong limit for the unprotected human body. Mars's atmospheric mass of 25 teratonnes compares to Earth's 5148 teratonnes; Mars has a scale height of 11.1 kilometres (6.9 mi) versus Earth's 8.5 kilometres (5.3 mi).

The Martian atmosphere consists of approximately 96% carbon dioxide, 1.9% argon, 1.9% nitrogen, and traces of free oxygen, carbon monoxide, water and methane, among other gases, for a mean molar mass of 43.34 g/mol. There has been renewed interest in its composition since the detection of traces of methane in 2003[174] that may indicate life but may also be produced by a geochemical process, volcanic or hydrothermal activity.

The atmosphere is quite dusty, giving the Martian sky a light brown or orange-red color when seen from the surface; data from the Mars Exploration Rovers indicate suspended particles of roughly 1.5 micrometres in diameter.

On 16 December 2014, NASA reported detecting an unusual increase, then decrease, in the amounts of methane in the atmosphere of the planet Mars. Organic chemicals have been detected in powder drilled from a rock by the *Curiosity* rover. Based on deuterium to hydrogen ratio studies, much of the water at Gale Crater on Mars was found to have been lost during ancient times, before the lakebed in the crater was formed; afterwards, large amounts of water continued to be lost.

On 18 March 2015, NASA reported the detection of an aurora that is not fully understood and an unexplained dust cloud in the atmosphere of Mars.

On 4 April 2015, NASA reported studies, based on measurements by the Sample Analysis at Mars (SAM) instrument on the *Curiosity* rover, of the Martian atmosphere using xenon and argon isotopes. Results provided support for a "vigorous" loss of atmosphere early in the history of Mars and were consistent with an atmospheric signature found in bits of atmosphere captured in some Martian meteorites found on Earth. This was further supported by results from the MAVEN orbiter circling Mars, that the solar wind is responsible for stripping away the atmosphere of Mars over the years.

In September 2017, NASA reported radiation levels on the surface of the planet Mars were temporarily doubled, and were associated with an aurora 25 times brighter than any observed earlier due to a massive, and unexpected, solar storm in the middle of the month.

On 1 June 2018, NASA scientists detected signs of a dust storm (see image) on the planet Mars which may affect the survivability of the solar-powered *Opportunity* rover since the dust may block the sunlight (see image) needed to operate; as of 12 June, the storm is the worst ever recorded at the surface of the planet, and spanned an area about the size of North America and Russia combined (about a quarter of the planet); as of 13 June, *Opportunity* was reported to be experiencing serious communication problem(s) due to the dust storm; a NASA teleconference about the dust storm was presented on 13 June 2018 at 01:30 pm/et/usa and is available for replay[175]. In July 2018, researchers reported that the largest single source of dust on the planet Mars comes from the Medusae Fossae Formation.

On 7 June 2018, NASA announced a cyclical seasonal variation in atmospheric methane.

Structure

Pressure comparison

Where	Pressure
Olympus Mons summit	0.03 kilopascals (0.0044 psi)
Mars average	0.6 kilopascals (0.087 psi)
Hellas Planitia bottom	1.16 kilopascals (0.168 psi)
Armstrong limit	6.25 kilopascals (0.906 psi)
Mount Everest summit	33.7 kilopascals (4.89 psi)
Earth sea level	101.3 kilopascals (14.69 psi)

Mars's atmosphere is composed of the following layers:

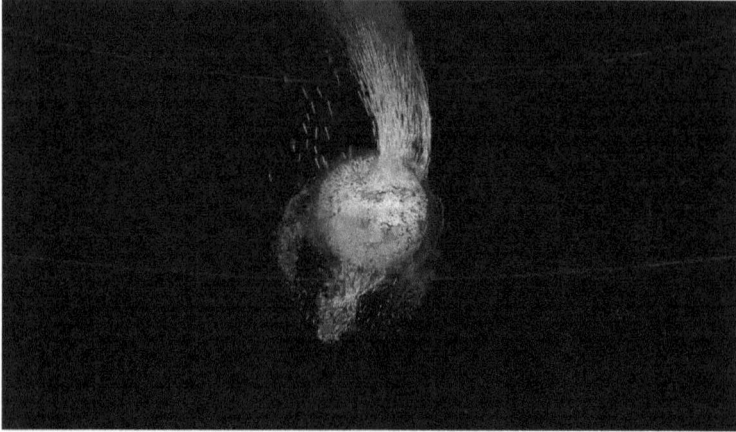

Figure 156:
The solar wind accelerates ions from the Mars upper atmosphere into space
(video (01:13); 5 November 2015)

- Exosphere: Typically stated to start at 200 km (120 mi) and higher, this region is where the last wisps of atmosphere merge into the vacuum of space. There is no distinct boundary where the atmosphere ends; it just tapers away.
- Upper atmosphere, or thermosphere: A region with very high temperatures, caused by heating from the Sun. Atmospheric gases start to separate from each other at these altitudes, rather than forming the even mix found in the lower atmospheric layers.
- Middle atmosphere: The region in which Mars's jetstream flows.
- Lower atmosphere: A relatively warm region affected by heat from airborne dust and from the ground.

There is also a complicated ionosphere,[176] and a seasonal ozone layer over the south pole.[177] The *MAVEN* spacecraft determined in 2015 that there is a substantial layered structure present in both neutral gases and ion densities.

Initial analyses by the *MAVEN* orbiter[178] and the ExoMars Trace Gas Orbiter, have shown high thermal and density variability in the atmosphere with a slightly lower average density than predicted by existing models.[179]

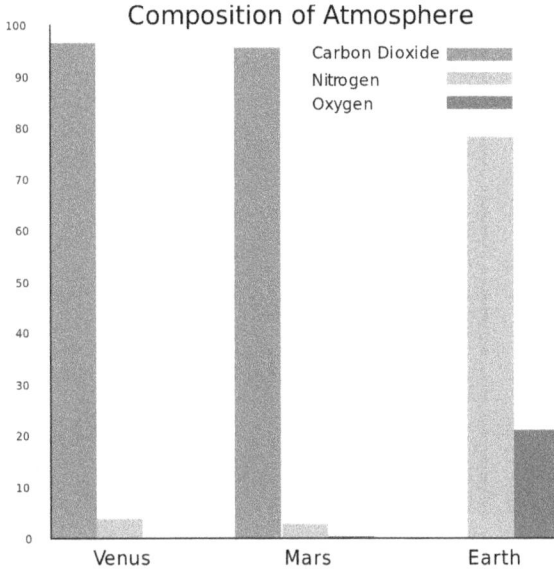

Figure 157: *Comparison of the atmospheric compositions of Venus, Mars, and Earth.*

Observations and measurement from Earth

In 1864, William Rutter Dawes observed "that the ruddy tint of the planet does not arise from any peculiarity of its atmosphere seems to be fully proved by the fact that the redness is always deepest near the centre, where the atmosphere is thinnest." Spectroscopic observations in the 1860s and 1870s[180] led many to think the atmosphere of Mars is similar to Earth's. In 1894, though, spectral analysis and other qualitative observations by William Wallace Campbell suggested Mars resembles the Moon, which has no appreciable atmosphere, in many respects.

In 1926, photographic observations by William Hammond Wright at the Lick Observatory allowed Donald Howard Menzel to discover quantitative evidence of Mars's atmosphere.

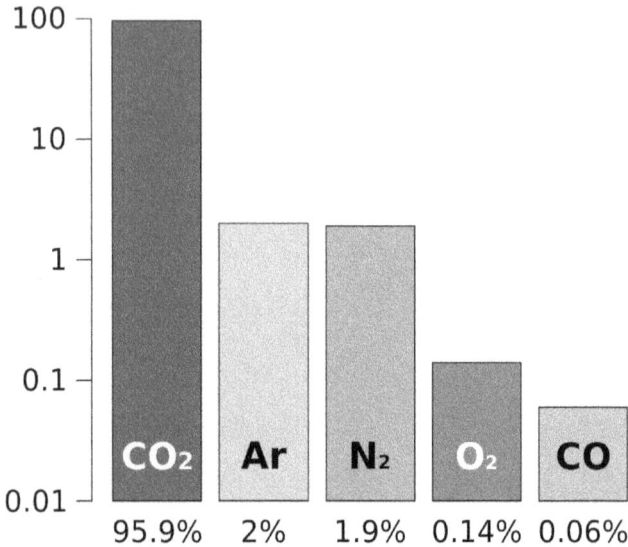

Figure 158: *Most abundant gases on Mars.*

Composition

Carbon dioxide

The main component of the atmosphere of Mars is carbon dioxide (CO_2) at 95.9%. Each pole is in continual darkness during its hemisphere's winter, and the surface gets so cold that as much as 25% of the atmospheric CO_2 condenses at the polar caps into solid CO_2 ice (dry ice). When the pole is again exposed to sunlight during summer, the CO_2 ice sublimes back into the atmosphere. This process leads to a significant annual variation in the atmospheric pressure and atmospheric composition around the Martian poles.

It has been suggested that Mars had a much thicker, warmer, and wetter atmosphere early in its history. Much of this early atmosphere would have consisted of carbon dioxide. Such an atmosphere would have raised the temperature, at least in some places, to above the freezing point of water. With the higher temperature, running water could have carved out the many channels and outflow valleys that are common on the planet. It also might have gathered to form lakes and maybe an ocean. Some researchers have suggested that the atmosphere of Mars may have been many times as thick as the present one of Earth; however, research published in fall 2015 advanced the idea that perhaps the early Martian atmosphere was not as thick as previously thought.[181]

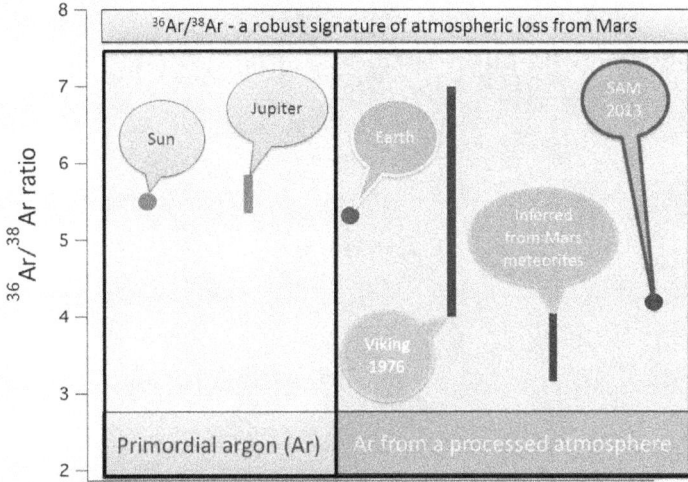

Figure 159: *Argon isotope ratios are a signature of atmospheric loss on Mars.*

Currently, the atmosphere is very thin. For many years, it was assumed that as with Earth, most of the early carbon dioxide was locked up in minerals, called carbonates. However, despite the use of many orbiting instruments that looked for carbonates, very few carbonate deposits have been found. Today, it is thought that much of the carbon dioxide in the Martian air was removed by the solar wind. Researchers have discovered a two-step process that sends the gas into space. Ultraviolet light from the sun could strike a carbon dioxide molecule, breaking it into carbon monoxide and oxygen. A second photon of ultraviolet light could subsequently break the carbon monoxide into oxygen and carbon which would receive enough energy to escape the planet. In this process the light isotope of carbon (^{12}C) is most likely to leave the atmosphere. Hence, the carbon dioxide left in the atmosphere would be enriched with the heavy isotope (^{13}C). This higher level of the heavy isotope is what was recently found by the *Curiosity* rover that sits on the surface of Mars.

Argon

The atmosphere of Mars is enriched considerably with the noble gas argon, in comparison to the atmosphere of the other planets within the Solar System. Unlike carbon dioxide, the argon content of the atmosphere does not condense, and hence the total amount of argon in the Mars atmosphere is constant. However, the relative concentration at any given location can change

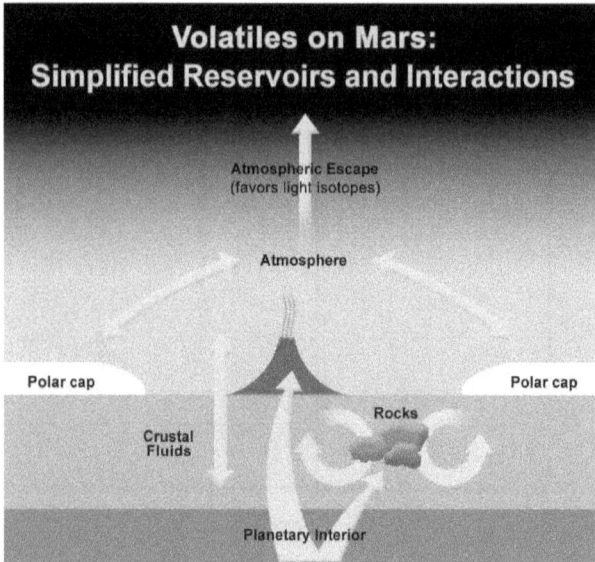

Figure 160: *Volatile gases on Mars.*

as carbon dioxide moves in and out of the atmosphere. Recent satellite data shows an increase in atmospheric argon over the southern pole during its autumn, which dissipates the following spring.

Water

Some aspects of the Martian atmosphere vary significantly. As carbon dioxide sublimes back into the atmosphere during the Martian summer, it leaves traces of water. Seasonal winds transport large amounts of dust and water vapor giving rise to Earth-like frost and large cirrus clouds. These clouds of water-ice were photographed by the *Opportunity* rover in 2004.[182] NASA scientists working on the *Phoenix* Mars mission confirmed on July 31, 2008 that they had indeed found subsurface water ice at Mars's northern polar region.

Methane

Trace amounts of methane (CH_4), at the level of several parts per billion (ppb), were first reported in Mars's atmosphere by a team at the NASA Goddard Space Flight Center in 2003. In March 2004, the Mars Express Orbiter and ground-based observations by three groups also suggested the presence of methane in the atmosphere at a concentration of about 10 ppb (parts per billion). Large differences in the abundances were measured between observations taken in 2003 and 2006, which suggested that the methane was locally concentrated and probably seasonal.[183]

Because methane on Mars would quickly break down due to ultraviolet radiation from the Sun and chemical reactions with other gases, its reported persistent presence in the atmosphere also implies the existence of a source to continually replenish the gas. Current photochemical models alone can not explain the rapid variability of the methane levels. It had been proposed that the methane might be replenished by meteorites entering the atmosphere of Mars, but researchers from Imperial College London found that the volumes of methane released this way are too low to sustain the measured levels of the gas.

Research suggests that the implied methane destruction lifetime is as long as $\sim$4 Earth years and as short as $\sim$0.6 Earth years. This lifetime is short enough for the atmospheric circulation to yield the observed uneven distribution of methane across the planet. In either case, the destruction lifetime for methane is much shorter than the timescale ($\sim$350 years) estimated for photochemical (UV radiation) destruction. The rapid destruction (or "sink") of methane suggests that another process must dominate removal of atmospheric methane on Mars, and it must be more efficient than destruction by light by a factor of 100 to 600. This unexplained fast destruction rate also suggests a very active replenishing source. In 2014 it was concluded that presence of strong methane sinks are not subject to atmospheric oxidation. A possibility is that the methane is not consumed at all, but rather condenses and evaporates seasonally from clathrates. Another possibility is that methane reacts with tumbling surface sand quartz (SiO

$_2$) and olivine to form covalent Si–CH

$_3$ bonds.

The principal candidates for the origin of Mars' methane include nonbiological processes such as water–rock reactions, radiolysis of water, and pyrite formation, all of which produce H_2 that could then generate methane and other hydrocarbons via Fischer–Tropsch synthesis with CO and CO_2. It has also been shown that methane could be produced by a process involving water, carbon dioxide, and the mineral olivine, which is known to be common on Mars. The required conditions for this reaction (i.e. high temperature and pressure) do not exist on the surface, but may exist within the crust.[184] A detection of the mineral by-product serpentinite would suggest that this process is occurring. An analog on Earth suggests that low-temperature production and exhalation of methane from serpentinized rocks may be possible on Mars. Another possible geophysical source could be ancient methane trapped in clathrate hydrates that may be released occasionally. Under the assumption

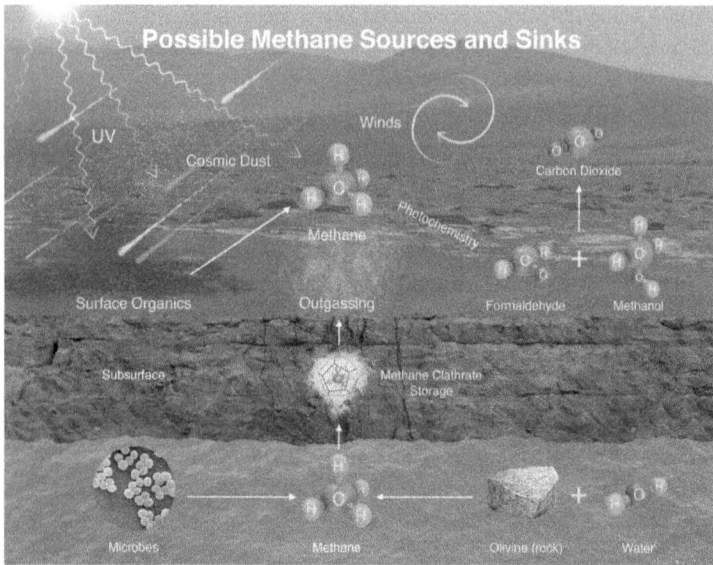

Figure 161: *Possible methane sources and sinks on Mars.*

of a cold early Mars environment, a cryosphere could trap such methane as clathrates in stable form at depth, that might exhibit sporadic release.

A group of Mexican scientists performed plasma experiments in a synthetic Mars atmosphere and found that bursts of methane can be produced when a discharge interacts with water ice. A potential source of the discharges can be the electrification of dust particles from sand storms and dust devils. The ice can be found in trenches or in the permafrost. The electrical discharge ionizes gaseous CO_2 and water molecules and their byproducts recombine to produce methane. The results obtained show that pulsed electrical discharges over ice samples in a Martian atmosphere produce about 1.41×10^{16} molecules of methane per joule of applied energy.

Living microorganisms, such as methanogens, are another possible source, but no evidence for the presence of such organisms has been found on Mars. In Earth's oceans, biological methane production tends to be accompanied by ethane, whereas volcanic methane is accompanied by sulfur dioxide. Several studies of trace gases in the Martian atmosphere have found no evidence for sulfur dioxide in the Martian atmosphere, which makes volcanism unlikely to be the source of methane.

In 2011, NASA scientists reported a comprehensive search using ground-based high-resolution infrared spectroscopy for trace species (including

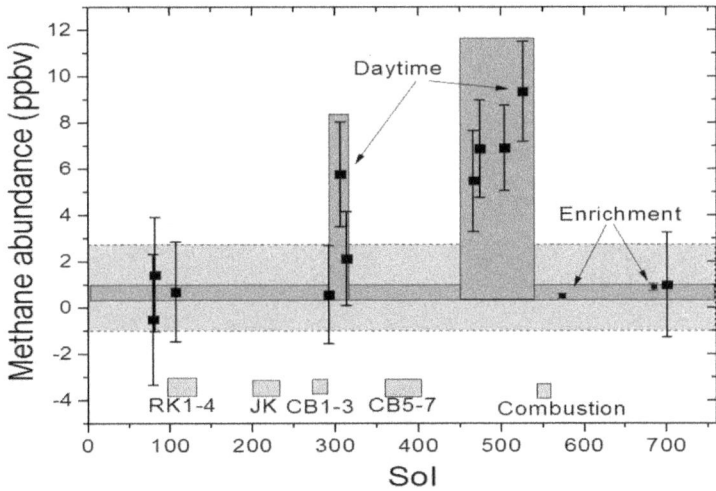

Figure 162: *Methane measurements in the atmosphere of Mars by the Curiosity rover.*

methane) on Mars, deriving sensitive upper limits for methane (<7 ppbv), ethane (<0.2 ppbv), methanol (<19 ppbv) and others (H_2CO, C_2H_2, C_2H_4, N_2O, NH_3, HCN, CH_3Cl, HCl, HO_2 – all limits at ppbv levels). The data were acquired over a period of 6 years and span different seasons and locations on Mars, suggesting that if organics are being released into the atmosphere, these events were extremely rare or currently non-existent, considering the expected long lifetimes for some of these species.

In August 2012, the *Curiosity* rover landed on Mars. The rover's instruments are capable of making precise abundance measurements, which can be used to distinguish between different isotopologues of methane. The first measurements with *Curiosity*'s Tunable Laser Spectrometer (TLS) in 2012 indicated that there was no methane or less than 5 ppb of methane at the landing site, later calculated to a baseline of 0.3 to 0.7 ppb.[185] On 2013, NASA scientists again reported no detection of methane beyond a baseline. But on 2014, NASA reported that the *Curiosity* rover detected a tenfold increase ('spike') in methane in the atmosphere around it in late 2013 and early 2014. Four measurements taken over two months in this period averaged 7.2 ppb, implying that Mars is episodically producing or releasing methane from an unknown source. Before and after that, readings averaged around one-tenth that level.

The Indian Mars Orbiter Mission, which entered orbit around Mars on 24 September 2014, is equipped with a Fabry–Pérot interferometer to measure

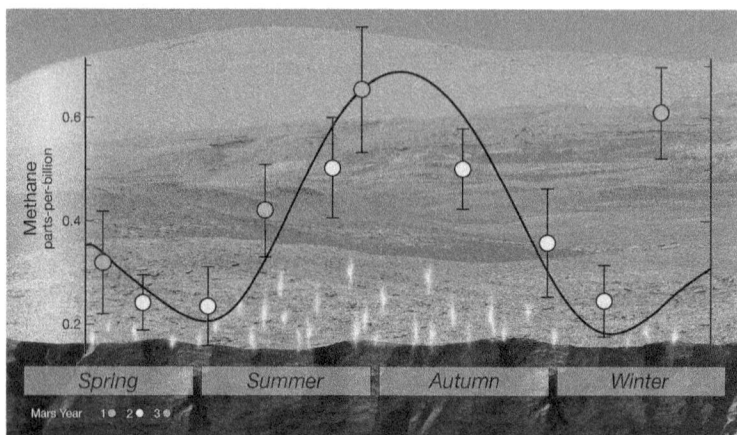

Figure 163:
Curiosity detected a cyclical seasonal variation in atmospheric methane.

atmospheric methane, but after entering Mars orbit it was determined that it was not capable of detecting methane.[186:57] The instrument was repurposed as an albedo mapper.[187] The ExoMars Trace Gas Orbiter, which entered orbit on 19 October 2016, will further study the methane, as well as its decomposition products such as formaldehyde and methanol starting on April 2018.[188]

Measuring the ratio of hydrogen and methane levels on Mars may help determine the likelihood of life on Mars. According to the scientists, "...low H_2/ CH_4 ratios (less than approximately 40) indicate that life is likely present and active."

On June 7, 2018, NASA announced a cyclical seasonal variation in atmospheric methane.

Sulfur dioxide

Sulfur dioxide in the atmosphere is thought to be a tracer of current volcanic activity. It has become especially interesting due to the long-standing controversy of methane on Mars. If methane on Mars were being produced by volcanoes (as it is in part on Earth) we would expect to find sulfur dioxide in large quantities. Several teams have searched for sulfur dioxide on Mars using the NASA Infrared Telescope Facility. No sulfur dioxide was detected in these studies, but they were able to place stringent upper limits on the atmospheric concentration of 0.2 ppb. In March 2013, a team led by scientists at NASA Goddard Space Flight Center reported a detection of SO_2 in Rocknest soil samples analyzed by the *Curiosity* rover.[189]

Figure 164: *Rotation of Mars near opposition. Ecliptic south is up.*

Ozone

As reported by the European Space Agency (ESA) on September 29, 2013,Wikipedia:Citation needed a new comparison of spacecraft data with computer models explains how global atmospheric circulation creates a layer of ozone (O
$_3$) above Mars's southern pole in winter. Ozone was most likely difficult to detect on Mars because its concentration is typically 300 times lower than on Earth, although it varies greatly with location and time. The SPICAM —an UV/IR spectrometer— on board Mars Express has shown the presence of two distinct ozone layers at low-to-mid latitudes. These comprise a persistent, near-surface layer below an altitude of 30 km, a separate layer that is only present in northern spring and summer with an altitude varying from 30 to 60 km, and another separate layer that exists 40–60 km above the southern pole in winter, with no counterpart above the Mars's north pole. This third ozone layer shows an abrupt decrease in elevation between 75 and 50 degrees south. SPICAM detected a gradual increase in ozone concentration at 50 km until midwinter, after which it slowly decreased to very low concentrations, with no layer detectable above 35 km. The reporting scientists think that the observed polar ozone layers are the result of the same atmospheric circulation pattern that creates a distinct oxygen emission identified in the polar night and also present in Earth's atmosphere. This circulation takes the form of a huge Hadley cell in which warmer air rises and travels toward the south pole before cooling and sinking at higher latitudes. Mars is on a quite elliptical orbit and has a large axial tilt, which causes extreme seasonal variations in temperature amongst the northern and southern hemispheres. Mars's temperature difference greatly influences the amount of water vapor in the atmosphere, because

warmer air can contain more moisture. This, in turn, affects the production of ozone-destroying hydrogen radicals.[190]

Oxygen

In early 2016, Stratospheric Observatory for Infrared Astronomy (SOFIA) detected atomic oxygen in the atmosphere of Mars.[191] This was the first time in forty years it was detected, the last time being the Viking and Mariner missions in the 1970s.

Potential for use by humans

The atmosphere of Mars is a resource of known composition available at any landing site on Mars. It has been proposed that human exploration of Mars could use carbon dioxide (CO_2) from the Martian atmosphere to make rocket fuel for the return mission. Mission studies that propose using the atmosphere in this way include the Mars Direct proposal of Robert Zubrin and the NASA Design reference mission study. Two major chemical pathways for use of the carbon dioxide are the Sabatier reaction, converting atmospheric carbon dioxide along with additional hydrogen (H_2), to produce methane (CH_4) and oxygen (O_2), and electrolysis, using a zirconia solid oxide electrolyte to split the carbon dioxide into oxygen (O_2) and carbon monoxide (CO).

History

Mars's atmosphere is thought to have changed over the course of the planet's lifetime, with evidence suggesting the possibility that Mars had large oceans a few billion years ago. As stated in the Mars ocean hypothesis, atmospheric pressure on the present-day Martian surface only exceeds that of the triple point of water (6.11 hectopascals (0.0886 psi)) in the lowest elevations; at higher elevations water can exist only in solid or vapor form. Annual mean temperatures at the surface are currently < 210 K (–63 °C; –82 °F), significantly lower than that needed to sustain liquid water. However, early in its history Mars may have had conditions more conducive to retaining liquid water at the surface. In 2013, scientists published that Mars once had "oxygen-rich" atmosphere billions of years ago.

Possible causes for the depletion of a previously thicker Martian atmosphere include:

- Gradual erosion of the atmosphere by solar wind. On 5 November 2015, NASA announced that data from MAVEN shows that the erosion of Mars' atmosphere increases significantly during solar storms. This shift took place between about 4.2 to 3.7 billion years ago, as the shielding effect of the global magnetic field was lost when the planet's internal dynamo cooled.
- Catastrophic collision by a body large enough to blow away a significant percentage of the atmosphere;
- Mars' low gravity allowing the atmosphere to "blow off" into space by Jeans escape.[192]

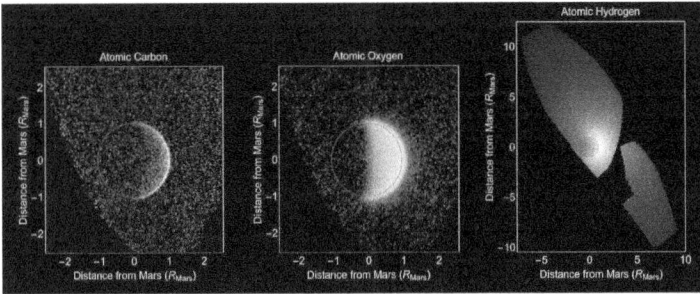

File:PIA18613-MarsMAVEN-Atmosphere-3UV-Views-20141014.jpg

Mars's escaping atmosphere—carbon, oxygen, hydrogen—made by *MAVEN* UV spectrograph).

Images

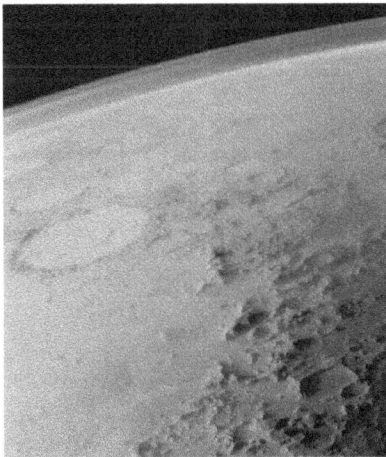

Figure 165: *Mars's thin atmosphere, visible on the horizon.*

Figure 166: *Mars Pathfinder – Martian sky with water ice clouds.*

Figure 167: *A storm front moves in*

File:MarsSunset losslesscrop.jpg

Martian sunset by Spirit rover at Gusev crater (May, 2005).

File:Mars sunset PIA01547.jpg

Martian sunset by Pathfinder at Ares Vallis (July, 1997).

Further reading

- "Mars Clouds Higher Than Any On Earth"[193]. *Space.com.*
- Mikulski, Lauren (2000). "Pressure on the Surface of Mars"[194]. *The Physics Factbook.*
- "Atmospheric pressure – summary of the Viking lander results"[195].
- Khan, Michael (4 December 2009). "The Low Down on Methane on Mars"[196]. Archived from the original[197] on 7 December 2009. Retrieved 8 December 2009.
- "A seasonal ozone layer over the Martian south pole"[198]. ESA. 29 September 2013. Retrieved 17 December 2013.

External links

- ☙ Media related to Atmosphere of Mars at Wikimedia Commons

Climate of Mars

Climate of Mars

The climate of the planet Mars has been a topic of scientific curiosity for centuries, in part because it is the only terrestrial planet whose surface can be directly observed in detail from the Earth with help from a telescope.

Although Mars is smaller than the Earth, at 11% of Earth's mass, and 50% farther from the Sun than the Earth, its climate has important similarities, such as the polar ice caps, seasonal changes and the observable presence of weather patterns. It has attracted sustained study from planetologists and climatologists. While Mars's climate has similarities to Earth's, including periodic ice ages, there are also important differences, such as much lower thermal inertia. Mars' atmosphere has a scale height of approximately 11 km (36,000 ft), 60% greater than that on Earth. The climate is of considerable relevance to the question of whether life is or was present on the planet. The climate briefly received more interest in the news due to NASA measurements indicating increased sublimation of one near-polar region leading to some popular press speculation that Mars was undergoing a parallel bout of global warming, although Mars' average temperature has actually cooled in recent decades, and the polar caps themselves are growing.

Mars has been studied by Earth-based instruments since the 17th century but it is only since the exploration of Mars began in the mid-1960s that close-range observation has been possible. Flyby and orbital spacecraft have provided data from above, while landers and rovers have measured atmospheric conditions directly. Advanced Earth orbital instruments today continue to provide some useful "big picture" observations of relatively large weather phenomena.

The first Martian flyby mission was Mariner 4 which arrived in 1965. That quick two-day pass (July 14–15, 1965) with crude instruments contributed little to the state of knowledge of Martian climate. Later Mariner missions

Figure 168: *Mars as seen by Rosetta in 2007*

(Mariner 6, and Mariner 7) filled in some of the gaps in basic climate information. Data-based climate studies started in earnest with the Viking program landers in 1975 and continue with such probes as the Mars Reconnaissance Orbiter.

This observational work has been complemented by a type of scientific computer simulation called the Mars general circulation model. Several different iterations of MGCM have led to an increased understanding of Mars as well as the limits of such models.

Historical climate observations

Giacomo Maraldi determined in 1704 that the southern cap is not centered on the rotational pole of Mars.[199] During the opposition of 1719, Maraldi observed both polar caps and temporal variability in their extent.

William Herschel was the first to deduce the low density of the Martian atmosphere in his 1784 paper entitled *On the remarkable appearances at the polar regions on the planet Mars, the inclination of its axis, the position of its poles, and its spheroidal figure; with a few hints relating to its real diameter and atmosphere.* When Mars appeared to pass close by two faint stars with no effect on their brightness, Herschel correctly concluded that this meant that there was little atmosphere around Mars to interfere with their light.

Honore Flaugergues 1809 discovery of "yellow clouds" on the surface of Mars is the first known observation of Martian dust storms.[200] Flaugergues also observed in 1813 significant polar ice waning during Martian springtime. His speculation that this meant that Mars was warmer than earth proved inaccurate.

Martian paleoclimatology

There are two dating systems now in use for Martian geological time. One is based on crater density and has three ages: Noachian, Hesperian, and Amazonian. The other is a mineralogical timeline, also having three ages: Phyllocian, Theikian, and Siderikian.

Recent observations and modeling are producing information not only about the present climate and atmospheric conditions on Mars but also about its past. The Noachian-era Martian atmosphere had long been theorized to be carbon dioxide–rich. Recent spectral observations of deposits of clay minerals on Mars and modeling of clay mineral formation conditions have found that there is little to no carbonate present in clay of that era. Clay formation in a carbon dioxide–rich environment is always accompanied by carbonate formation, although the carbonate may later be dissolved by volcanic acidity.

The discovery of water-formed minerals on Mars including hematite and jarosite, by the Opportunity rover and goethite by the Spirit rover, has led to the conclusion that climatic conditions in the distant past allowed for free-flowing water on Mars. The morphology of some crater impacts on Mars indicate that the ground was wet at the time of impact. Geomorphic observations of both landscape erosion rates and Martian valley networks also strongly imply warmer, wetter conditions on Noachian-era Mars (earlier than about 4 billion years ago). However, chemical analysis of Martian meteorite samples suggests that the ambient near-surface temperature of Mars has most likely been below 0 °C for the last four billion years.

Some scientists maintain that the great mass of the Tharsis volcanoes has had a major influence on Mars's climate. Erupting volcanoes give off great amounts of gas, mainly water vapor and CO_2. Enough gas may have been released by volcanoes to have made the earlier Martian atmosphere thicker than Earth's. The volcanoes could also have emitted enough H_2O to cover the whole Martian surface to a depth of 120 m (390 ft). CO_2 is a greenhouse gas that raises the temperature of a planet: it traps heat by absorbing infrared radiation. So Tharsis volcanoes, by giving off CO_2, could have made Mars more Earth-like in the past. Mars may have once had a much thicker and warmer atmosphere,

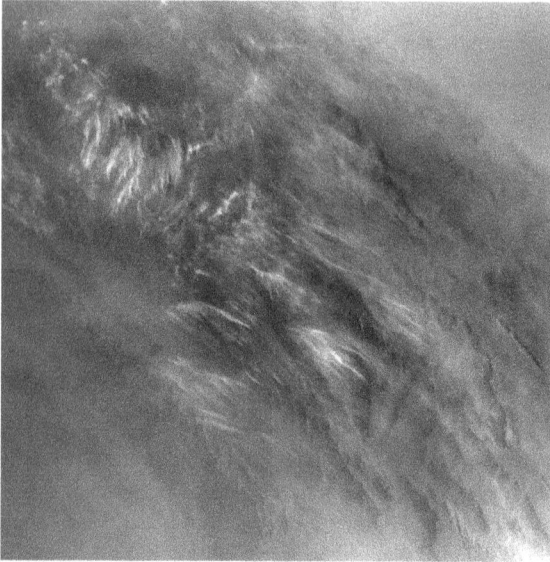

Figure 169: *Martian morning clouds (Viking Orbiter 1, 1976)*

and oceans and/or lakes may have been present.[201] It has, however, proven extremely difficult to construct convincing global climate models for Mars which produce temperatures above 0 °C at any point in its history, although this may simply reflect problems in accurately calibrating such models.

Weather

Mars' temperature and circulation vary every Martian year (as expected for any planet with an atmosphere). Mars lacks oceans, a source of much inter-annual variation on Earth.Wikipedia:Please clarify Mars Orbiter Camera data beginning in March 1999 and covering 2.5 Martian years show that Martian weather tends to be more repeatable and hence more predictable than that of Earth. If an event occurs at a particular time of year in one year, the available data (sparse as it is) indicate that it is fairly likely to repeat the next year at nearly the same location, give or take a week.

On September 29, 2008, the Phoenix lander took pictures of snow falling from clouds 4.5 km above its landing site near Heimdal Crater. The precipitation vaporized before reaching the ground, a phenomenon called virga.

Figure 170: *Ice clouds moving above the Phoenix landing site over a period of 10 min (August 29, 2008)*

Clouds

Mars' dust storms can kick up fine particles in the atmosphere around which clouds can form. These clouds can form very high up, up to 100 km (62 mi) above the planet. The clouds are very faint and can only be seen reflecting sunlight against the darkness of the night sky. In that respect, they look similar to the mesospheric clouds, also known as noctilucent clouds on Earth, which occur about 80 km (50 mi) above our planet.

Temperature

Measurements of Martian temperature predate the Space Age. However, early instrumentation and techniques of radio astronomy produced crude, differing results. Early flyby probes (Mariner 4) and later orbiters used radio occultation to perform aeronomy. With chemical composition already deduced from spectroscopy, temperature and pressure could then be derived. Nevertheless, flyby occultations can only measure properties along two transects, at their trajectories' entries and exits from Mars' disk as seen from Earth. This results in weather "snapshots" at a particular area, at a particular time. Orbiters then increase the number of radio transects. Later missions, starting with the dual

Mariner 6 and 7 flybys, plus the Soviet Mars 2 and 3, carried infrared detectors to measure radiant energy. Mariner 9 was the first to place an infrared radiometer and spectrometer in Mars orbit in 1971, along with its other instruments and radio transmitter. Viking 1 and 2 followed, with not merely Infrared Thermal Mappers (IRTM). The missions could also corroborate these remote sensing datasets with not only their *in situ* lander metrology booms, but with higher-altitude temperature and pressure sensors for their descent.

Differing *in situ* values have been reported for the average temperature on Mars, with a common value being –55 °C (218 K; –67 °F). Surface temperatures may reach a high of about 20 °C (293 K; 68 °F) at noon, at the equator, and a low of about –153 °C (120 K; –243 °F) at the poles. Actual temperature measurements at the Viking landers' site range from –17.2 °C (256.0 K; 1.0 °F) to –107 °C (166 K; –161 °F). The warmest soil temperature estimated by the Viking Orbiter was 27 °C (300 K; 81 °F).[202] The Spirit rover recorded a maximum daytime air temperature in the shade of 35 °C (308 K; 95 °F), and regularly recorded temperatures well above 0 °C (273 K; 32 °F), except in winter.[203]

It has been reported that "On the basis of the nightime air temperature data, every northern spring and early northern summer yet observed were identical to within the level of experimental error (to within ±1 °C)" but that the "daytime data, however, suggest a somewhat different story, with temperatures varying from year-to-year by up to 6 °C in this season. This day-night discrepancy is unexpected and not understood". In southern spring and summer, variance is dominated by dust storms which increase the value of the night low temperature and decrease the daytime peak temperature.[204] This results in a small (20 °C) decrease in average surface temperature, and a moderate (30 °C) increase in upper atmosphere temperature.

Before and after the Viking missions, newer, more advanced Martian temperatures were determined from Earth via microwave spectroscopy. As the microwave beam, of under 1 arcminute, is larger than the disk of the planet, the results are global averages. Later, the Mars Global Surveyor's Thermal Emission Spectrometer and to a lesser extent 2001 Mars Odyssey's THEMIS could not merely reproduce infrared measurements but intercompare lander, rover, and Earth microwave data. The Mars Reconnaissance Orbiter's Mars Climate Sounder can similarly derive atmospheric profiles. The datasets "suggest generally colder atmospheric temperatures and lower dust loading in recent decades on Mars than during the Viking Mission," although Viking data had previously been revised downward. The TES data indicates "Much colder (10–20 K) global atmospheric temperatures were observed during the 1997 versus 1977 perihelion periods" and "that the global aphelion atmosphere of Mars is colder, less dusty, and cloudier than indicated by the established Viking climatology," again, taking into account the Wilson and Richardson revisions to Viking data.

A later comparison, while admitting "it is the microwave record of air temperatures which is the most representative," attempted to merge the discontinuous spacecraft record. No measurable trend in global average temperature between Viking IRTM and MGS TES was visible. "Viking and MGS air temperatures are essentially indistinguishable for this period, suggesting that the Viking and MGS eras are characterized by essentially the same climatic state." It found "a strong dichotomy" between the northern and southern hemispheres, a "very asymmetric paradigm for the Martian annual cycle: a northern spring and summer which is relatively cool, not very dusty, and relatively rich in water vapor and ice clouds; and a southern summer rather similar to that observed by Viking with warmer air temperatures, less water vapor and water ice, and higher levels of atmospheric dust."

The Mars Reconnaissance Orbiter MCS (Mars Climate Sounder) instrument was, upon arrival, able to operate jointly with MGS for a brief period; the less-capable Mars Odyssey THEMIS and Mars Express SPICAM datasets may also be used to span a single, well-calibrated record. While MCS and TES temperatures are generally consistent, investigators report possible cooling below the analytical precision. "After accounting for this modeled cooling, MCS MY 28 temperatures are an average of 0.9 (daytime) and 1.7 K (night-time) cooler than TES MY 24 measurements."

It has been suggested that Mars had a much thicker, warmer atmosphere early in its history. Much of this early atmosphere would have consisted of carbon dioxide. Such an atmosphere would have raised the temperature, at least in some places, to above the freezing point of water. With the higher temperature running water could have carved out the many channels and outflow valleys that are common on the planet. It also may have gathered together to form lakes and maybe an ocean. Some researchers have suggested that the atmosphere of Mars may have been many times as thick as the Earth's; however research published in September 2015 advanced the idea that perhaps the early Martian atmosphere was not as thick as previously thought.[205]

Currently, the atmosphere is very thin. For many years, it was assumed that as with the Earth, most of the early carbon dioxide was locked up in minerals, called carbonates. However, despite the use of many orbiting instruments that looked for carbonates, very few carbonate deposits have been found. Today, it is thought that much of the carbon dioxide in the Martian air was removed by the solar wind. Researchers have discovered a two-step process that sends the gas into space. Ultraviolet light from the Sun could strike a carbon dioxide molecule, breaking it into carbon monoxide and oxygen. A second photon of ultraviolet light could subsequently break the carbon monoxide into oxygen and

carbon which would get enough energy to escape the planet. In this process the light isotope of carbon (C
12) would be most likely to leave the atmosphere. Hence, the carbon dioxide left in the atmosphere would be enriched with the heavy isotope (C
13).[206] This higher level of the heavy isotope is what was found by the *Curiosity* rover on Mars.

Climate data for Gale Crater (2012–2015)

Month	Jan	Feb	Mar	Apr	May	Jun	Jul	Aug	Sep	Oct	Nov	Dec	Year
Record high °C (°F)	6 (43)	6 (43)	1 (34)	0 (32)	7 (45)	14 (57)	20 (68)	19 (66)	7 (45)	7 (45)	8 (46)	8 (46)	20 (68)
Average high °C (°F)	-7 (19)	-18 (0)	-23 (-9)	-20 (-4)	-4 (25)	0.0 (32)	2 (36)	1 (34)	1 (34)	4 (39)	-1 (30)	-3 (27)	-5.7 (21.7)
Average low °C (°F)	-82 (-116)	-86 (-123)	-88 (-126)	-87 (-125)	-85 (-121)	-78 (-108)	-76 (-105)	-69 (-92)	-68 (-90)	-73 (-99)	-73 (-99)	-77 (-107)	-78.5 (-109.3)
Record low °C (°F)	-95 (-139)	-127 (-197)	-114 (-173)	-97 (-143)	-98 (-144)	-125 (-193)	-84 (-119)	-80 (-112)	-78 (-108)	-79 (-110)	-83 (-117)	-110 (-166)	-127 (-197)

Source: Centro de Astrobiología, Mars Weather, NASA Quest, SpaceDaily

Atmospheric properties and processes

Low atmospheric pressure

The Martian atmosphere is composed mainly of carbon dioxide and has a mean surface pressure of about 600 pascals (Pa), much lower than the Earth's 101,000 Pa. One effect of this is that Mars' atmosphere can react much more quickly to a given energy input than that of Earth's atmosphere. As a consequence, Mars is subject to strong thermal tides produced by solar heating rather than a gravitational influence. These tides can be significant, being up to 10% of the total atmospheric pressure (typically about 50 Pa). Earth's atmosphere experiences similar diurnal and semidiurnal tides but their effect is less noticeable because of Earth's much greater atmospheric mass.

Although the temperature on Mars can reach above freezing (0 °C (273 K; 32 °F)), liquid water is unstable over much of the planet, as the atmospheric pressure is below water's triple point and water ice sublimes into water vapor. Exceptions to this are the low-lying areas of the planet, most notably in the Hellas Planitia impact basin, the largest such crater on Mars. It is so deep that the atmospheric pressure at the bottom reaches 1155 Pa, which is above the triple point, so if the temperature exceeded 0 °C liquid water could exist there.Wikipedia:Citation needed

Wind

The surface of Mars has a very low thermal inertia, which means it heats quickly when the sun shines on it. Typical daily temperature swings, away from the polar regions, are around 100 K. On Earth, winds often develop in areas where thermal inertia changes suddenly, such as from sea to land. There are no seas on Mars, but there are areas where the thermal inertia of the soil changes, leading to morning and evening winds akin to the sea breezes on Earth. The Antares project "Mars Small-Scale Weather" (MSW) has recently identified some minor weaknesses in current global climate models (GCMs) due to the GCMs' more primitive soil modeling "heat admission to the ground and back is quite important in Mars, so soil schemes have to be quite accurate. "[209] Those weaknesses are being corrected and should lead to more accurate future assessments, but make continued reliance on older predictions of modeled Martian climate somewhat problematic.

At low latitudes the Hadley circulation dominates, and is essentially the same as the process which on Earth generates the trade winds. At higher latitudes a series of high and low pressure areas, called baroclinic pressure waves, dominate the weather. Mars is drier and colder than Earth, and in consequence dust raised by these winds tends to remain in the atmosphere longer than on Earth

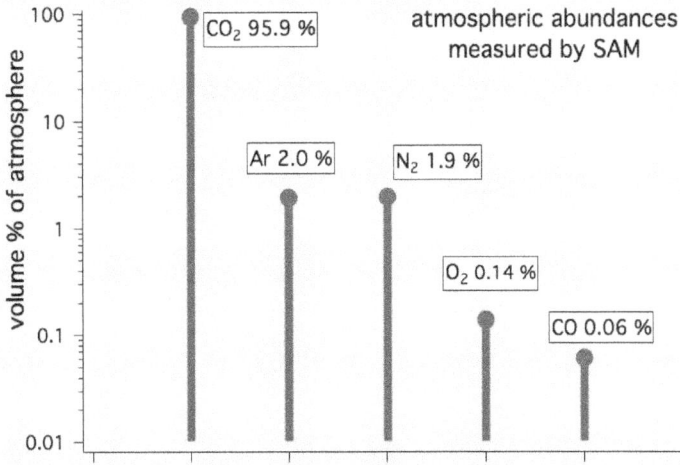

Figure 171: *Planet Mars – most abundant gases – (Curiosity rover, Sample Analysis at Mars device, October 2012).*

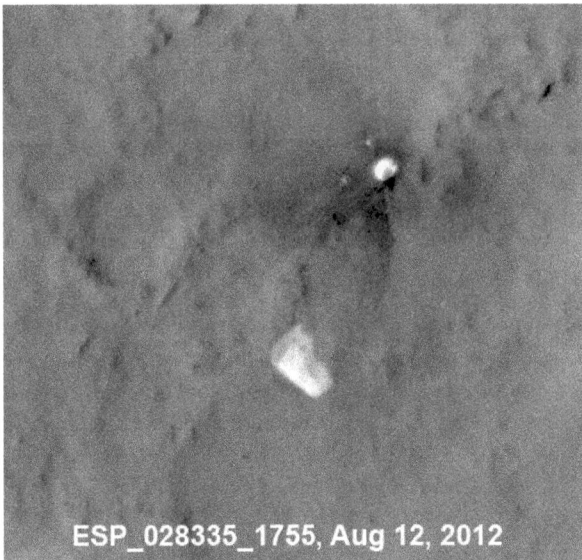

Figure 172: *Curiosity rover's parachute flapping in the Martian wind (HiRISE/MRO) (August 12, 2012 to January 13, 2013).*

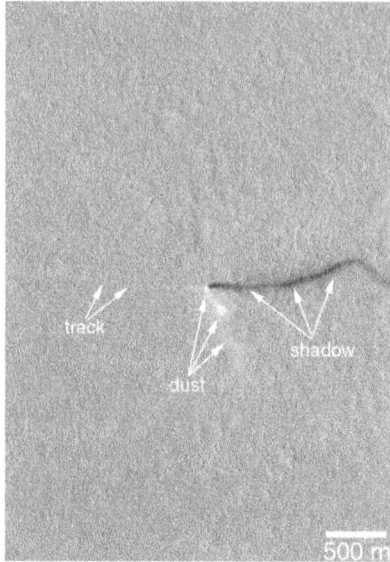

Figure 173: *Martian Dust Devil – in Amazonis Planitia (April 10, 2001) (also*[207]*) (video (02:19)*[208]*).*

as there is no precipitation to wash it out (excepting CO_2 snowfall). One such
cyclonic storm was recently captured by the Hubble space telescope (pictured
below).

One of the major differences between Mars' and Earth's Hadley circulations is
their speed which is measured on an overturning timescale. The overturning
timescale on Mars is about 100 Martian days while on Earth, it is over a year.

Effect of dust storms

When the Mariner 9 probe arrived at Mars in 1971, the world expected to see
crisp new pictures of surface detail. Instead they saw a near planet-wide dust
storm with only the giant volcano Olympus Mons showing above the haze.
The storm lasted for a month, an occurrence scientists have since learned is
quite common on Mars. Using data from Mariner 9, James B. Pollack et
al. proposed a *mechanism for Mars dust storms* in 1973.[210] <templatestyles
src="Multiple_image/styles.css" />

Dust storms on Mars.

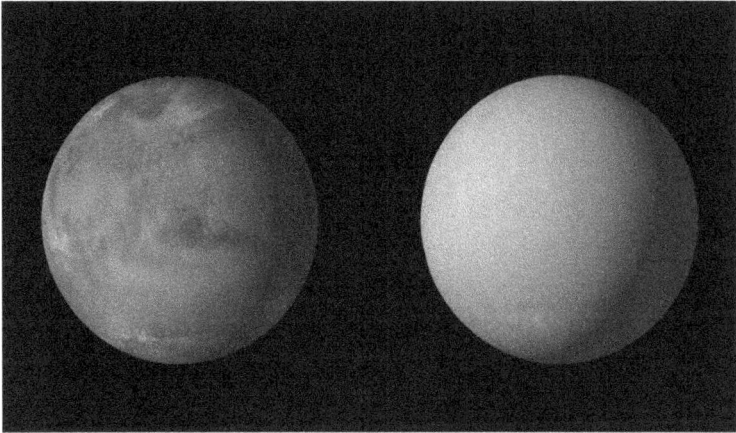

Figure 174:
Mars (before/after) dust storm
(July 2018)

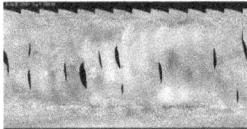

June 6, 2018.

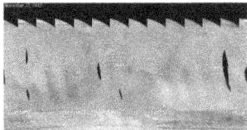

November 25, 2012

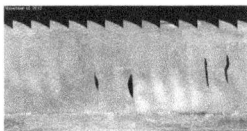

November 18, 2012

Locations of *Opportunity* and *Curiosity* rovers are noted (MRO).

As observed by the Viking spacecraft from the surface, "during a global dust storm the diurnal temperature range narrowed sharply, from 50° Celsius to only about ten degrees, and the wind speeds picked up considerably—indeed, within only an hour of the storm's arrival they had increased to 17 m/s (61 km/

Figure 175: *2001 Hellas Basin dust storm*

Figure 176: *Time-lapse composite of Martian horizon over 30 Martian days shows how much sunlight the July 2007 dust storms blocked; Tau of 4.7 indicates 99% blocked.*

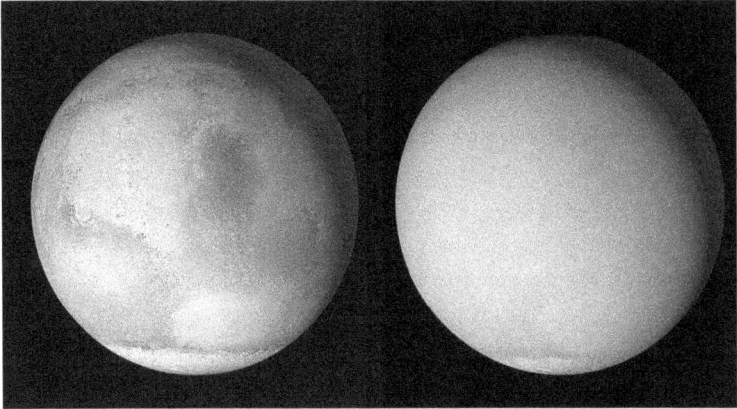

Figure 177: *Mars without a dust storm on June 2001 (on left) and with a global dust storm on July 2001 (on right), as seen by Mars Global Surveyor*

h), with gusts up to 26 m/s (94 km/h). Nevertheless, no actual transport of material was observed at either site, only a gradual brightening and loss of contrast of the surface material as dust settled onto it." On June 26, 2001, the Hubble Space Telescope spotted a dust storm brewing in Hellas Basin on Mars (pictured right). A day later the storm "exploded" and became a global event. Orbital measurements showed that this dust storm reduced the average temperature of the surface and raised the temperature of the atmosphere of Mars by 30 K. The low density of the Martian atmosphere means that winds of 18 to 22 m/s (65 to 79 km/h) are needed to lift dust from the surface, but since Mars is so dry, the dust can stay in the atmosphere far longer than on Earth, where it is soon washed out by rain. The season following that dust storm had daytime temperatures 4 K below average. This was attributed to the global covering of light-colored dust that settled out of the dust storm, temporarily increasing Mars' albedo.

In mid-2007 a planet-wide dust storm posed a serious threat to the solar-powered *Spirit* and *Opportunity* Mars Exploration Rovers by reducing the amount of energy provided by the solar panels and necessitating the shut-down of most science experiments while waiting for the storms to clear. Following the dust storms, the rovers had significantly reduced power due to settling of dust on the arrays.

Dust storms are most common during perihelion, when the planet receives 40 percent more sunlight than during aphelion. During aphelion water ice clouds form in the atmosphere, interacting with the dust particles and affecting the temperature of the planet.

A large intensifying dust storm began in late-May 2018 and had persisted as of mid-June. By 10 June 2018, as observed at the location of the rover *Opportunity*, the storm was more intense than the 2007 dust storm endured by *Opportunity*.[211] On 20 June 2018, NASA reported that the dust storm had grown to completely cover the entire planet.

It has been suggested that dust storms on Mars could play a role in storm formation similar to that of water clouds on Earth.Wikipedia:Citation needed Observation since the 1950s has shown that the chances of a planet-wide dust storm in a particular Martian year are approximately one in three.

Dust storms contribute to water loss on Mars. A study of dust storms with the Mars Reconnaissance Orbiter suggested that 10 percent of the water loss from Mars may have been caused by dust storms. Instruments on board the Mars Reconnaissance Orbiter detected observed water vapor at very high altitudes during global dust storms. Ultraviolet light from the sun can then break the water apart into hydrogen and oxygen. The Hydrogen from the water molecule then escapes into space.[212,213,214]

Saltation

The process of geological saltation is quite important on Mars as a mechanism for adding particulates to the atmosphere. Saltating sand particles have been observed on the MER *Spirit* rover.[215] Theory and real world observations have not agreed with each other, classical theory missing up to half of real-world saltating particles. A new model more closely in accord with real world observations demonstrates that saltating particles create an electrical field that increases the saltation effect. Mars grains saltate in 100 times higher and longer trajectories and reach 5–10 times higher velocities than Earth grains do.

Repeating northern annular cloud

A large doughnut shaped cloud appears in North polar region of Mars around the same time every Martian year and of about the same size. It forms in the morning and dissipates by the Martian afternoon. The outer diameter of the cloud is roughly 1,600 km (1,000 mi), and the inner hole or eye is 320 km (200 mi) across. The cloud is thought to be composed of water-ice, so it is white in color, unlike the more common dust storms.

It looks like a cyclonic storm, similar to a hurricane, but it does not rotate. The cloud appears during the northern summer and at high latitude. Speculation is that this is due to unique climate conditions near the northern pole. Cyclone-like storms were first detected during the Viking orbital mapping program, but the northern annular cloud is nearly three times larger. The cloud has also

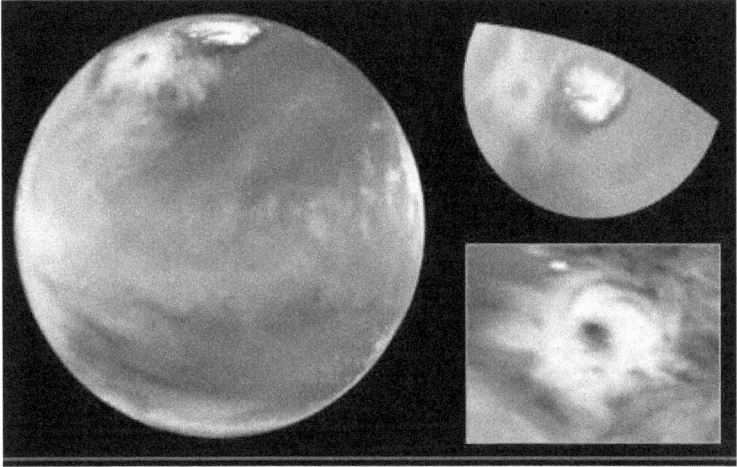

Figure 178: *Hubble view of the colossal polar cloud on Mars*

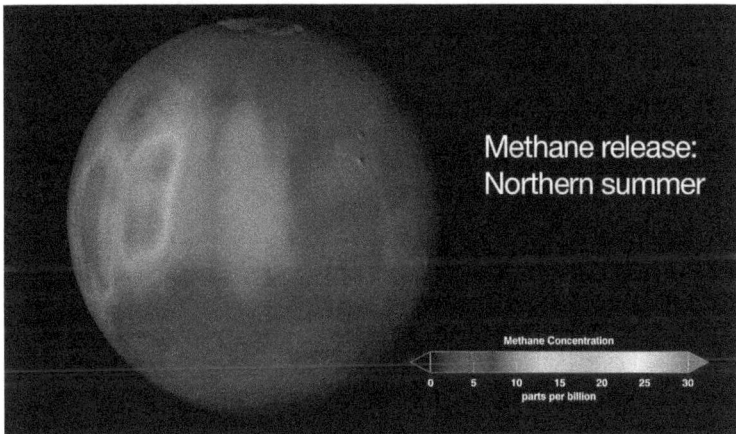

Figure 179: *Methane map*

been detected by various probes and telescopes including the Hubble and Mars Global Surveyor.

Other repeating events are dust storms and dust devils.

Methane presence

Although methane is a greenhouse gas on Earth, the small amounts that have been claimed to be present on Mars would have little effect on the Martian

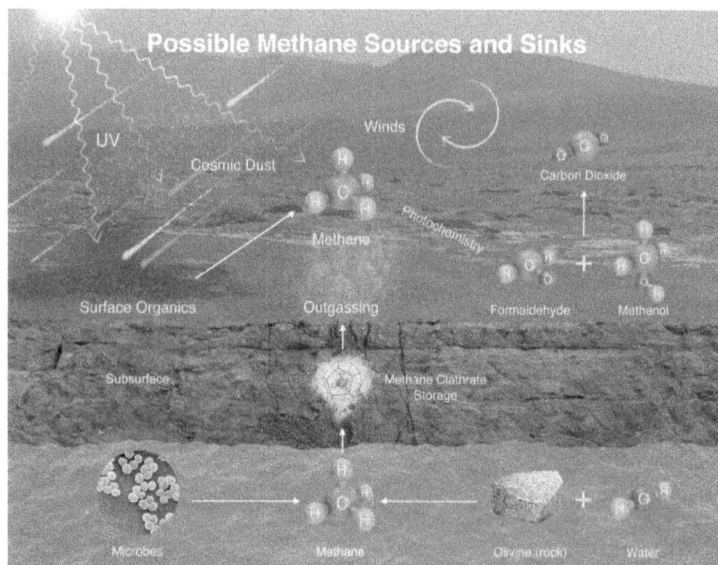

Figure 180: *Methane (CH₄) on Mars: potential sources and sinks*

global climate. Trace amounts of methane (CH_4) at concentration of several parts per billion (ppb), were first reported in the atmosphere of Mars by a team at the NASA Goddard Space Flight Center in 2003.[216]

In March 2004 the Mars Express Orbiter and ground-based observations from Canada-France-Hawaii Telescope also suggested the presence of methane in the atmosphere with a mole fraction of about 10 nmol/mol. However, the complexity of these observations has sparked discussion as to the reliability of the results. Since breakup of that much methane by ultraviolet light would only take 350 years under current Martian conditions, if methane is present some sort of active source must be replenishing the gas. Clathrate hydrates, or water-rock reactions could be possible geological sources of methane but there is presently no consensus on the source or existence of Martian methane.

The Curiosity rover landed on Mars in August 2012. It is able to make precise abundance measurements and also distinguish between different isotopologues of methane. The first measurements with the Tunable Laser Spectrometer (TLS) indicate that there is less than 5 ppb of methane at the landing site. On September 19, 2013 NASA scientists used further measurements from Curiosity to report a non-detection of atmospheric methane with a measured value of 0.18±0.67 ppbv corresponding to an upper limit of 1.3 ppbv (95% confidence limit).

On 16 December 2014, NASA reported the *Curiosity* rover detected a "tenfold spike", likely localized, in the amount of methane in the Martian atmosphere. Sample measurements taken "a dozen times over 20 months" showed increases in late 2013 and early 2014, averaging "7 parts of methane per billion in the atmosphere." Before and after that, readings averaged around one-tenth that level.

The Indian Mars Orbiter Mission, launched in November 5, 2013, will attempt to detect and map the sources of methane, if they exist. The ExoMars Trace Gas Orbiter planned to launch in 2016 would further study the methane, as well as its decomposition products such as formaldehyde and methanol.

Carbon dioxide carving

Mars Reconnaissance Orbiter images suggest an unusual erosion effect occurs based on Mars' unique climate. Spring warming in certain areas leads to CO_2 ice subliming and flowing upwards, creating highly unusual erosion patterns called "spider gullies". Translucent CO_2 ice forms over winter and as the spring sunlight warms the surface, it vaporizes the CO_2 to gas which flows uphill under the translucent CO_2 ice. Weak points in that ice lead to CO_2 geysers.

Mountains

Martian storms are significantly affected by Mars' large mountain ranges. Individual mountains like record holding Olympus Mons (26 km (85,000 ft)) can affect local weather but larger weather effects are due to the larger collection of volcanoes in the Tharsis region.

One unique repeated weather phenomenon involving mountains is a spiral dust cloud that forms over Arsia Mons. The spiral dust cloud over Arsia Mons can tower 15 to 30 km (49,000 to 98,000 ft) above the volcano. Clouds are present around Arsia Mons throughout the Martian year, peaking in late summer.

Clouds surrounding mountains display a seasonal variability. Clouds at Olympus Mons and Ascreaus Mons appear in northern hemisphere spring and summer, reaching a total maximum area of approximately 900,000 km^2 and 1,000,000 km^2 respectively in late spring. Clouds around Alba Patera and Pavonis Mons show an additional, smaller peak in late summer. Very few clouds were observed in winter. Predictions from the Mars General Circulation Model are consistent with these observations.

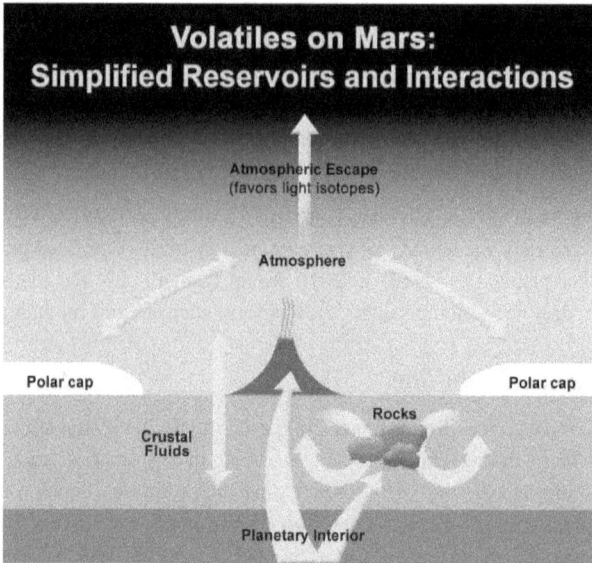

Figure 181: *Planet Mars' volatile gases* (*Curiosity rover, October 2012*)

Polar caps

Mars has ice caps at its north pole and south pole, which mainly consist of water ice; however, there is frozen carbon dioxide (dry ice) present on their surfaces. Dry ice accumulates in the north polar region (Planum Boreum) in winter only, subliming completely in summer, while the south polar region additionally has a permanent dry ice cover up to eight meters (25 feet) thick. This difference is due to the higher elevation of the south pole.

So much of the atmosphere can condense at the winter pole that the atmospheric pressure can vary by up to a third of its mean value. This condensation and evaporation will cause the proportion of the noncondensable gases in the atmosphere to change inversely. The eccentricity of Mars's orbit affects this cycle, as well as other factors. In the spring and autumn wind due to the carbon dioxide sublimation process is so strong that it can be a cause of the global dust storms mentioned above.

The northern polar cap has a diameter of approximately 1,000 km during the northern Mars summer, and contains about 1.6 million cubic kilometres of ice, which if spread evenly on the cap would be 2 km thick. (This compares to a volume of 2.85 million cubic kilometres for the Greenland ice sheet.) The southern polar cap has a diameter of 350 km and a maximum thickness of 3 km. Both polar caps show spiral troughs, were initially thought to form as

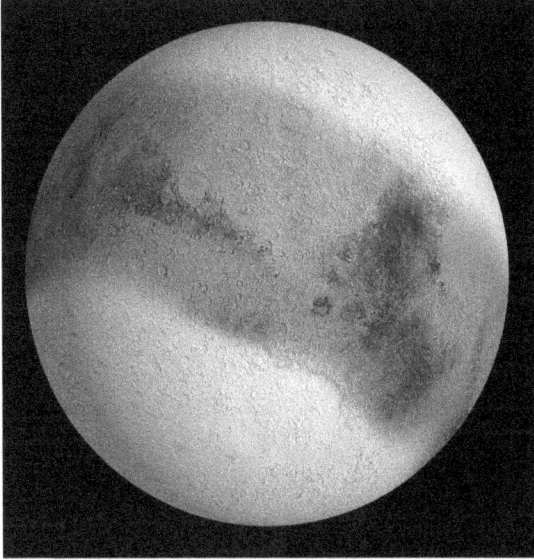

Figure 182: *How Mars might have looked during an ice age between 2.1 million and 400,000 years ago, when Mars's axial tilt is thought to have been larger than today.*

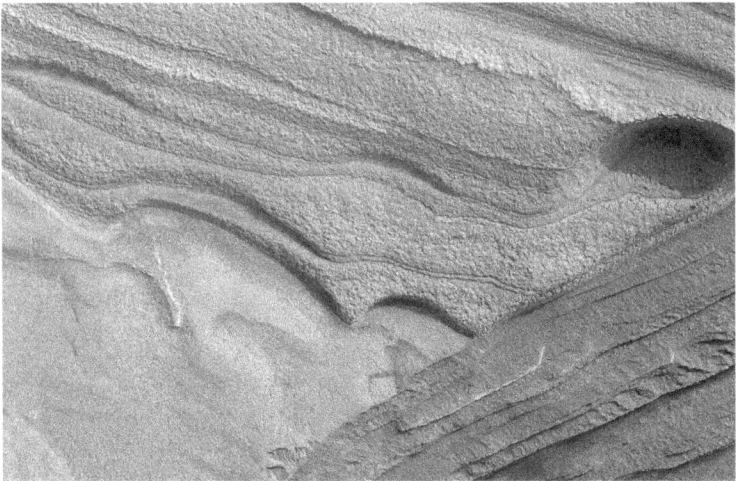

Figure 183: *HiRISE view of Olympia Rupes in Planum Boreum, one of many exposed water ice layers found in the polar regions of Mars. Depicted width: 1.3 km (0.8 miles)*

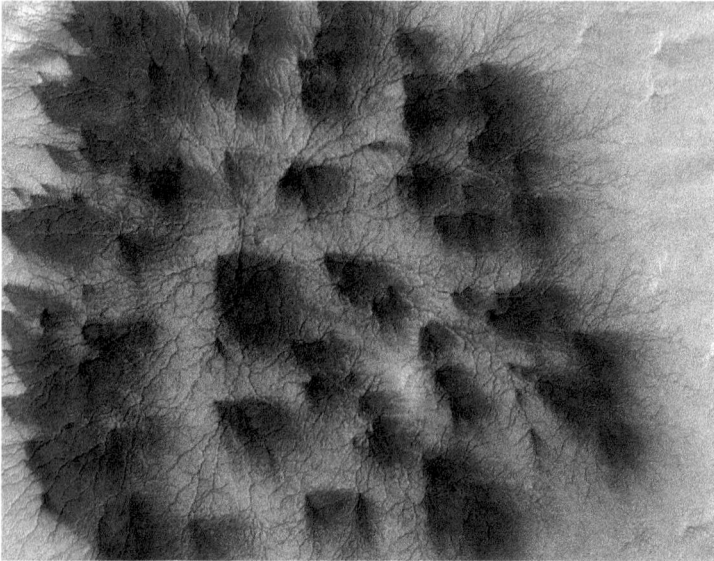

Figure 184: *HiRISE image of "dark dune spots" and fans formed by eruptions of CO_2 gas geysers on Mars' south polar ice sheet.*

a result of differential solar heating, coupled with the sublimation of ice and condensation of water vapor. Recent analysis of ice penetrating radar data from SHARAD has demonstrated that the spiral troughs are formed from a unique situation in which high density katabatic winds descend from the polar high to transport ice and create large wavelength bedforms. The spiral shape comes from Coriolis effect forcing of the winds, much like winds on earth spiral to form a hurricane. The troughs did not form with either ice cap, instead they began to form between 2.4 million and 500,000 years ago, after three fourths of the ice cap was in place. This suggests that a climatic shift allowed for their onset. Both polar caps shrink and regrow following the temperature fluctuation of the Martian seasons; there are also longer-term trends that are better understood in the modern era.

During the southern hemisphere spring, solar heating of dry ice deposits at the south pole leads in places to accumulation of pressurized CO_2 gas below the surface of the semitransparent ice, warmed by absorption of radiation by the darker substrate. After attaining the necessary pressure, the gas bursts through the ice in geyser-like plumes. While the eruptions have not been directly observed, they leave evidence in the form of "dark dune spots" and lighter fans atop the ice, representing sand and dust carried aloft by the eruptions, and a spider-like pattern of grooves created below the ice by the outrushing gas. (see

Geysers on Mars.) Eruptions of nitrogen gas observed by *Voyager 2* on Triton are thought to occur by a similar mechanism.

Both polar caps are currently accumulating, confirming predicted Milankovich cycling on timescales of $\sim$400,000 and $\sim$4,000,000 years. Soundings by the Mars Reconnaissance Orbiter SHARAD indicate total cap growth of $\sim$0.24 km3/year. Of this, 92%, or $\sim$0.86 mm/year, is going to the north, as Mars' offset Hadley circulation acts as a nonlinear pump of volatiles northward.

Solar wind

Mars lost most of its magnetic field about four billion years ago. As a result, solar wind and cosmic radiation interacts directly with the Martian ionosphere. This keeps the atmosphere thinner than it would otherwise be by solar wind action constantly stripping away atoms from the outer atmospheric layer. Most of the historical atmospheric loss on Mars can be traced back to this solar wind effect. Current theory posits a weakening solar wind and thus today's atmosphere stripping effects are much less than those in the past when the solar wind was stronger.Wikipedia:Citation needed

Seasons

Mars has an axial tilt of 25.2°. This means that there are seasons on Mars, just as on Earth. The eccentricity of Mars' orbit is 0.1, much greater than the Earth's present orbital eccentricity of about 0.02. The large eccentricity causes the insolation on Mars to vary as the planet orbits the Sun. (The Martian year lasts 687 days, roughly 2 Earth years.) As on Earth, Mars' obliquity dominates the seasons but, because of the large eccentricity, winters in the southern hemisphere are long and cold while those in the North are short and warm.

It is now thought that ice accumulated when Mars' orbital tilt was very different from what it is now. (The axis the planet spins on has considerable "wobble," meaning its angle changes over time.)[217,218,219] A few million years ago, the tilt of the axis of Mars was 45 degrees instead of its present 25 degrees. Its tilt, also called obliquity, varies greatly because its two tiny moons cannot stabilize it like our moon.

Many features on Mars, especially in the Ismenius Lacus quadrangle, are thought to contain large amounts of ice. The most popular model for the origin of the ice is climate change from large changes in the tilt of the planet's rotational axis. At times the tilt has even been greater than 80 degrees Large changes in the tilt explains many ice-rich features on Mars.

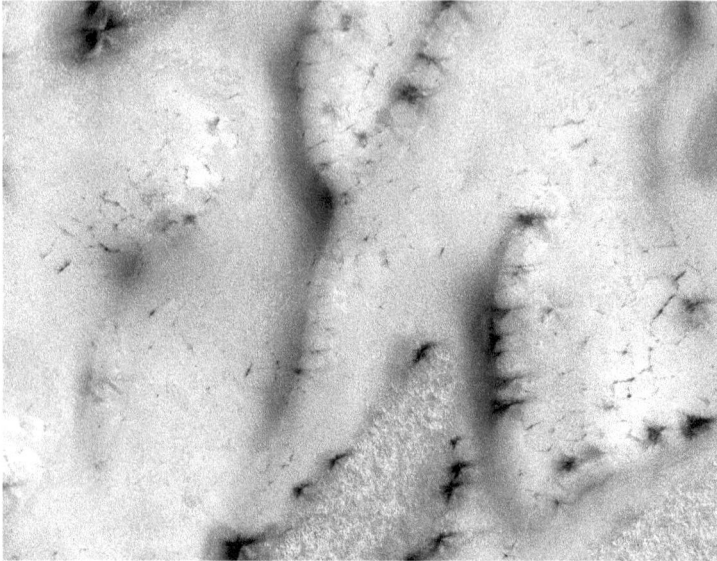

Figure 185: *In spring, sublimation of ice causes sand from below the ice layer to form fan-shaped deposits on top of the seasonal ice.*

Studies have shown that when the tilt of Mars reaches 45 degrees from its current 25 degrees, ice is no longer stable at the poles. Furthermore, at this high tilt, stores of solid carbon dioxide (dry ice) sublimate, thereby increasing the atmospheric pressure. This increased pressure allows more dust to be held in the atmosphere. Moisture in the atmosphere will fall as snow or as ice frozen onto dust grains. Calculations suggest this material will concentrate in the mid-latitudes.[220] General circulation models of the Martian atmosphere predict accumulations of ice-rich dust in the same areas where ice-rich features are found. When the tilt begins to return to lower values, the ice sublimates (turns directly to a gas) and leaves behind a lag of dust. The lag deposit caps the underlying material so with each cycle of high tilt levels, some ice-rich mantle remains behind.[221] Note, that the smooth surface mantle layer probably represents only relative recent material. Below are images of layers in this smooth mantle that drops from the sky at times.

Present unequal lengths of the seasons

Season	Mars' Sols	Earth Days
Northern Spring, Southern Autumn	193.30	92.764
Northern Summer, Southern Winter	178.64	93.647
Northern Autumn, Southern Spring	142.70	89.836
Northern Winter, Southern Summer	153.95	88.997

Precession in the alignment of the obliquity and eccentricity lead to global warming and cooling ('great' summers and winters) with a period of 170,000 years.

Like Earth, the obliquity of Mars undergoes periodic changes which can lead to long-lasting changes in climate. Once again, the effect is more pronounced on Mars because it lacks the stabilizing influence of a large moon. As a result, the obliquity can alter by as much as 45°. Jacques Laskar, of France's National Centre for Scientific Research, argues that the effects of these periodic climate changes can be seen in the layered nature of the ice cap at the Martian north pole. Current research suggests that Mars is in a warm interglacial period which has lasted more than 100,000 years.

Because the Mars Global Surveyor was able to observe Mars for 4 Martian years, it was found that Martian weather was similar from year to year. Any differences were directly related to changes in the solar energy that reached Mars. Scientists were even able to accurately predict dust storms that would occur during the landing of Beagle 2. Regional dust storms were discovered to be closely related to where dust was available.[222]

Evidence for recent climatic change

There have been regional changes around the south pole (Planum Australe) over the past few Martian years. In 1999 the Mars Global Surveyor photographed pits in the layer of frozen carbon dioxide at the Martian south pole. Because of their striking shape and orientation these pits have become known as swiss cheese features. In 2001 the craft photographed the same pits again and found that they had grown larger, retreating about 3 meters in one Martian year. These features are caused by the sublimation of the dry ice layer, thereby exposing the inert water ice layer. More recent observations indicate that the ice at Mars' south pole is continuing to sublimate. The pits in the ice continue to grow by about 3 meters per Martian year. Malin states that conditions on Mars are not currently conducive to the formation of new ice. A NASA press

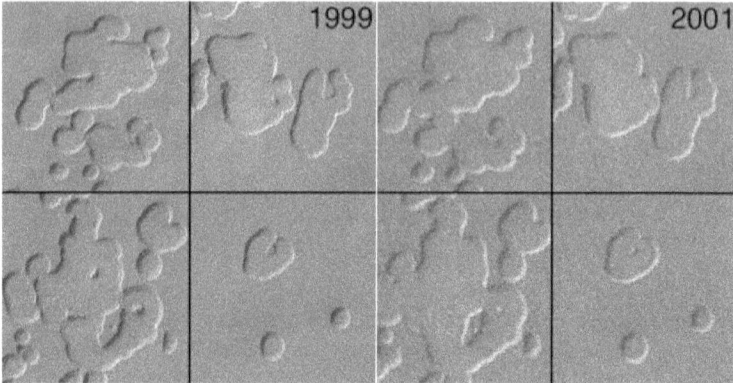

Figure 186: *Pits in south polar ice cap (MGS 1999, NASA)*

release indicates that "climate change [is] in progress" on Mars. In a summary of observations with the Mars Orbiter Camera, researchers speculated that some dry ice may have been deposited between the Mariner 9 and the Mars Global Surveyor mission. Based on the current rate of loss, the deposits of today may be gone in a hundred years.

Elsewhere on the planet, low latitude areas have more water ice than they should have given current climatic conditions.[223] Mars Odyssey "is giving us indications of recent global climate change in Mars," said Jeffrey Plaut, project scientist for the mission at NASA's Jet Propulsion Laboratory, in non-peer reviewed published work in 2003.

Attribution theories

Polar changes

Colaprete et al. conducted simulations with the Mars General Circulation Model which show that the local climate around the Martian south pole may currently be in an unstable period. The simulated instability is rooted in the geography of the region, leading the authors to speculate that the sublimation of the polar ice is a local phenomenon rather than a global one. The researchers showed that even with a constant solar luminosity the poles were capable of jumping between states of depositing or losing ice. The trigger for a change of states could be either increased dust loading in the atmosphere or an albedo change due to deposition of water ice on the polar cap. This theory is somewhat problematic due to the lack of ice deposition after the 2001 global dust storm. Another issue is that the accuracy of the Mars General Circulation Model decreases as the scale of the phenomenon becomes more local.

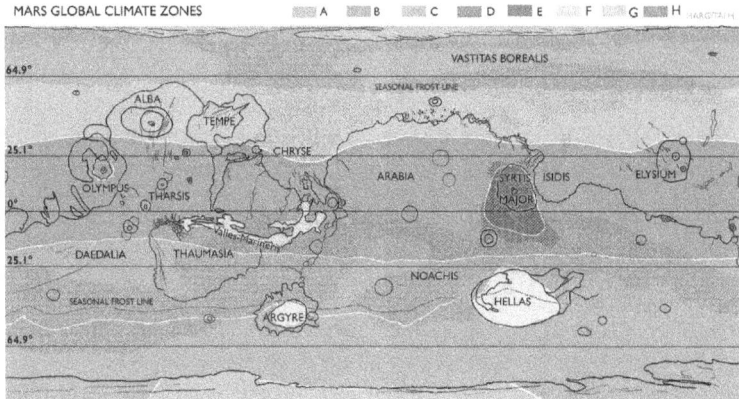

Figure 187: *Mars Global Climate Zones (based on temperature, modified by topography, albedo, actual solar radiation)*

It has been argued that "observed regional changes in south polar ice cover are almost certainly due to a regional climate transition, not a global phenomenon, and are demonstrably unrelated to external forcing." Writing in a *Nature* news story, Chief News and Features Editor Oliver Morton said "The warming of other solar bodies has been seized upon by climate sceptics. On Mars, the warming seems to be down to dust blowing around and uncovering big patches of black basaltic rock that heat up in the day."

Solar irradiance

K. I. Abdusamatov has proposed that "parallel global warmings" observed simultaneously on Mars and on Earth can only be a consequence of the same factor: a long-time change in solar irradiance." While some individuals who reject the science of global warming take this as proof that humans are not causing climate change, Abdusamatov's hypothesis has not been accepted by the scientific community. His assertions have not been published in the peer-reviewed literature, and have been dismissed by other scientists, who have stated that "the idea just isn't supported by the theory or by the observations" and that it "doesn't make physical sense." Other scientists have proposed that the observed variations are caused by irregularities in the orbit of Mars or a possible combination of solar and orbital effects.

Climate zones

Terrestrial Climate zones first have been defined by Wladimir Köppen based on the distribution of vegetation groups. Climate classification is furthermore based on temperature, rainfall, and subdivided based upon differences in the seasonal distribution of temperature and precipitation; and a separate group exists for extrazonal climates like in high altitudes. Mars has neither vegetation nor rainfall, so any climate classification could be only based upon temperature; a further refinement of the system may be based on dust distribution, water vapor content, occurrence of snow. Solar Climate Zones can also be easily defined for Mars.

Current missions

The 2001 Mars Odyssey is currently orbiting Mars and taking global atmospheric temperature measurements with the TES instrument. The Mars Reconnaissance Orbiter is currently taking daily weather and climate related observations from orbit. One of its instruments, the Mars climate sounder is specialized for climate observation work. The MSL was launched in November 2011 and landed on Mars on August 6, 2012. Orbiters MAVEN, Mangalyaan, and TGO are currently orbiting Mars and studying its atmosphere. <templatestyles src="Multiple_image/styles.css" />

Curiosity rover – Temperature, Pressure, Humidity
at Gale Crater on Mars (August 2012 – February 2013)

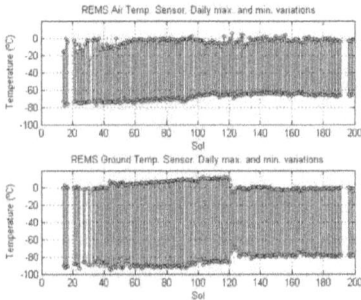

Temperature

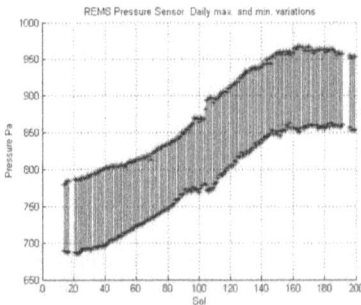

Pressure

ATMOSPHERE RELATIVE HUMIDITY

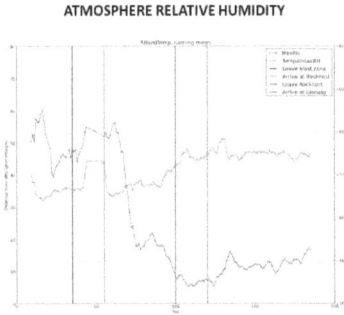

Humidity

Future missions

• ExoMars Trace Gas Orbiter was launched in 2016

Further reading

• Jakosky, Bruce M.; Phillips, Roger J. (2001). "Mars' volatile and climate history". *Nature*. **412** (6843): 237–244. doi: 10.1038/35084184[224]. PMID 11449285[225]. review article

External links

• Nature study explains mystery of Mars icecaps.[226]
• Mars could be undergoing major global warming[227]
• Mars Global Surveyor MOC2-1151 Release[228]
• Global warming on Mars?[229]
• Images of melting ice cap: Evidence for Recent Climate Change on Mars[230]
• Article from National Geographic on the issue of Martian Global Warming[231]
• Weather Reports[232] from the Curiosity Rover (REMS)[233]
• HRSC – Clouds[234]

Orbit of Mars

Orbit of Mars

Mars has an orbit with a semimajor axis of 1.524 astronomical units (228 million kilometers), and an eccentricity of 0.0934.[235] The planet orbits the Sun in 687 days[236] and travels 9.55 AU in doing so,[237] making the average orbital speed 24 km/s.

The eccentricity is greater than that of every other planet except Mercury, and this causes a large difference between the aphelion and perihelion distances—they are 1.6660 and 1.3814 AU.[238] Wikipedia:Citation needed

Changes in the orbit

Mars is in the midst of a long-term increase in eccentricity.Wikipedia:Please clarify It reached a minimum of 0.079 about 19 millennia ago, and will peak at about 0.105 after about 24 millennia from now (and with perihelion distances a mere 1.3621 astronomical units). The orbit is at times near circular: it was 0.002 1.35 million years ago, and will be about 0.01 a million years into the future. The maximum eccentricity between those two minima is 0.12.[239]

Oppositions

Mars reaches opposition when there is a 180° difference between the geocentric longitudes of it and the Sun. At a time near opposition (within 8½ days) the Earth–Mars distance is as small as it will get during that 780-day synodic period.[240] Every opposition has some significance because Mars is visible from Earth all night, high and fully lit, but the ones of special interest happen when Mars is near perihelion, because this is when Mars is also nearest to Earth. One perihelic opposition is followed by another either 15 or 17 years later. In fact every opposition is followed by a similar one 7 or 8 synodic periods later,

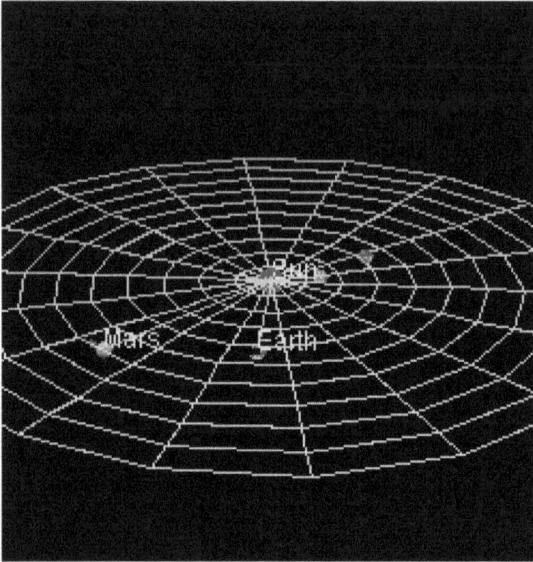

Figure 188: *Orbit of Mars relative to the orbits of inner Solar system planets*

and by a very similar one 37 synodic periods (79 years) later.[241] In the so-called perihelic opposition Mars is closest to the Sun and is particularly close to Earth: Oppositions range from about 0.68 AU when Mars is near aphelion to only about 0.37 AU when Mars is near perihelion.

Close approaches to Earth

Mars comes closer to Earth than any other planet save Venus at its nearest—56 versus 40 million km. The distances have been declining over the years, and in 2003 the minimum distance was 55.76 Gm, nearer than any such encounter in almost 60,000 years (57617 BC). This modern record will be beaten in 2287, and the record before 3000 will be set in 2729 at 55.65. By 4000, the record will stand at 55.44. The distances will continue to decrease for about 24,000 years.[242]

Historical importance

Until the work of Johannes Kepler (1571–1630), a German astronomer, it was believed, or assumed, that planets traveled in circular orbits around the Sun. When Kepler studied Danish astronomer Tycho Brahe's careful observations of Mars's position in the sky on many nights, Kepler realized that Mars's orbit could not be a circle. After considerable analysis, Kepler discovered that Mars's orbit was an ellipse, with the Sun occupying one of the elliptical orbit's two focal points. This, in turn, led to Kepler's discovery that all planets orbit the Sun in elliptical orbits, with the Sun at one of the two focal points. This became the first of Kepler's three laws of planetary motion.[243]

Accuracy/predictability

From the perspective of all but the most demanding, the path of Mars is simple. An equation in *Astronomical Algorithms* that assumes an unperturbed elliptical orbit predicts the perihelion and aphelion times with an error of "a few hours".[244] Using orbital elements to calculate those distances agrees to actual averages to at least five significant figures. Formulas for computing position straight from orbital elements typically do not provide or need corrections for the effects of other planets.[245]

For a higher level of accuracy the perturbations of planets are required. These are well known, and are believed to be modeled well enough to achieve high accuracy. These are all of the bodies that need to be considered for even many demanding problems. When Aldo Vitagliano calculated the date of close Martian approaches in the distant past or future, he tested the potential effect caused by the uncertainties of the asteroid belt models by running the simulations both with and without the biggest three asteroids, and found the effects were negligible.

Observations are much better now, and space age technology has replaced the older techniques. E. Myles Standish wrote: "Classical ephemerides over the past centuries have been based entirely upon optical observations:almost exclusively, meridian circle transit timings. With the advent of planetary radar, spacecraft missions, VLBI, etc., the situation for the four inner planets has changed dramatically." (8.5.1 page 10) For DE405, created in 1995, optical observations were dropped and as he wrote "initial conditions for the inner four planets were adjusted to ranging data primarily..."[246] The error in DE 405 is known to be about 2 km and is now sub-kilometer.[247]

Although the perturbations on Mars by asteroids have caused problems, they have also been used to estimate the masses of certain asteroids.[248] But improving the model of the asteroid belt is of great concern to those requiring or attempting to provide the highest-accuracy ephemerides.[249]

Orbital parameters

No more than five significant figures are presented in the following table of Mars's orbital elements. To this level of precision, the numbers match very well the VSOP87 elements and calculations derived from them, as well as Standish's (of JPL) 250-year best fit, and calculations using the actual positions of Mars over time.

Distances and eccentricity	(AU)	(Gm)
Semimajor axis	1.5237	227.9
Perihelion	1.3814	206.7
Aphelion	1.6660	249.2
Average^{250}	1.5303	228.9
Circumference	9.553	1429
Closest approach to Earth	0.3727	55.76
Farthest distance from Earth	2.675	400.2
Eccentricity	0.0934	
Angles	(°)	
Inclination	1.850	
Period	(days)	(years)
Orbital	687.0	1.881
Synodic	779.9	2.135
Speed	(km/s)	
Average	24.1	
Maximum	26.5	
Minimum	22.0	

Habitability

Life on Mars

This article is one of a series on:

Life in the Universe

Astrobiology

Habitability in the Solar System

- Habitability of Venus
- Life on Earth
- Habitability of Mars
- Habitability of Enceladus
- Habitability of Europa
- Habitability of Titan

Life outside the Solar System

- Circumstellar habitable zone
- Exoplanetology
- Planetary habitability
- SETI

- $\underline{v}$
- $\underline{t}$
- $\underline{e}^{251}$

The possibility of life on Mars is a subject of significant interest to astrobiology due to its proximity and similarities to Earth. To date, no proof has been found of past or present life on Mars. Cumulative evidence shows that during the ancient Noachian time period, the surface environment of Mars had liquid water and may have been habitable for microorganisms. The existence of habitable conditions does not necessarily indicate the presence of life.

Scientific searches for evidence of life began in the 19th century, and they continue today via telescopic investigations and deployed probes. While early work focused on phenomenology and bordered on fantasy, the modern scientific inquiry has emphasized the search for water, chemical biosignatures in the soil and rocks at the planet's surface, and biomarker gases in the atmosphere.

Mars is of particular interest for the study of the origins of life because of its similarity to the early Earth. This is especially so since Mars has a cold climate and lacks plate tectonics or continental drift, so it has remained almost unchanged since the end of the Hesperian period. At least two-thirds of Mars's surface is more than 3.5 billion years old, and Mars may thus hold the best record of the prebiotic conditions leading to abiogenesis, even if life does not or has never existed there.

Following the confirmation of the past existence of surface liquid water, the *Curiosity* and *Opportunity* rovers started searching for evidence of past life, including a past biosphere based on autotrophic, chemotrophic, or chemolithoautotrophic microorganisms, as well as ancient water, including fluvio-lacustrine environments (plains related to ancient rivers or lakes) that may have been habitable. The search for evidence of habitability, taphonomy (related to fossils), and organic compounds on Mars is now a primary NASA and ESA objective.

The findings of organic compounds inside sedimentary rocks and of boron on Mars are of interest as they are precursors for prebiotic chemistry. Such findings, along with previous discoveries that liquid water was clearly present on ancient Mars, further supports the possible early habitability of Gale Crater on Mars. Currently, the surface of Mars is bathed with radiation, and when reacting with the perchlorates on the surface, it may be more toxic to microorganisms than thought earlier. Therefore, the consensus is that if life exists —or existed— on Mars, it could be found or is best preserved in the subsurface, away from present-day harsh surface processes.

In June 2018 NASA announced the detection of seasonal variation of methane levels on Mars. Methane could be produced by microorganisms or by geological means. The European ExoMars Trace Gas Orbiter started mapping the atmospheric methane in April 2018, and the 2020 ExoMars rover will drill and analyze subsurface samples, while the NASA Mars 2020 rover will cache dozens of drill samples for their potential transport to Earth laboratories in the late 2020s or 2030s.

Early speculation

<templatestyles src="Multiple_image/styles.css" />

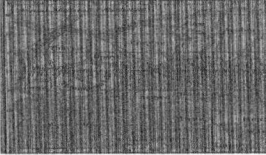

Historical map of Mars from Giovanni Schiaparelli

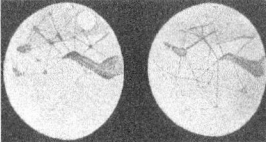

Mars canals illustrated by astronomer Percival Lowell, 1898

Mars' polar ice caps were discovered in the mid-17th century. In the late 18th century, William Herschel proved they grow and shrink alternately, in the summer and winter of each hemisphere. By the mid-19th century, astronomers knew that Mars had certain other similarities to Earth, for example that the length of a day on Mars was almost the same as a day on Earth. They also knew that its axial tilt was similar to Earth's, which meant it experienced seasons just as Earth does — but of nearly double the length owing to its much longer year. These observations led to increase in speculation that the darker albedo features were water and the brighter ones were land, whence followed speculation on whether Mars may be inhabited by some form of life.

In 1854, William Whewell, a fellow of Trinity College, Cambridge, who popularized the word *scientist,* theorized that Mars had seas, land and possibly life forms. Speculation about life on Mars exploded in the late 19th century, following telescopic observation by some observers of apparent Martian canals — which were later found to be optical illusions. Despite this, in 1895, American astronomer Percival Lowell published his book *Mars,* followed by *Mars and its Canals* in 1906, proposing that the canals were the work of a long-gone civilization. This idea led British writer H. G. Wells to write *The War of the Worlds* in 1897, telling of an invasion by aliens from Mars who were fleeing the planet's desiccation.

Spectroscopic analysis of Mars' atmosphere began in earnest in 1894, when U.S. astronomer William Wallace Campbell showed that neither water nor oxygen were present in the Martian atmosphere. By 1909, better telescopes and the best perihelic opposition of Mars since 1877 conclusively put an end to the canal hypothesis.

Habitability

Chemical, physical, geological, and geographic attributes shape the environments on Mars. Isolated measurements of these factors may be insufficient to deem an environment habitable, but the sum of measurements can help predict locations with greater or lesser habitability potential. The two current ecological approaches for predicting the potential habitability of the Martian surface use 19 or 20 environmental factors, with an emphasis on water availability, temperature, the presence of nutrients, an energy source, and protection from solar ultraviolet and galactic cosmic radiation.

Scientists do not know the minimum number of parameters for determination of habitability potential, but they are certain it is greater than one or two of the factors in the table below. Similarly, for each group of parameters, the habitability threshold for each is to be determined. Laboratory simulations show that whenever multiple lethal factors are combined, the survival rates plummet quickly. There are no full-Mars simulations published yet that include all of the biocidal factors combined.

Habitability factors	
Water	· liquid water activity (a_w) · Past/future liquid (ice) inventories · Salinity, pH, and Eh of available water
Chemical environment	**Nutrients:** · C, H, N, O, P, S, essential metals, essential micronutrients · Fixed nitrogen · Availability/mineralogy **Toxin abundances and lethality:** · Heavy metals (e.g., Zn, Ni, Cu, Cr, As, Cd, etc., some essential, but toxic at high levels) · Globally distributed oxidizing soils
Energy for metabolism	**Solar** (surface and near-surface only) **Geochemical** (subsurface) · Oxidants · Reductants · Redox gradients

Conducive physical conditions	· Temperature · Extreme diurnal temperature fluctuations · Low pressure (Is there a low-pressure threshold for terrestrial anaerobes?) · Strong ultraviolet germicidal irradiation · Galactic cosmic radiation and solar particle events (long-term accumulated effects) · Solar UV-induced volatile oxidants, e.g., O_2^-, O^-, H_2O_2, O_3 · Climate/variability (geography, seasons, diurnal, and eventually, obliquity variations) · Substrate (soil processes, rock microenvironments, dust composition, shielding) · High CO_2 concentrations in the global atmosphere · Transport (aeolian, groundwater flow, surface water, glacial)

Past

Recent models have shown that, even with a dense CO_2 atmosphere, early Mars was colder than Earth has ever been. Transiently warm conditions related to impacts or volcanism could have produced conditions favoring the formation of the late Noachian valley networks, even though the mid-late Noachian global conditions were probably icy. Local warming of the environment by volcanism and impacts would have been sporadic, but there should have been many events of water flowing at the surface of Mars. Both the mineralogical and the morphological evidence indicates a degradation of habitability from the mid Hesperian onward. The exact causes are not well understood but may be related to a combination of processes including loss of early atmosphere, or impact erosion, or both.

The loss of the Martian magnetic field strongly affected surface environments through atmospheric loss and increased radiation; this change significantly degraded surface habitability. When there was a magnetic field, the atmosphere would have been protected from erosion by the solar wind, which would ensure the maintenance of a dense atmosphere, necessary for liquid water to exist on the surface of Mars. The loss of the atmosphere was accompanied by decreasing temperatures. Part of the liquid water inventory sublimed and was transported to the poles, while the rest became trapped in permafrost, a subsurface ice layer.

Observations on Earth and numerical modeling have shown that a crater-forming impact can result in the creation of a long-lasting hydrothermal system when ice is present in the crust. For example, a 130 km large crater could sustain an active hydrothermal system for up to 2 million years, that is, long enough for microscopic life to emerge, but unlikely to have progressed any further down the evolutionary path.[252]

Soil and rock samples studied in 2013 by NASA's *Curiosity* rover's onboard instruments brought about additional information on several habitability factors. The rover team identified some of the key chemical ingredients for life in

Figure 189: *Alga crater is thought to have deposits of impact glass that may have preserved ancient biosignatures, if present during the impact.*

this soil, including sulfur, nitrogen, hydrogen, oxygen, phosphorus and possibly carbon, as well as clay minerals, suggesting a long-ago aqueous environment — perhaps a lake or an ancient streambed — that had neutral acidity and not too salty. On December 9, 2013, NASA reported that, based on evidence from *Curiosity* studying Aeolis Palus, Gale Crater contained an ancient freshwater lake which could have been a hospitable environment for microbial life. The confirmation that liquid water once flowed on Mars, the existence of nutrients, and the previous discovery of a past magnetic field that protected the planet from cosmic and solar radiation, together strongly suggest that Mars could have had the environmental factors to support life. The assessment of past habitability is not in itself evidence that Martian life has ever actually existed. If it did, it was probably microbial, existing communally in fluids or on sediments, either free-living or as biofilms, respectively.

Impactite, shown to preserve signs of life on Earth, was discovered on Mars and could contain signs of ancient life, if life ever existed on the planet.

On June 7, 2018, NASA announced that the *Curiosity* rover had discovered organic molecules in sedimentary rocks dating to three billion years old. The detection of organic molecules in rocks indicate that some of the building blocks for life were present.

Present

Conceivably, if life exists (or existed) on Mars, evidence of that life could be found, or is best preserved, in the subsurface, away from present-day harsh surface conditions. Present-day life on Mars, or its biosignatures, could occur kilometers below the surface, or in subsurface geothermal hot spots, or it could occur a few meters below the surface. The permafrost layer on Mars is only a couple of centimeters below the surface, and salty brines can be liquid a few centimeters below that but not far down. Water is close to its boiling point even at the deepest points in the Hellas basin, and so cannot remain liquid for long on the surface of Mars in its present state, except when covered in ice or after a sudden release of underground water.

So far, NASA has pursued a "follow the water" strategy on Mars and has not searched for biosignatures for life there directly since the *Viking* mission. As of 2017, the consensus by astrobiologists at NASA is that it may be necessary to access the Martian subsurface to find currently habitable environments.

Dormant subsurface life

Curiosity measured ionizing radiation levels of 76 mGy a year. This level of ionizing radiation is sterilizing for dormant life on the surface of Mars. It varies considerably in habitability depending on its orbital eccentricity and the tilt of its axis. If the surface life has been reanimated as recently as 450,000 years ago, then rovers on Mars could find dormant but still viable life at a depth of one meter below the surface, according to an estimate.

Cosmic radiation

In 1965, the Mariner 4 probe discovered that Mars had no global magnetic field that would protect the planet from potentially life-threatening cosmic radiation and solar radiation; observations made in the late 1990s by the Mars Global Surveyor confirmed this discovery. Scientists speculate that the lack of magnetic shielding helped the solar wind blow away much of Mars's atmosphere over the course of several billion years. As a result, the planet has been vulnerable to radiation from space for about 4 billion years.

Recent *in-situ* data from *Curiosity* rover indicates that ionizing radiation from galactic cosmic rays (GCR) and solar particle events (SPE) may not be a limiting factor in habitability assessments for present-day surface life on Mars. The level of 76 mGy per year measured by *Curiosity* is similar to levels inside the ISS. In the 2014 Findings of the Second MEPAG Special Regions Science Analysis Group, their conclusion was:

"From MSL RAD measurements, ionizing radiation from galactic cosmic rays (GCR) at Mars is so low as to be negligible. Intermittent Solar particle events (SPE) can increase the atmospheric ionization down to ground level and increase the total dose, but these events are sporadic and last at most a few (2–5) days. These facts are not used to distinguish Special Regions on Mars." A Special Region is defined as a region on the Mars surface where Earth life could potentially survive.

Cumulative effects

Even the hardiest cells known could not possibly survive the cosmic radiation near the surface of Mars since Mars lost its protective magnetosphere and atmosphere. After mapping cosmic radiation levels at various depths on Mars, researchers have concluded that over time, any life within the first several meters of the planet's surface would be killed by lethal doses of cosmic radiation. The team calculated that the cumulative damage to DNA and RNA by cosmic radiation would limit retrieving viable dormant cells on Mars to depths greater than 7.5 meters below the planet's surface. Even the most radiation-tolerant Earthly bacteria would survive in dormant spore state only 18,000 years at the surface; at 2 meters —the greatest depth at which the ExoMars rover will be capable of reaching— survival time would be 90,000 to half a million years, depending on the type of rock.

Data collected by the Radiation assessment detector (RAD) instrument on board the *Curiosity* rover revealed that the absorbed dose measured is 76 mGy/ year at the surface, and that "ionizing radiation strongly influences chemical compositions and structures, especially for water, salts, and redox-sensitive components such as organic molecules." Regardless of the source of Martian organic compounds (meteoric, geological, or biological), its carbon bonds are susceptible to breaking and reconfiguring with surrounding elements by ionizing charged particle radiation. These improved subsurface radiation estimates give insight into the potential for the preservation of possible organic biosignatures as a function of depth as well as survival times of possible microbial or bacterial life forms left dormant beneath the surface. The report concludes that the *in situ* "surface measurements —and subsurface estimates— constrain the preservation window for Martian organic matter following exhumation and exposure to ionizing radiation in the top few meters of the Martian surface."

In September 2017, NASA reported radiation levels on the surface of the planet Mars were temporarily doubled and were associated with an aurora 25 times brighter than any observed earlier, due to a major, and unexpected, solar storm in the middle of the month.

UV radiation

On UV radiation, a 2014 report concludes that "[T]he Martian UV radiation environment is rapidly lethal to unshielded microbes but can be attenuated by global dust storms and shielded completely by < 1 mm of regolith or by other organisms." In addition, laboratory research published in July 2017 demonstrated that UV irradiated perchlorates cause a 10.8-fold increase in cell death when compared to cells exposed to UV radiation after 60 seconds of exposure. The penetration depth of UV radiation into soils is in the sub-millimeter to millimeter range and depends on the properties of the soil.[253]

Perchlorates

The Martian regolith is known to contain a maximum of 0.5% (w/v) perchlorate (ClO_4^-) that is toxic for most living organisms,[254] but since they drastically lower the freezing point of water and few extremophiles can use it as an energy source (see Perchlorates - Biology), it has prompted speculation of what their influence would be on habitability.[255]

Research published in July 2017 shows that when irradiated with a simulated Martian UV flux, perchlorates become even more lethal to bacteria (bactericide). Even dormant spores lost viability within minutes. In addition, two other compounds of the Martian surface, iron oxides and hydrogen peroxide, act in synergy with irradiated perchlorates to cause a 10.8-fold increase in cell death when compared to cells exposed to UV radiation after 60 seconds of exposure. It was also found that abraded silicates (quartz and basalt) lead to the formation of toxic reactive oxygen species.[256] The researchers concluded that "the surface of Mars is lethal to vegetative cells and renders much of the surface and near-surface regions uninhabitable."[257] This research demonstrates that the present-day surface is more uninhabitable than previously thought,[258] and reinforces the notion to inspect at least a few meters into the ground to ensure the levels of radiation would be relatively low.[259]

Recurrent slope lineae

Recurrent slope lineae (RSL) features form on Sun-facing slopes at times of the year when the local temperatures reach above the melting point for ice. The streaks grow in spring, widen in late summer and then fade away in autumn. This is hard to model in any other way except as involving liquid water in some form, though the streaks themselves are thought to be a secondary effect and not a direct indication of the dampness of the regolith. Although these features are now confirmed to involve liquid water in some form, the water could be either too cold or too salty for life. At present they are treated as potentially habitable, as "Uncertain Regions, to be treated as Special Regions".

The 'Special Regions assessment' says of them: "Although no single model currently proposed for the origin of RSL adequately explains all observations, they are currently best interpreted as being due to the seepage of water at > 250 K, with a_w (water activity) unknown and perhaps variable. As such they meet the criteria for Uncertain Regions, to be treated as Special Regions. There are other features on Mars with characteristics similar to RSL, but their relationship to possible liquid water is much less likely." They were first reported in 2011. They were already suspected as involving flowing brines back then, as all the other models available involved liquid water in some form.

The thermodynamic availability of water (water activity) strictly limits microbial propagation on Earth, particularly in hypersaline environments, and there are indications that the brine ionic strength is a barrier to the habitability of Mars. Experiments show that high ionic strength, driven to extremes on Mars by the ubiquitous occurrence of divalent ions, "renders these environments uninhabitable despite the presence of biologically available water."[260]

Nitrogen fixation

After carbon, nitrogen is arguably the most important element needed for life. Thus, measurements of nitrate over the range of 0.1% to 5% are required to address the question of its occurrence and distribution. There is nitrogen (as N_2) in the atmosphere at low levels, but this is not adequate to support nitrogen fixation for biological incorporation. Nitrogen in the form of nitrate could be a resource for human exploration both as a nutrient for plant growth and for use in chemical processes. On Earth, nitrates correlate with perchlorates in desert environments, and this may also be true on Mars. Nitrate is expected to be stable on Mars and to have formed by thermal shock from impact or volcanic plume lightning on ancient Mars.

On 24 March 2015, NASA reported that the SAM instrument on the *Curiosity* rover detected nitrates by heating surface sediments. The nitrogen in nitrate is in a "fixed" state, meaning that it is in an oxidized form that can be used by living organisms. The discovery supports the notion that ancient Mars may have been hospitable for life. It is suspected that all nitrate on Mars is a relic, with no modern contribution.[261] Nitrate abundance ranges from non-detection to 681 ± 304 mg/kg in the samples examined until late 2017. Modeling indicates that the transient condensed water films on the surface should be transported to lower depths (≈ 10 m) potentially transporting nitrates, where subsurface microorganisms could thrive.

In contrast, phosphate, one of the chemical nutrients thought to be essential for life, is readily available on Mars.[262]

Low pressure

Further complicating estimates of the habitability of the Martian surface is the fact that very little is known about the growth of microorganisms at pressures close to those on the surface of Mars. Some teams determined that some bacteria may be capable of cellular replication down to 25 mbar, but that is still above the atmospheric pressures found on Mars (range 1–14 mbar). In another study, twenty-six strains of bacteria were chosen based on their recovery from spacecraft assembly facilities, and only *Serratia liquefaciens* strain ATCC 27592 exhibited growth at 7 mbar, 0 °C, and CO_2-enriched anoxic atmospheres.

Liquid water

Liquid water is a necessary but not sufficient condition for life as we know it, as habitability is a function of a multitude of environmental parameters. Liquid water cannot exist on the surface of Mars except at the lowest elevations for minutes or hours. Liquid water does not appear at the surface itself, but it could form in minuscule amounts around dust particles in snow heated by the Sun. Also, the ancient equatorial ice sheets beneath the ground may slowly sublimate or melt, accessible from the surface via caves. <templatestyles src="Multiple_image/styles.css" />

Mars - Utopia Planitia

Scalloped terrain led to the discovery of a large amount of underground ice

enough water to fill Lake Superior (November 22, 2016)

Martian terrain

Figure 190: *A series of artist's conceptions of past water coverage on Mars*

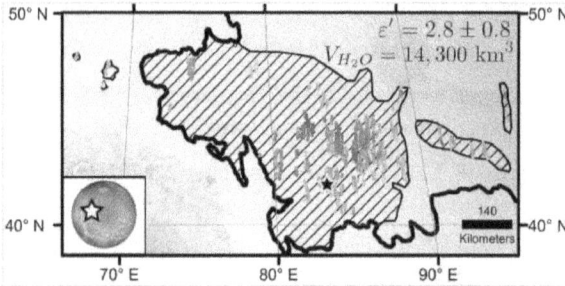

Map of terrain

Water on Mars exists almost exclusively as water ice, located in the Martian polar ice caps and under the shallow Martian surface even at more temperate latitudes. A small amount of water vapor is present in the atmosphere. There are no bodies of liquid water on the Martian surface because its atmospheric pressure at the surface averages 600 pascals (0.087 psi)—about 0.6% of Earth's mean sea level pressure—and because the temperature is far too low, (210 K (−63 °C)) leading to immediate freezing. Despite this, about 3.8 billion years ago, there was a denser atmosphere, higher temperature, and vast amounts of liquid water flowed on the surface, including large oceans.

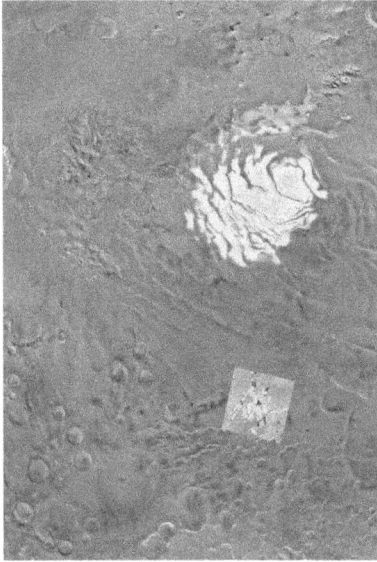

Figure 191:
Mars SouthPole
Site of Subglacial Water
(25 July 2018)

It has been estimated that the primordial oceans on Mars would have covered between 36% and 75% of the planet. On November 22, 2016, NASA reported finding a large amount of underground ice in the Utopia Planitia region of Mars. The volume of water detected has been estimated to be equivalent to the volume of water in Lake Superior. Analysis of Martian sandstones, using data obtained from orbital spectrometry, suggests that the waters that previously existed on the surface of Mars would have had too high a salinity to support most Earth-like life. Tosca *et al.* found that the Martian water in the locations they studied all had water activity, $a_w \leq 0.78$ to 0.86—a level fatal to most Terrestrial life. Haloarchaea, however, are able to live in hypersaline solutions, up to the saturation point.

In June 2000, possible evidence for current liquid water flowing at the surface of Mars was discovered in the form of flood-like gullies. Additional similar images were published in 2006, taken by the Mars Global Surveyor, that suggested that water occasionally flows on the surface of Mars. The images showed changes in steep crater walls and sediment deposits, providing the strongest evidence yet that water coursed through them as recently as several years ago.

Figure 192: *The silica-rich patch discovered by Spirit rover*

There is disagreement in the scientific community as to whether or not the recent gully streaks were formed by liquid water. Some suggest the flows were merely dry sand flows. Others suggest it may be liquid brine near the surface, but the exact source of the water and the mechanism behind its motion are not understood.

In July 2018, scientists reported the discovery of a subglacial lake on Mars, 1.5 km (0.93 mi) below the southern polar ice cap, and extending sideways about 20 km (12 mi), the first known stable body of water on the planet. The lake was discovered using the MARSIS radar on board the *Mars Express* orbiter, and the profiles were collected between May 2012 and December 2015.[263] The lake is centered at 193°E, 81°S, a flat area that does not exhibit any peculiar topographic characteristics but is surrounded by higher ground, except on its eastern side, where there is a depression.

Silica

In May 2007, the *Spirit* rover disturbed a patch of ground with its inoperative wheel, uncovering an area 90% rich in silica. The feature is reminiscent of the effect of hot spring water or steam coming into contact with volcanic rocks. Scientists consider this as evidence of a past environment that may have been favorable for microbial life and theorize that one possible origin for the silica may have been produced by the interaction of soil with acid vapors produced by volcanic activity in the presence of water.

Based on Earth analogs, hydrothermal systems on Mars would be highly attractive for their potential for preserving organic and inorganic biosignatures. For this reason, hydrothermal deposits are regarded as important targets in the exploration for fossil evidence of ancient Martian life.

Possible biosignatures

In May 2017, evidence of the earliest known life on land on Earth may have been found in 3.48-billion-year-old geyserite and other related mineral deposits (often found around hot springs and geysers) uncovered in the Pilbara Craton of Western Australia. These findings may be helpful in deciding where best to search for early signs of life on the planet Mars.

Methane

Possible trace amounts of methane in the atmosphere of Mars were first discovered in 2003 with Earth-based telescopes and fully verified in 2004 by the ESA Mars Express spacecraft in orbit around Mars. As methane is an unstable gas, its presence indicates that there must be an active source on the planet in order to keep such levels in the atmosphere. It is estimated that Mars must produce 270 ton/year of methane, but asteroid impacts account for only 0.8% of the total methane production. Although geologic sources of methane such as serpentinization are possible, the lack of current volcanism, hydrothermal activity or hotspots are not favorable for geologic methane. It has been suggested that the methane was produced by chemical reactions in meteorites, driven by the intense heat during entry through the atmosphere. Although research published in December 2009 ruled out this possibility, research published in 2012 suggest that a source may be organic compounds on meteorites that are converted to methane by ultraviolet radiation.

The existence of life in the form of microorganisms such as methanogens is among possible but as yet unproven sources. Methanogens do not require oxygen or organic nutrients, are non-photosynthetic, use hydrogen as their energy source and carbon dioxide (CO_2) as their carbon source, so they could exist in subsurface environments on Mars. If microscopic Martian life is producing the methane, it probably resides far below the surface, where it is still warm enough for liquid water to exist.

Since the 2003 discovery of methane in the atmosphere, some scientists have been designing models and *in vitro* experiments testing the growth of methanogenic bacteria on simulated Martian soil, where all four methanogen

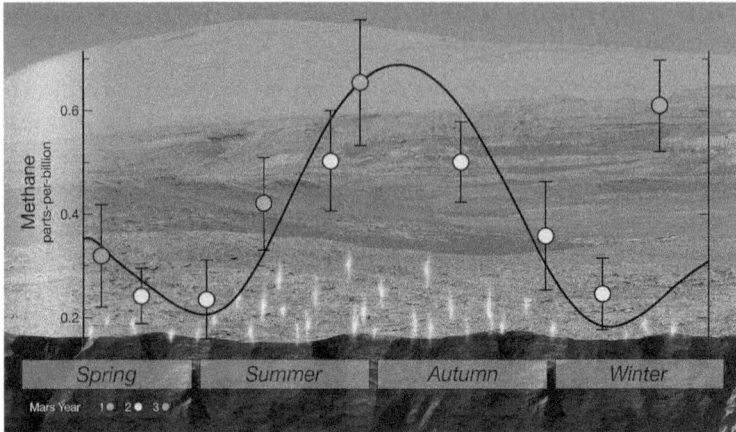

Figure 193:
Curiosity detected a cyclical seasonal varia-
tion in atmospheric methane (June 7, 2018).

strains tested produced substantial levels of methane, even in the presence of 1.0wt% perchlorate salt.

A team led by Levin suggested that both phenomena—methane production and degradation—could be accounted for by an ecology of methane-producing and methane-consuming microorganisms.

Research at the University of Arkansas presented in June 2015 suggested that some methanogens could survive on Mars's low pressure. Rebecca Mickol found that in her laboratory, four species of methanogens survived low-pressure conditions that were similar to a subsurface liquid aquifer on Mars. The four species that she tested were *Methanothermobacter wolfeii*, *Methanosarcina barkeri*, *Methanobacterium formicicum*, and *Methanococcus maripaludis*. In June 2012, scientists reported that measuring the ratio of hydrogen and methane levels on Mars may help determine the likelihood of life on Mars. According to the scientists, "...low H_2/CH_4 ratios (less than approximately 40) indicate that life is likely present and active." Other scientists have recently reported methods of detecting hydrogen and methane in extraterrestrial atmospheres.

The *Curiosity* rover, which landed on Mars in August 2012, is able to make measurements that distinguish between different isotopologues of methane, and in 2014, Curiosity detected a "tenfold spike" in the level of methane in the Martian atmosphere compared to the usual background readings. Even if the mission is to determine that microscopic Martian life is the seasonal source of

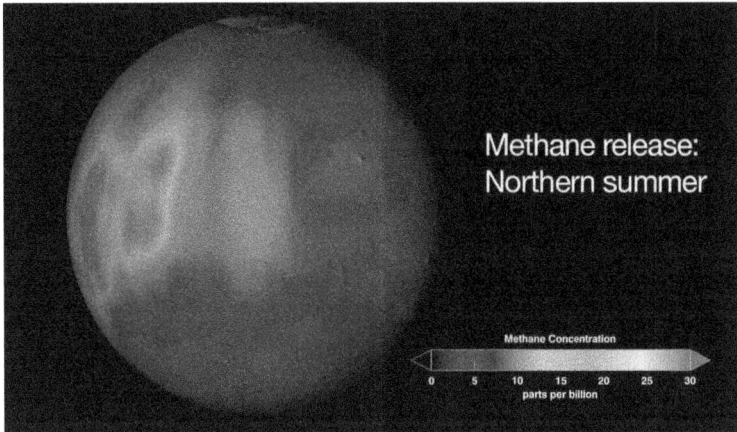

Figure 194:
*Distribution of methane in the atmosphere of
Mars in the Northern Hemisphere during summer*

the methane, the life forms probably reside far below the surface, outside of the rover's reach. The first measurements with the Tunable Laser Spectrometer (TLS) in the *Curiosity* rover indicated that there is less than 5 ppb of methane at the landing site at the point of the measurement. On July 19, 2013, NASA scientists published the results of a new analysis of the atmosphere of Mars, reporting a lack of methane around the landing site of the *Curiosity* rover. On September 19, 2013, NASA again reported no detection of atmospheric methane with a measured value of 0.18±0.67 ppbv corresponding to an upper limit of only 1.3 ppbv (95% confidence limit) and, as a result, concluded that the probability of current methanogenic microbial activity on Mars is reduced. On 16 December 2014, NASA reported that Curiosity had detected a tenfold increase ('spike') in methane in the atmosphere around it in late 2013 and early 2014. Four measurements taken over two months in this period averaged 7 ppb, suggesting that methane is released at intervals.

On June 7, 2018, NASA announced it has detected a seasonal variation of methane levels on Mars. Chris Webster, a senior researcher at Jet Propulsion Laboratory says, "Not only have we got this wonderful repeatability, but the seasonal cycle changes by a factor of three. That's a huge change, completely unexpected. And what it does, it gives us a key to unlocking the mysteries associated with Mars methane because now we have something to test our models and our understanding against." Methane lasts only a few centuries before it is broken down. He also says "...so if we see methane in the martian atmosphere, that means something is happening today, it's being released or it's being created."

The ExoMars Trace Gas Orbiter, launched in March 2016, began on 21 April 2018 to map the concentration and sources of methane in the atmosphere, as well as its decomposition products such as formaldehyde and methanol.

Formaldehyde

In February 2005, it was announced that the Planetary Fourier Spectrometer (PFS) on the European Space Agency's Mars Express Orbiter had detected traces of formaldehyde in the atmosphere of Mars. Vittorio Formisano, the director of the PFS, has speculated that the formaldehyde could be the byproduct of the oxidation of methane and, according to him, would provide evidence that Mars is either extremely geologically active or harboring colonies of microbial life. NASA scientists consider the preliminary findings well worth a follow-up but have also rejected the claims of life.

Viking lander biological experiments

The 1970s Viking program placed two identical landers on the surface of Mars tasked to look for biosignatures of microbial life on the surface. Of the four experiments performed by each Viking lander, only the 'Labeled Release' (LR) experiment gave a positive result for metabolism, while the other three did not detect organic compounds. The LR was a specific experiment designed to test only a narrowly defined critical aspect of the theory concerning the possibility of life on Mars; therefore, the overall results were declared inconclusive. No Mars lander mission has found meaningful traces of biomolecules or biosignatures. The claim of extant microbial life on Mars is based on old data collected by the Viking landers, currently reinterpreted as sufficient evidence of life, mainly by Gilbert Levin, Joseph D. Miller, Navarro, Giorgio Bianciardi and Patricia Ann Straat, that the Viking LR experiments detected extant microbial life on Mars.

Assessments published in December 2010 by Rafael Navarro–Gonzáles indicate that organic compounds "could have been present" in the soil analyzed by both Viking 1 and 2. The study determined that perchlorate —discovered in 2008 by Phoenix lander— can destroy organic compounds when heated, and produce chloromethane and dichloromethane as a byproduct, the identical chlorine compounds discovered by both Viking landers when they performed the same tests on Mars. Because perchlorate would have broken down any Martian organics, the question of whether or not Viking found organic compounds is still wide open.

The Labeled Release evidence was not generally accepted initially, and, to this day lacks the consensus of the scientific community.

Curiosity rover sediment sampling

In June 2018, NASA reported that the *Curiosity* rover had found evidence of complex organic compounds from mudstone rocks aged approximately 3.5 billion years old, sampled from two distinct sites in a dry lake in the Pahrump Hills of the Gale crater. The rock samples, when pyrolyzed via the *Curiosity*'s Sample Analysis at Mars instrument, released an array of organic molecules; these include sulfur-containing thiophenes, aromatic compounds such as benzene and toluene, and aliphatic compounds such as propane and butene. The concentration of organic compounds is 100-fold higher than earlier measurements. The authors speculate that the presence of sulfur may have helped preserve them. The products resemble those obtained from the breakdown of kerogen, a precursor to oil and natural gas on Earth. NASA stated that these findings are not evidence that life existed on the planet, but that the organic compounds needed to sustain microscopic life were present and there may be deeper sources of organic compounds on the planet.

Meteorites

There are 34 known Martian meteorites (some of which were found in several fragments). These are valuable because they are the only physical samples of Mars available to Earth-bound laboratories. Some researchers have argued that microscopic morphological features found in ALH84001 are biomorphs, however this interpretation has been highly controversial and is not supported by the majority of researchers in the field.

Seven criteria have been established for the recognition of past life within terrestrial geologic samples. Those criteria are:

1. Is the geologic context of the sample compatible with past life?
2. Is the age of the sample and its stratigraphic location compatible with possible life?
3. Does the sample contain evidence of cellular morphology and colonies?
4. Is there any evidence of biominerals showing chemical or mineral disequilibria?
5. Is there any evidence of stable isotope patterns unique to biology?
6. Are there any organic biomarkers present?
7. Are the features indigenous to the sample?

For general acceptance of past life in a geologic sample, essentially most or all of these criteria must be met. All seven criteria have not yet been met for any of the Martian samples.

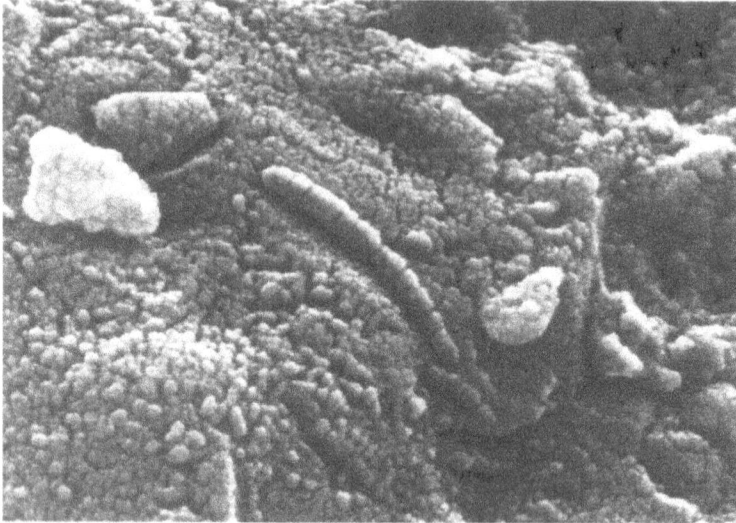

Figure 195: *An electron microscope reveals bacteria-*
like structures in meteorite fragment ALH84001

ALH84001

In 1996, the Martian meteorite ALH84001, a specimen that is much older than the majority of Martian meteorites that have been recovered so far, received considerable attention when a group of NASA scientists led by David S. McKay reported microscopic features and geochemical anomalies that they considered to be best explained by the rock having hosted Martian bacteria in the distant past. Some of these features resembled terrestrial bacteria, aside from their being much smaller than any known form of life. Much controversy arose over this claim, and ultimately all of the evidence McKay's team cited as evidence of life was found to be explainable by non-biological processes. Although the scientific community has largely rejected the claim ALH 84001 contains evidence of ancient Martian life, the controversy associated with it is now seen as a historically significant moment in the development of exobiology.

Nakhla

The Nakhla meteorite fell on Earth on June 28, 1911, on the locality of Nakhla, Alexandria, Egypt.

In 1998, a team from NASA's Johnson Space Center obtained a small sample for analysis. Researchers found preterrestrial aqueous alteration phases and

Figure 196: *Nakhla meteorite*

objects of the size and shape consistent with Earthly fossilized nanobacteria. Analysis with gas chromatography and mass spectrometry (GC-MS) studied its high molecular weight polycyclic aromatic hydrocarbons in 2000, and NASA scientists concluded that as much as 75% of the organic compounds in Nakhla "may not be recent terrestrial contamination".

This caused additional interest in this meteorite, so in 2006, NASA managed to obtain an additional and larger sample from the London Natural History Museum. On this second sample, a large dendritic carbon content was observed. When the results and evidence were published in 2006, some independent researchers claimed that the carbon deposits are of biologic origin. It was remarked that since carbon is the fourth most abundant element in the Universe, finding it in curious patterns is not indicative or suggestive of biological origin.[264]

Shergotty

The Shergotty meteorite, a 4 kg Martian meteorite, fell on Earth on Shergotty, India on August 25, 1865, and was retrieved by witnesses almost immediately. It is composed mostly of pyroxene and thought to have undergone preterrestrial aqueous alteration for several centuries. Certain features in its interior suggest remnants of a biofilm and its associated microbial communities. Work is in progress on searching for magnetites within alteration phases.

Yamato 000593

Yamato 000593 is the second largest meteorite from Mars found on Earth. Studies suggest the Martian meteorite was formed about 1.3 billion years ago from a lava flow on Mars. An impact occurred on Mars about 12 million years ago and ejected the meteorite from the Martian surface into space. The meteorite landed on Earth in Antarctica about 50,000 years ago. The mass of the meteorite is 13.7 kg (30 lb) and it has been found to contain evidence of past water movement. At a microscopic level, spheres are found in the meteorite that are rich in carbon compared to surrounding areas that lack such spheres. The carbon-rich spheres may have been formed by biotic activity according to NASA scientists.

Geysers on Mars

<templatestyles src="Multiple_image/styles.css" />

Artist concept showing sand-laden jets erupt from geysers on Mars.

Close up of dark dune spots, probably created by cold geyser-like eruptions.

The seasonal frosting and defrosting of the southern ice cap results in the formation of spider-like radial channels carved on 1-meter thick ice by sunlight. Then, sublimed CO_2 – and probably water – increase pressure in their interior producing geyser-like eruptions of cold fluids often mixed with dark basaltic sand or mud. This process is rapid, observed happening in the space of a few days, weeks or months, a growth rate rather unusual in geology – especially for Mars.

A team of Hungarian scientists propose that the geysers' most visible features, dark dune spots and spider channels, may be colonies of photosynthetic Martian microorganisms, which over-winter beneath the ice cap, and as the sunlight returns to the pole during early spring, light penetrates the ice, the microorganisms photosynthesize and heat their immediate surroundings. A pocket of liquid water, which would normally evaporate instantly in the thin Martian atmosphere, is trapped around them by the overlying ice. As this ice layer thins, the microorganisms show through grey. When the layer has completely melted, the microorganisms rapidly desiccate and turn black, surrounded by a grey aureole. The Hungarian scientists believe that even a complex sublimation process is insufficient to explain the formation and evolution of the dark dune spots in space and time.[265] Since their discovery, fiction writer Arthur C. Clarke promoted these formations as deserving of study from an astrobiological perspective.

A multinational European team suggests that if liquid water is present in the spiders' channels during their annual defrost cycle, they might provide a niche where certain microscopic life forms could have retreated and adapted while sheltered from solar radiation. A British team also considers the possibility that organic matter, microbes, or even simple plants might co-exist with these inorganic formations, especially if the mechanism includes liquid water and a geothermal energy source. They also remark that the majority of geological structures may be accounted for without invoking any organic "life on Mars" hypothesis. It has been proposed to develop the Mars Geyser Hopper lander to study the geysers up close.

Forward contamination

Planetary protection of Mars aims to prevent biological contamination of the planet. A major goal is to preserve the planetary record of natural processes by preventing human-caused microbial introductions, also called forward contamination. There is abundant evidence as to what can happen when organisms from regions on Earth that have been isolated from one another for significant periods of time are introduced into each other's environment. Species that are constrained in one environment can thrive – often out of control – in another environment much to the detriment of the original species that were present. In some ways, this problem could be compounded if life forms from one planet were introduced into the totally alien ecology of another world.

The prime concern of hardware contaminating Mars derives from incomplete spacecraft sterilization of some hardy terrestrial bacteria (extremophiles) despite best efforts. Hardware includes landers, crashed probes, end-of-mission disposal of hardware, and the hard landing of entry, descent, and landing systems. This has prompted research on survival rates of radiation-resistant microorganisms including the species *Deinococcus radiodurans* and genera *Brevundimonas*, *Rhodococcus*, and *Pseudomonas* under simulated Martian conditions. Results from one of these experimental irradiation experiments, combined with previous radiation modeling, indicate that *Brevundimonas* sp. MV. 7 emplaced only 30 cm deep in Martian dust could survive the cosmic radiation for up to 100,000 years before suffering 10^6 population reduction. The diurnal Mars-like cycles in temperature and relative humidity affected the viability of *Deinococcus radiodurans* cells quite severely. In other simulations, *Deinococcus radiodurans* also failed to grow under low atmospheric pressure, under 0 °C, or in the absence of oxygen.

Survival under simulated Martian conditions

On 26 April 2012, scientists reported that an extremophile lichen survived and showed remarkable results on the adaptation capacity of photosynthetic activity within the simulation time of 34 days under Martian conditions in the Mars Simulation Laboratory (MSL) maintained by the German Aerospace Center (DLR). The ability to survive in an environment is not the same as the ability to thrive, reproduce, and evolve in that same environment, necessitating further study.

Although numerous studies point to resistance to some of Mars conditions, they do so separately, and none has considered the full range of Martian surface conditions, including temperature, pressure, atmospheric composition, radiation, humidity, oxidizing regolith, and others, all at the same time and in combination.[266] Laboratory simulations show that whenever multiple lethal factors are combined, the survival rates plummet quickly.

Missions

Mars-2

Mars-1 was the first spacecraft launched to Mars in 1962, but communication was lost while en route to Mars. With Mars-2 and Mars-3 in 1971-1972, information was obtained on the nature of the surface rocks and altitude profiles of the surface density of the soil, its thermal conductivity, and thermal anomalies detected on the surface of Mars. The program found that its northern polar cap has a temperature below -110 °C and that the water vapor content in the atmosphere of Mars is five thousand times less than on Earth. No signs of life were found.

Mariner 4

<templatestyles src="Multiple_image/styles.css" />

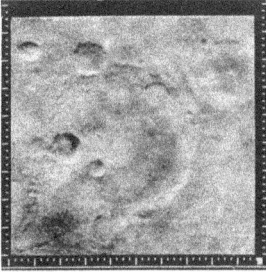

Mariner Crater, as seen by Mariner 4 in 1965. Pictures like this suggested that Mars is too dry for any kind of life.

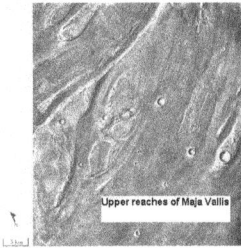

Streamlined Islands seen by Viking orbiter showed that large floods occurred on Mars. The image is located in Lunae Palus quadrangle.

Mariner 4 probe performed the first successful flyby of the planet Mars, returning the first pictures of the Martian surface in 1965. The photographs showed an arid Mars without rivers, oceans, or any signs of life. Further, it revealed that the surface (at least the parts that it photographed) was covered in craters, indicating a lack of plate tectonics and weathering of any kind for the last 4 billion years. The probe also found that Mars has no global magnetic field that would protect the planet from potentially life-threatening cosmic rays. The probe was able to calculate the atmospheric pressure on the planet to be about 0.6 kPa (compared to Earth's 101.3 kPa), meaning that liquid water could not exist on the planet's surface. After Mariner 4, the search for life on Mars changed to a search for bacteria-like living organisms rather than for multicellular organisms, as the environment was clearly too harsh for these.

Figure 197: *Carl Sagan poses next to a replica of the Viking landers.*

Viking orbiters

Liquid water is necessary for known life and metabolism, so if water was present on Mars, the chances of it having supported life may have been determinant. The Viking orbiters found evidence of possible river valleys in many areas, erosion and, in the southern hemisphere, branched streams.

Viking experiments

The primary mission of the Viking probes of the mid-1970s was to carry out experiments designed to detect microorganisms in Martian soil because the favorable conditions for the evolution of multicellular organisms ceased some four billion years ago on Mars. The tests were formulated to look for microbial life similar to that found on Earth. Of the four experiments, only the Labeled Release (LR) experiment returned a positive result,Wikipedia:Accuracy dispute#Disputed statement showing increased $^{14}CO_2$ production on first exposure of soil to water and nutrients. All scientists agree on two points from the Viking missions: that radiolabeled $^{14}CO_2$ was evolved in the Labeled Release experiment, and that the GCMS detected no organic molecules. There are vastly different interpretations of what those results imply.

A 2011 astrobiology textbook notes that the GCMS was the decisive factor due to which "For most of the Viking scientists, the final conclusion was that the *Viking* missions failed to detect life in the Martian soil."

One of the designers of the Labeled Release experiment, Gilbert Levin, believes his results are a definitive diagnostic for life on Mars. Levin's interpretation is disputed by many scientists. A 2006 astrobiology textbook noted that "With unsterilized Terrestrial samples, though, the addition of more nutrients after the initial incubation would then produce still more radioactive gas as the dormant bacteria sprang into action to consume the new dose of food. This was not true of the Martian soil; on Mars, the second and third nutrient injections did not produce any further release of labeled gas." Other scientists argue that superoxides in the soil could have produced this effect without life being present. An almost general consensus discarded the Labeled Release data as evidence of life, because the gas chromatograph and mass spectrometer, designed to identify natural organic matter, did not detect organic molecules. More recently, high levels of organic chemicals, particularly chlorobenzene, were detected in powder drilled from one of the rocks, named "Cumberland", analyzed by the *Curiosity* rover. The results of the Viking mission concerning life are considered by the general expert community as inconclusive.

In 2007, during a Seminar of the Geophysical Laboratory of the Carnegie Institution (Washington, D.C., US), Gilbert Levin's investigation was assessed once more. Levin still maintains that his original data were correct, as the positive and negative control experiments were in order. Moreover, Levin's team, on 12 April 2012, reported a statistical speculation, based on old data —reinterpreted mathematically through cluster analysis— of the Labeled Release experiments, that may suggest evidence of "extant microbial life on Mars." Critics counter that the method has not yet been proven effective for differentiating between biological and non-biological processes on Earth so it is premature to draw any conclusions.

A research team from the National Autonomous University of Mexico headed by Rafael Navarro-González concluded that the GCMS equipment (TV-GC-MS) used by the Viking program to search for organic molecules, may not be sensitive enough to detect low levels of organics. Klaus Biemann, the principal investigator of the GCMS experiment on *Viking* wrote a rebuttal. Because of the simplicity of sample handling, TV–GC–MS is still considered the standard method for organic detection on future Mars missions, so Navarro-González suggests that the design of future organic instruments for Mars should include other methods of detection.

After the discovery of perchlorates on Mars by the Phoenix lander, practically the same team of Navarro-González published a paper arguing that the Viking GCMS results were compromised by the presence of perchlorates. A 2011 astrobiology textbook notes that "while perchlorate is too poor an oxidizer to reproduce the LR results (under the conditions of that experiment perchlorate does not oxidize organics), it does oxidize, and thus destroy, organics at the

Figure 198: *An artist's concept of the Phoenix spacecraft*

higher temperatures used in the Viking GCMS experiment." Biemann has written a commentary critical of this Navarro-González paper as well, to which the latter have replied; the exchange was published in December 2011.

Phoenix lander, 2008

The Phoenix mission landed a robotic spacecraft in the polar region of Mars on May 25, 2008 and it operated until November 10, 2008. One of the mission's two primary objectives was to search for a "habitable zone" in the Martian regolith where microbial life could exist, the other main goal being to study the geological history of water on Mars. The lander has a 2.5 meter robotic arm that was capable of digging shallow trenches in the regolith. There was an electrochemistry experiment which analysed the ions in the regolith and the amount and type of antioxidants on Mars. The Viking program data indicate that oxidants on Mars may vary with latitude, noting that Viking 2 saw fewer oxidants than Viking 1 in its more northerly position. Phoenix landed further north still. Phoenix's preliminary data revealed that Mars soil contains perchlorate, and thus may not be as life-friendly as thought earlier. The pH and salinity level were viewed as benign from the standpoint of biology. The analysers also indicated the presence of bound water and CO_2. A recent analysis of Martian meteorite EETA79001 found 0.6 ppm ClO_4^-, 1.4 ppm ClO_3^-, and 16 ppm NO_3^-, most likely of Martian origin. The ClO_3^- suggests presence of other highly oxidizing oxychlorines such as ClO_2^- or ClO, produced both by

Figure 199: *Curiosity rover self-portrait.*

UV oxidation of Cl and X-ray radiolysis of ClO_4^-. Thus only highly refractory and/or well-protected (sub-surface) organics are likely to survive.[267] In addition, recent analysis of the Phoenix WCL showed that the $Ca(ClO_4)_2$ in the Phoenix soil has not interacted with liquid water of any form, perhaps for as long as 600 Myr. If it had, the highly soluble $Ca(ClO_4)_2$ in contact with liquid water would have formed only $CaSO_4$. This suggests a severely arid environment, with minimal or no liquid water interaction.

Mars Science Laboratory

The Mars Science Laboratory mission is a NASA project that launched on November 26, 2011, the *Curiosity* rover, a nuclear-powered robotic vehicle, bearing instruments designed to assess past and present habitability conditions on Mars. The *Curiosity* rover landed on Mars on Aeolis Palus in Gale Crater, near Aeolis Mons (a.k.a. Mount Sharp), on August 6, 2012.

On 16 December 2014, NASA reported the *Curiosity* rover detected a "tenfold spike", likely localized, in the amount of methane in the Martian atmosphere. Sample measurements taken "a dozen times over 20 months" showed increases in late 2013 and early 2014, averaging "7 parts of methane per billion in the atmosphere." Before and after that, readings averaged around one-tenth that level. In addition, low levels of chlorobenzene (C

$_6$H

$_5$Cl), were detected in powder drilled from one of the rocks, named "Cumberland", analyzed by the Curiosity rover.

<templatestyles src="Multiple_image/styles.css" />

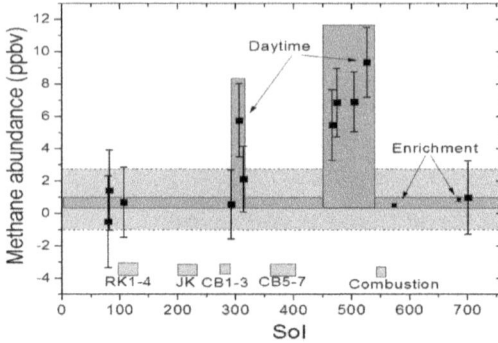

Methane measurements in the atmosphere of Mars

by the *Curiosity* rover (August 2012 to September 2014).

Methane (CH_4) on Mars - potential sources and sinks.

<templatestyles src="Multiple_image/styles.css" />

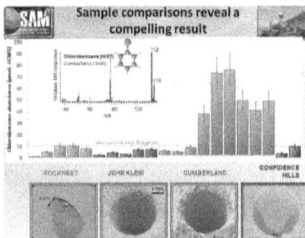

Comparison of organic compounds in Martian rocks - Chlorobenzene levels were much higher in the "Cumberland" rock sample.

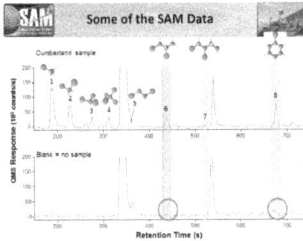

Detection of organic compounds in the "Cumberland" rock sample.

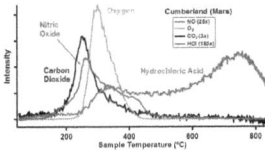

Sample Analysis at Mars (SAM) of "Cumberland" rock.

Future astrobiology missions

- ExoMars is a European-led multi-spacecraft programme currently under development by the European Space Agency (ESA) and the Russian
  Federal Space Agency for launch in 2016 and 2020. Its primary scientific mission will be to search for possible biosignatures on Mars, past
  or present. A rover with a 2 m (6.6 ft) core drill will be used to sample
  various depths beneath the surface where liquid water may be found
  and where microorganisms (or organic biosignaturesWikipedia:Citation
  needed) might survive cosmic radiation.
- Mars 2020 – The *Mars 2020* rover is a Mars planetary rover mission by
  NASA with a planned launch in 2020. It is intended to investigate an
  astrobiologically relevant ancient environment on Mars, investigate its
  surface geological processes and history, including the assessment of its
  past habitability and potential for preservation of biosignatures within
  accessible geological materials.
- Mars Sample Return Mission — The best life detection experiment proposed is the examination on Earth of a soil sample from Mars. However,
  the difficulty of providing and maintaining life support over the months
  of transit from Mars to Earth remains to be solved. Providing for still
  unknown environmental and nutritional requirements is daunting. Should

dead organisms be found in a sample, it would be difficult to conclude that those organisms were alive when obtained.

Human colonization of Mars

Some of the main reasons for colonizing Mars include economic interests, long-term scientific research best carried out by humans as opposed to robotic probes, and sheer curiosity. Surface conditions and the presence of water on Mars make it arguably the most hospitable of the planets in the Solar System, other than Earth. Human colonization of Mars would require *in situ* resource utilization (ISRU); A NASA report states that "applicable frontier technologies include robotics, machine intelligence, nanotechnology, synthetic biology, 3-D printing/additive manufacturing, and autonomy. These technologies combined with the vast natural resources should enable, pre- and post-human arrival ISRU to greatly increase reliability and safety and reduce cost for human colonization of Mars."

External links

- Study Reveals Young Mars Was A Wet World[268]
- NASA – The Mars Exploration Program[269]
- Scientists have discovered that Mars once had saltwater oceans[270]
- BBC News: Ammonia on Mars could mean life[271]
- Scientist says that life on Mars is likely today[272]
- Ancient salty sea on Mars wins as the most important scientific achievement of 2004 – Journal Science[273]
- Mars meteor found on Earth provides evidence that suggests microbial life once existed on Mars[274]
- Scientific American Magazine (November 2005 Issue) Did Life Come from Another World?[275]
- Audio interview about "Dark Dune Spots"[276]

Colonization of Mars

Mars is the focus of much scientific study about possible human colonization. Mars' surface conditions and past presence of water make it arguably the most hospitable planet in the Solar System besides Earth. Mars requires less energy per unit mass (delta-v) to reach from Earth than any planet, except Venus.

Permanent human habitation on other planets, including Mars, is one of science fiction's most prevalent themes. As technology advances, and concerns about humanity's future on Earth increase, arguments favoring space colonization gain momentum. Other reasons for colonizing space include economic interests, long-term scientific research best carried out by humans as opposed to robotic probes, and sheer curiosity.

More reasons include planting crops to provide for persons back on Earth in order to stop any future food shortages, or killings from fears of future food shortages or from irrational fears of overpopulation, or famines. In this scenario, once crops, food and goods are maintained well enough to stop aforementioned food shortages or fear of these food shortages happening, based off of the agriculture on Mars being large and stable, then large, very stable colonies and/or cities can be set up. Various asteroids, moons, and places on Mars can be mined while crops are initially being planted. This method will save lives on Earth, add potentially trillions of dollars to the world economy annually, and provide a stable **colony on Mars**. Some have also stated that persons should stay on Earth, but get food and resources from Mars and surrounding asteroids, moons and structures.

Both private and public organizations have made commitments to researching the viability of long-term colonization efforts and to taking steps toward a permanent human presence on Mars. Space agencies engaged in research or mission planning include NASA, Roscosmos, and the China National Space Administration. Private organizations include Mars One, SpaceX, Lockheed Martin, and Boeing.

Mission concepts and timelines

All of the early human mission concepts to Mars as conceived by national governmental space programs—such as those being tentatively planned by NASA, Rocosmos and ESA—would not be direct precursors to colonization. They are intended solely as exploration missions, as the *Apollo* missions to the Moon were not planned to be sites of a permanent base.

Colonization requires the establishment of permanent habitats that have potential for self-expansion and self-suteinance. Two early proposals for building habitats on Mars are the Mars Direct and the Semi-Direct concepts, advocated by Robert Zubrin.

Figure 200: *An artist's conception of a Mars habitat, with a 3D printed dome made of water ice, air-lock, and pressurized rover designs on Mars*[277]

Figure 201: *An artist's conception of a human Mars base, with a cutaway revealing an interior horticultural area*

Figure 202: *Various components of a human Mars surface mission*

SpaceX (colonization)

As of 2019, SpaceX owner Elon Musk is funding and developing a series of Mars-bound cargo flights with the BFR rocket and spaceship system as early as 2022, followed by the first crewed flight to Mars on the next launch window in 2024. During the first phase, the goal will be to launch several BFRs to transport and assemble a methane/oxygen propellant plant and to build up a base in preparation for an expanded surface presence.[278] A successful colonization would ultimately involve many more economic actors—whether individuals, companies, or governments—to facilitate the growth of the human presence on Mars over many decades.

Relative similarity to Earth

Space colonization

- Solar System
  - Inner
    - Mercury
    - Venus
    - Earth
      - Moon
      - Lagrange points
    - Mars
      - Phobos
      - Deimos
  - Asteroid mining
  - Free space
  - Outer
    - Jupiter
      - Io
      - Europa
      - Ganymede
      - Callisto
    - Saturn
      - Titan
    - Uranus
    - Neptune
    - Trans-Neptunian objects

This box:
- view
- talk
- edit[279]

Earth is similar to Venus in bulk composition, size and surface gravity, but
Mars's similarities to Earth are more compelling when considering colonization. These include:

- The Martian day (or **sol**) is very close in duration to Earth's. A solar day
  on Mars is 24 hours, 39 minutes and 35.244 seconds.[280]
- Mars has a surface area that is 28.4% of Earth's, only slightly less than the
  amount of dry land on Earth (which is 29.2% of Earth's surface). Mars
  has half the radius of Earth and only one-tenth the mass. This means that
  it has a smaller volume ($\sim$15%) and lower average density than Earth.
- Mars has an axial tilt of 25.19°, similar to Earth's 23.44°. As a result,
  Mars has seasons much like Earth, though they last nearly twice as long
  because the Martian year is about 1.88 Earth years. The Martian north
  pole currently points at Cygnus, not Ursa Minor like Earth's.
- Recent observations by NASA's *Mars Reconnaissance Orbiter*, ESA's
  Mars Express and NASA's *Phoenix* Lander confirm the presence of water
  ice on Mars.

Differences from Earth

Atmospheric pressure comparison

Location	Pressure
Olympus Mons summit	0.03 kPa (0.0044 psi)
Mars average	0.6 kPa (0.087 psi)
Hellas Planitia bottom	1.16 kPa (0.168 psi)
Armstrong limit	6.25 kPa (0.906 psi)
Mount Everest summit	33.7 kPa (4.89 psi)
Earth sea level	101.3 kPa (14.69 psi)

- Although there are some extremophile organisms that survive in hostile conditions on Earth, including simulations that approximate Mars, plants and animals generally cannot survive the ambient conditions present on the surface of Mars.
- Surface gravity of Mars is 38% that of Earth. Although microgravity is known to cause health problems such as muscle loss and bone demineralization, it is not known if Martian gravity would have a similar effect. The Mars Gravity Biosatellite was a proposed project designed to learn more about what effect Mars's lower surface gravity would have on humans, but it was cancelled due to a lack of funding.
- Mars is much colder than Earth, with mean surface temperatures between 186 and 268 K (–87 and –5 °C; –125 and 23 °F) (depending on position). The lowest temperature ever recorded on Earth was 180 K (–89.2 °C, –128.6 °F) in Antarctica.
- Surface water on Mars may occur transiently, but only under certain conditions.
- Because Mars is about 52% farther from the Sun, the amount of solar energy entering its upper atmosphere per unit area (the solar constant) is only around 43.3% of what reaches the Earth's upper atmosphere. However, due to the much thinner atmosphere, a higher fraction of the solar energy reaches the surface.Wikipedia:Verifiability The maximum solar irradiance on Mars is about 590 W/m^2 compared to about 1000 W/m^2 at the Earth's surface. Also, year-round dust storms on Mars may block sunlight for weeks at a time.
- Mars's orbit is more eccentric than Earth's, increasing temperature and solar constant variations.
- Due to the lack of a magnetosphere, solar particle events and cosmic rays can easily reach the Martian surface.

Figure 203: *Expedition style crewed mission would operate on the surface, but for limited amounts of time*

- The atmospheric pressure on Mars is far below the Armstrong limit at which people can survive without pressure suits. Since terraforming cannot be expected as a near-term solution, habitable structures on Mars would need to be constructed with pressure vessels similar to spacecraft, capable of containing a pressure between 30 and 100 kPa. See Atmosphere of Mars.
- The Martian atmosphere is 95% carbon dioxide, 3% nitrogen, 1.6% argon, and traces of other gases including oxygen totaling less than 0.4%.
- The thin atmosphere does not filter out ultraviolet sunlight.

Conditions for human habitation

Conditions on the surface of Mars are closer to the conditions on Earth in terms of temperature and sunlight than on any other planet or moon, except for the cloud tops of Venus. However, the surface is not hospitable to humans or most known life forms due to greatly reduced air pressure, and an atmosphere with only 0.1% oxygen.

In 2012, it was reported that some lichen and cyanobacteria survived and showed remarkable adaptation capacity for photosynthesis after 34 days in simulated Martian conditions in the Mars Simulation Laboratory (MSL) maintained by the German Aerospace Center (DLR). Some scientists think that

Figure 204: *Dust is one concern for Mars missions*

cyanobacteria could play a role in the development of self-sustainable crewed outposts on Mars. They propose that cyanobacteria could be used directly for various applications, including the production of food, fuel and oxygen, but also indirectly: products from their culture could support the growth of other organisms, opening the way to a wide range of life-support biological processes based on Martian resources.

Humans have explored parts of Earth that match some conditions on Mars. Based on NASA rover data, temperatures on Mars (at low latitudes) are similar to those in Antarctica. The atmospheric pressure at the highest altitudes reached by piloted balloon ascents (35 km (114,000 feet) in 1961, 38 km in 2012) is similar to that on the surface of Mars.

Human survival on Mars would require complex life-support measures and living in artificial environments.

Potential access to in-situ water (frozen or otherwise) via drilling has been investigated by NASA.

Effects on human health

Mars presents a hostile environment for human habitation. Different technologies have been developed to assist long-term space exploration and may be adapted for habitation on Mars. The existing record for the longest consecutive space flight is 438 days by cosmonaut Valeri Polyakov, and the most accrued time in space is 878 days by Gennady Padalka. The longest time spent outside the protection of the Earth's Van Allen radiation belt is about 12 days for the Apollo 17 moon landing. This is minor in comparison to the 1100 day journey planned by NASA as soon as the year 2028. Scientists have also hypothesized that many different biological functions can be negatively affected by the environment of Mars colonies. Due to higher levels of radiation, there are a multitude of physical side-effects that must be mitigated.

Physical effects

The difference in gravity will negatively affect human health by weakening bones and muscles. There is also risk of osteoporosis and cardiovascular problems. Current rotations on the International Space Station put astronauts in zero gravity for six months, a comparable length of time to a one-way trip to Mars. This gives researchers the ability to better understand the physical state that astronauts going to Mars will arrive in. Once on Mars, surface gravity is only 38% of that on Earth. Upon return to Earth, recovery from bone loss and atrophy is a long process and the effects of microgravity may never fully reverse.

There are also severe radiation risks on Mars that can influence cognitive processes, deteriorate cardiovascular health, inhibit reproduction, and cause cancer.

Psychological effects

Due to the communication delays, new protocols need to be developed in order to assess crew members' psychological health. Researchers have developed a Martian simulation called HI-SEAS (Hawaii Space Exploration Analog and Simulation) that places scientists in a simulated Martian laboratory to study the psychological effects of isolation, repetitive tasks, and living in close-quarters with other scientists for up to a year at a time. Computer programs are being developed to assist crews with personal and interpersonal issues in absence of direct communication with professionals on earth. Current suggestions for Mars exploration and colonization are to select individuals who have passed psychological screenings. Psychosocial sessions for the return home are also suggested in order to reorient people to society.

Figure 205: *Artist's conception of the process of terraforming Mars.*

Terraforming

There is much discussion regarding the possibility of terraforming Mars to allow a wide variety of life forms, including humans, to survive unaided on Mars's surface, including the technologies needed to do so.

Radiation

Mars has no global magnetosphere as Earth does. Combined with a thin atmosphere, this permits a significant amount of ionizing radiation to reach the Martian surface. The Mars Odyssey spacecraft carries an instrument, the Mars Radiation Environment Experiment (MARIE), to measure the radiation. MARIE found that radiation levels in orbit above Mars are 2.5 times higher than at the International Space Station. The average daily dose was about 220 µGy (22 mrad) – equivalent to 0.08 Gy per year. A three-year exposure to such levels would be close to the safety limits currently adopted by NASA.Wikipedia:Citation needed Levels at the Martian surface would be somewhat lower and might vary significantly at different locations depending on altitude and local magnetic fields. Building living quarters underground (possibly in Martian lava tubes which are already present) would significantly lower the colonists' exposure to radiation. Occasional solar proton events (SPEs) produce much higher doses.

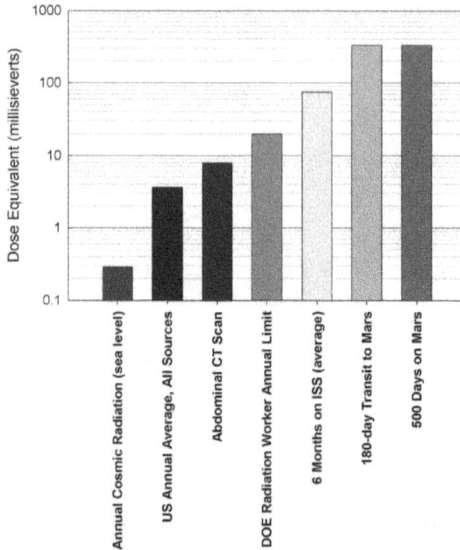

Figure 206: *Comparison of radiation doses – includes the amount detected on the trip from Earth to Mars by the RAD on the MSL (2011–2013).*

Much remains to be learned about space radiation. In 2003, NASA's Lyndon B. Johnson Space Center opened a facility, the NASA Space Radiation Laboratory, at Brookhaven National Laboratory, that employs particle accelerators to simulate space radiation. The facility studies its effects on living organisms, as well as experimenting with shielding techniques. Initially, there was some evidence that this kind of low level, chronic radiation is not quite as dangerous as once thought; and that radiation hormesis occurs. However, results from a 2006 study indicated that protons from cosmic radiation may cause twice as much serious damage to DNA as previously estimated, exposing astronauts to greater risk of cancer and other diseases. As a result of the higher radiation in the Martian environment, the summary report of the Review of U.S. Human Space Flight Plans Committee released in 2009 reported that "Mars is not an easy place to visit with existing technology and without a substantial investment of resources." NASA is exploring a variety of alternative techniques and technologies such as deflector shields of plasma to protect astronauts and spacecraft from radiation.

In September 2017, NASA reported radiation levels on the surface of the planet Mars were temporarily doubled, and were associated with an aurora 25-times brighter than any observed earlier, due to a massive, and unexpected, solar storm in the middle of the month.

Figure 207: *Rendezvous, an interplanetary
stage and lander stage come together over Mars*

Transportation

Interplanetary spaceflight

Mars requires less energy per unit mass (delta V) to reach from Earth than any planet except Venus. Using a Hohmann transfer orbit, a trip to Mars requires approximately nine months in space. Modified transfer trajectories that cut the travel time down to four to seven months in space are possible with incrementally higher amounts of energy and fuel compared to a Hohmann transfer orbit, and are in standard use for robotic Mars missions. Shortening the travel time below about six months requires higher delta-v and an exponentially-Wikipedia:Please clarifyWikipedia:Cleanup increasing amount of fuel, and is difficult with chemical rockets. It could be feasible with advanced spacecraft propulsion technologies, some of which have already been tested to varying levels, such as Variable Specific Impulse Magnetoplasma Rocket, and nuclear rockets. In the former case, a trip time of forty days could be attainable, and in the latter, a trip time down to about two weeks. In 2016, a University of California scientist said they could further reduce travel time for an robotic probe to Mars down to "as little as 72 hours" with the use of a "photonic propulsion" system instead of the fuel-based rocket propulsion system.

During the journey the astronauts would be subject to radiation, which would require a means to protect them. Cosmic radiation and solar wind cause DNA

Figure 208: *Mars (Viking 1, 1980)*

damage, which increases the risk of cancer significantly. The effect of long-term travel in interplanetary space is unknown, but scientists estimate an *added* risk of between 1% and 19% (one estimate is 3.4%) for men to die of cancer because of the radiation during the journey to Mars and back to Earth. For women the probability is higher due to generally larger glandular tissues.

Landing on Mars

Mars has a surface gravity 0.38 times that of Earth, and the density of its atmosphere is about 0.6% of that on Earth. The relatively strong gravity and the presence of aerodynamic effects make it difficult to land heavy, crewed spacecraft with thrusters only, as was done with the Apollo Moon landings, yet the atmosphere is too thin for aerodynamic effects to be of much help in aerobraking and landing a large vehicle. Landing piloted missions on Mars would require braking and landing systems different from anything used to land crewed spacecraft on the Moon or robotic missions on Mars.

If one assumes carbon nanotube construction material will be available with a strength of 130 GPa then a space elevator could be built to land people and material on Mars. A space elevator on Phobos has also been proposed.

Figure 209: *Painting of a landing on Mars (1986)*

Equipment needed for colonization

Colonization of Mars will require a wide variety of equipment—both equipment to directly provide services to humans and production equipment used to produce food, propellant, water, energy and breathable oxygen—in order to support human colonization efforts. Required equipment will include:

- Habitats
- Storage facilities
- Shop workspaces
- Airlock, for pressurization and dust management
- Resource extraction equipment—initially for water and oxygen, later for a wider cross section of minerals, building materials, etc.
- Equipment for energy production and energy storage, some solar and perhaps nuclear as well

- Food production spaces and equipment.
- Propellant production equipment, generally thought to be hydrogen and methane through the Sabatier reaction for fuel—with oxygen oxidizer—for chemical rocket engines
- Fuels or other energy source for use with surface transportation. Carbon monoxide/oxygen (CO/O_2) engines have been suggested for early surface

Figure 210: *Mars greenhouses feature in many colonization designs, especially for food production and other purposes*

Figure 211: *Various technologies and devices for
Mars are shown in the illustration of a Mars base*

transportation use as both carbon monoxide and oxygen can be straight-forwardly produced by zirconium dioxide electrolysis from the Martian atmosphere without requiring use of any of the Martian water resources to obtain hydrogen.

- Communication equipment
- Equipment for moving over the surface - Mars suit, crewed rovers and possibly even Mars aircraft.

According to Elon Musk, "even at a million people [working on Mars] you're assuming an incredible amount of productivity per person, because you would need to recreate the entire industrial base on Mars... You would need to mine and refine all of these different materials, in a much more difficult environment than Earth".[281]

Communication

Communications with Earth are relatively straightforward during the half-sol when Earth is above the Martian horizon. NASA and ESA included communications relay equipment in several of the Mars orbiters, so Mars already has communications satellites. While these will eventually wear out, additional orbiters with communication relay capability are likely to be launched before any colonization expeditions are mounted.

The one-way communication delay due to the speed of light ranges from about 3 minutes at closest approach (approximated by perihelion of Mars minus aphelion of Earth) to 22 minutes at the largest possible superior conjunction (approximated by aphelion of Mars plus aphelion of Earth). Real-time communication, such as telephone conversations or Internet Relay Chat, between Earth and Mars would be highly impractical due to the long time lags involved. NASA has found that direct communication can be blocked for about two weeks every synodic period, around the time of superior conjunction when the Sun is directly between Mars and Earth, although the actual duration of the communications blackout varies from mission to mission depending on various factors—such as the amount of link margin designed into the communications system, and the minimum data rate that is acceptable from a mission standpoint. In reality most missions at Mars have had communications blackout periods of the order of a month.

A satellite at the L_4 or L_5 Earth–Sun Lagrangian point could serve as a relay during this period to solve the problem; even a constellation of communications satellites would be a minor expense in the context of a full colonization program. However, the size and power of the equipment needed for these distances make the L4 and L5 locations unrealistic for relay stations, and the inherent stability of these regions, although beneficial in terms of station-keeping, also attracts dust and asteroids, which could pose a risk. Despite that

Figure 212: *Astronauts approach Viking 2 lander probe*

concern, the STEREO probes passed through the L4 and L5 regions without damage in late 2009.

Recent work by the University of Strathclyde's Advanced Space Concepts Laboratory, in collaboration with the European Space Agency, has suggested an alternative relay architecture based on highly non-Keplerian orbits. These are a special kind of orbit produced when continuous low-thrust propulsion, such as that produced from an ion engine or solar sail, modifies the natural trajectory of a spacecraft. Such an orbit would enable continuous communications during solar conjunction by allowing a relay spacecraft to "hover" above Mars, out of the orbital plane of the two planets. Such a relay avoids the problems of satellites stationed at either L4 or L5 by being significantly closer to the surface of Mars while still maintaining continuous communication between the two planets.

Robotic precursors

The path to a human colony could be prepared by robotic systems such as the Mars Exploration Rovers *Spirit*, *Opportunity* and *Curiosity*. These systems could help locate resources, such as ground water or ice, that would help a colony grow and thrive. The lifetimes of these systems would be measured in years and even decades, and as recent developments in commercial spaceflight have shown, it may be that these systems will involve private as well as government ownership. These robotic systems also have a reduced cost compared with early crewed operations, and have less political risk.

Wired systems might lay the groundwork for early crewed landings and bases, by producing various consumables including fuel, oxidizers, water, and construction materials. Establishing power, communications, shelter, heating, and manufacturing basics can begin with robotic systems, if only as a prelude to crewed operations.

Mars Surveyor 2001 Lander MIP (Mars ISPP Precursor) was to demonstrate manufacture of oxygen from the atmosphere of Mars,[282] and test solar cell technologies and methods of mitigating the effect of Martian dust on the power systems.[283]Wikipedia:Manual of Style/Dates and numbers#Chronological items

Before any people are transported to Mars on the notional 2030s Interplanetary Transport System envisioned by SpaceX, a number of robotic cargo missions would be undertaken first in order to transport the requisite equipment, habitats and supplies. Equipment that would be necessary would include "machines to produce fertilizer, methane and oxygen from Mars' atmospheric nitrogen and carbon dioxide and the planet's subsurface water ice" as well as construction materials to build transparent domes for initial agricultural areas.

Economics

As with early colonies in the New World, economics would be a crucial aspect to a colony's success. The reduced gravity well of Mars and its position in the Solar System may facilitate Mars–Earth trade and may provide an economic rationale for continued settlement of the planet. Given its size and resources, this might eventually be a place to grow food and produce equipment to mine the asteroid belt.

A major economic problem is the enormous up-front investment required to establish the colony and perhaps also terraform the planet.

Some early Mars colonies might specialize in developing local resources for Martian consumption, such as water and/or ice. Local resources can also be used in infrastructure construction. One source of Martian ore currently known to be available is metallic iron in the form of nickel–iron meteorites. Iron in this form is more easily extracted than from the iron oxides that cover the planet.

Another main inter-Martian trade good during early colonization could be manure.[284] Assuming that life doesn't exist on Mars, the soil is going to be very poor for growing plants, so manure and other fertilizers will be valued highly in

Figure 213: *Iron–nickel meteorite found on Mars's surface (Heat Shield Rock)*

any Martian civilization until the planet changes enough chemically to support growing vegetation on its own.

Solar power is a candidate for power for a Martian colony. Solar insolation (the amount of solar radiation that reaches Mars) is about 42% of that on Earth, since Mars is about 52% farther from the Sun and insolation falls off as the square of distance. But the thin atmosphere would allow almost all of that energy to reach the surface as compared to Earth, where the atmosphere absorbs roughly a quarter of the solar radiation. Sunlight on the surface of Mars would be much like a moderately cloudy day on Earth.

Economic drivers

Space colonization on Mars can roughly be said to be possible when the necessary methods of space colonization become cheap enough (such as space access by cheaper launch systems) to meet the cumulative funds that have been gathered for the purpose.

Although there are no immediate prospects for the large amounts of money required for any space colonization to be available given traditional launch costs,Wikipedia:Citing sources#What information to include there is some prospect of a radical reduction to launch costs in the 2020s, which would consequently lessen the cost of any efforts in that direction. With a published

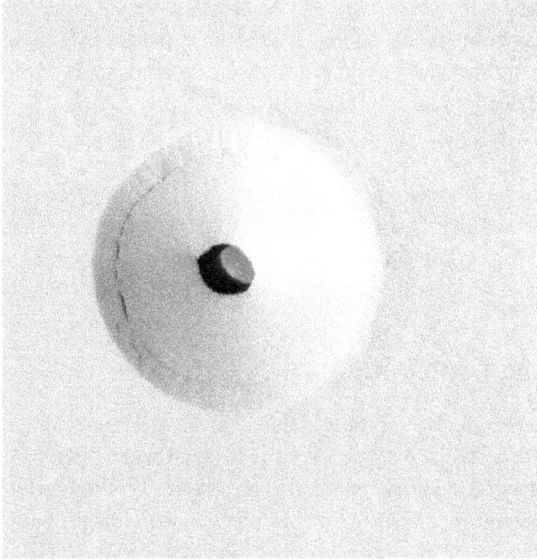

Figure 214: *Cropped version of a HiRISE image of a lava tube skylight entrance on the Martian volcano Pavonis Mons.*

price of US$62 million per launch of up to 22,800 kg (50,300 lb) payload to low Earth orbit or 4,020 kg (8,860 lb) to Mars, SpaceX Falcon 9 rockets are already the "cheapest in the industry". SpaceX's reusable plans include Falcon Heavy and future methane-based launch vehicles including the Interplanetary Transport System. If SpaceX is successful in developing the reusable technology, it would be expected to "have a major impact on the cost of access to space", and change the increasingly competitive market in space launch services.

Alternative funding approaches might include the creation of inducement prizes. For example, the 2004 President's Commission on Implementation of United States Space Exploration Policy suggested that an inducement prize contest should be established, perhaps by government, for the achievement of space colonization. One example provided was offering a prize to the first organization to place humans on the Moon and sustain them for a fixed period before they return to Earth.

Possible locations for settlements

Equatorial regions

Mars Odyssey found what appear to be natural caves near the volcano Arsia Mons. It has been speculated that settlers could benefit from the shelter that these or similar structures could provide from radiation and micrometeoroids. Geothermal energy is also suspected in the equatorial regions.

Lava tubes

Several possible Martian lava tube skylights have been located on the flanks of Arsia Mons. Earth based examples indicate that some should have lengthy passages offering complete protection from radiation and be relatively easy to seal using on-site materials, especially in small subsections.

Planetary protection

Robotic spacecraft to Mars are required to be sterilized, to have at most 300,000 spores on the exterior of the craft—and more thoroughly sterilized if they contact "special regions" containing water,[285,286] otherwise there is a risk of contaminating not only the life-detection experiments but possibly the planet itself.

It is impossible to sterilize human missions to this level, as humans are host to typically a hundred trillion microorganisms of thousands of species of the human microbiome, and these cannot be removed while preserving the life of the human. Containment seems the only option, but it is a major challenge in the event of a hard landing (i.e. crash).[287] There have been several planetary workshops on this issue, but with no final guidelines for a way forward yet.[288] Human explorers would also be vulnerable to back contamination to Earth if they become carriers of microorganisms.[289]

Ethical, political and legal challenges

One possible ethical challenge that space travelers might face is that of pregnancy during the trip. According to NASA's policies, it is forbidden for members of the crew to engage in sex in space. NASA wants its crewmembers to treat each other like coworkers would in a professional environment. A pregnant member on a spacecraft is dangerous to all those aboard. The pregnant woman and child would most likely need additional nutrition from the rations aboard, as well as special treatment and care. At some point during the trip, the pregnancy would most likely impede on the pregnant crew member's duties and abilities. It is still not fully known how the environment in a spacecraft would affect the development of a child aboard. It is known however that an unborn child in space would be more susceptible to solar radiation, which would likely have a negative effect on its cells and genetics.[290] During a long trip to Mars it is likely that members of craft may engage in sex due to their stressful and isolated environment.[291]

It is unforeseen how the first human landing on Mars will change the current policies regarding the exploration of space and occupancy of celestial bodies. In the 1967, United Nations Treaty on Principles Governing the Activities of States in the Exploration and Use of Outer Space, Including the Moon and Other Celestial Bodies, it was determined that no country may take claim to space or its inhabitants. Since the planet Mars offers a challenging environment and dangerous obstacles for humans to overcome, the laws and culture on the planet will most likely be very different from those on Earth.[292] With Elon Musk announcing his plans for travel to Mars, it is uncertain how the dynamic of a private company possibly being the first to put a human on Mars will play out on a national and global scale.[293] NASA had to deal with several cuts in funding. During the presidency of Barack Obama, the objective for NASA to reach Mars was pushed to the background.[294] In 2017, president Donald Trump promised to return humans to the Moon and eventually Mars,[295] effectively taking action by increasing NASA budget with $1.1 billion,[296] and mostly focus on the development of the new Space Launch System.

Advocacy

Mars colonization is advocated by several non-governmental groups for a range of reasons and with varied proposals. One of the oldest groups is the Mars Society who promote a NASA program to accomplish human exploration of Mars and have set up Mars analog research stations in Canada and the United States. Mars to Stay advocates recycling emergency return vehicles into permanent settlements as soon as initial explorers determine permanent habitation is possible. Mars One, which went public in June 2012, aims to establish a fully operational permanent human colony on Mars by 2027 with funding coming from a reality TV show and other commercial exploitation, although this approach has been widely criticized as unrealistic and infeasible.

Elon Musk founded SpaceX with the long-term goal of developing the technologies that will enable a self-sustaining human colony on Mars. In 2015 he stated "I think we've got a decent shot of sending a person to Mars in 11 or 12 years". Richard Branson, in his lifetime, is "determined to be a part of starting a population on Mars. I think it is absolutely realistic. It will happen... I think over the next 20 years, we will take literally hundreds of thousands of people to space and that will give us the financial resources to do even bigger things".

In June 2013, Buzz Aldrin, American engineer and former astronaut, and the second person to walk on the Moon, wrote an opinion, published in *The New*

Figure 215: *Buzz Aldrin, the 2nd human to set foot on the Moon, has recommended human Mars missions*

York Times, supporting a human mission to Mars and viewing the Moon "not as a destination but more a point of departure, one that places humankind on a trajectory to homestead Mars and become a two-planet species." In August 2015, Aldrin, in association with the Florida Institute of Technology, presented a "master plan", for NASA consideration, for astronauts, with a "tour of duty of ten years", to colonize Mars before the year 2040.

In fiction

A few instances in fiction provide detailed descriptions of Mars colonization. They include:

- *The Space Between Us* (2016 film), by Peter Chelsom
- *The Martian* (2011), by Andy Weir (and the film (2015), directed by Ridley Scott)
- *Mars (miniseries)* (2016) by National Geographic (U.S. TV channel)
- *Mr. Nobody* (2009), by Jaco Van Dormael
- *Aria* (2002–2008), by Kozue Amano
- *Red Faction* (2001), developed by Volition, published by THQ
- *First Landing* (2002), by Robert Zubrin
- *Mars Diaries* (2000), by Sigmund Brouwer

- *Mars Underground* (1997), by William K. Hartmann
- *Climbing Olympus* (1994), by Kevin J. Anderson
- *The Martian* (1992) and *Return to Mars* (1999), by Ben Bova
- *Total Recall* (1990), by Paul Verhoeven
- *Icehenge* (1985), the Mars trilogy (*Red Mars, Green Mars, Blue Mars*, 1992–96), and *The Martians* (1999), by Kim Stanley Robinson
- *Man Plus* (1976), by Frederik Pohl
- *The Destruction of Faena* (1974), by Alexander Kazantsev
- *We Can Remember It for You Wholesale* (1966), by Philip K. Dick
- *The Sands of Mars* (1951), by Arthur C. Clarke
- *The Martian Chronicles* (1950), by Ray Bradbury
- *John Carter (film)* (2012), by Mark Andrews

Further reading

- Robert Zubrin, *The Case for Mars: The Plan to Settle the Red Planet and Why We Must*, Simon & Schuster/Touchstone, 1996, ISBN 0-684-83550-9
- Frank Crossman and Robert Zubrin, editors, *On to Mars: Colonizing a New World*. Apogee Books Space Series, 2002, ISBN 1-896522-90-4.
- Frank Crossman and Robert Zubrin, editors, *On to Mars 2: Exploring and Settling a New World*. Apogee Books Space Series, 2005, ISBN 978-1-894959-30-8.
- Resource Utilization Concepts for MoonMars[297]; ByIris Fleischer, Olivia Haider, Morten W. Hansen, Robert Peckyno, Daniel Rosenberg and Robert E. Guinness; 30 September 2003; IAC Bremen, 2003 (29 Sept – 03 Oct 2003) and MoonMars Workshop (26–28 Sept 2003, Bremen). Accessed on 18 January 2010
- MARTIAN OUTPOST: The Challenges of Establishing a Human Settlement on Mars[298]; by Erik Seedhouse; Praxis Publishing; 2009; ISBN 978-0-387-98190-1. Also see[299],[300]
- Ice, mineral-rich soil could support human outpost on Mars[301]; by Sharon Gaudin; 27 June 2008; IDG News Service

External links

Wikimedia Commons has media related to *Colonization of Mars*.

Wikibooks has a book on the topic of: *Colonising Mars*

- Mars Society[302]
- The Planetary Society: Mars Millennium Project[303]
- 4Frontiers Corporation[304]
- The Mars Foundation[305]
- Making Mars the New Earth – National Geographic[306]
- Should we colonize Mars? – Wikidebate in Wikiversity

Moons

Moons of Mars

<templatestyles src="Multiple_image/styles.css" />

Color image of Phobos (MRO, 23 March 2008)

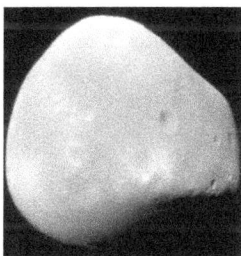

Color image of Deimos (MRO, 21 February 2009)

The **two moons of Mars** are Phobos and Deimos. Both were discovered by Asaph Hall in August 1877 and are named after the Greek mythological twin characters Phobos (panic/fear) and Deimos (terror/dread) who accompanied their father Ares, god of war, into battle. Ares was known as Mars to the Romans.

Phobos orbits closer to Mars, with a semi-major axis of 9,377 km (5,827 mi) to Deimos' 23,460 km (14,580 mi).

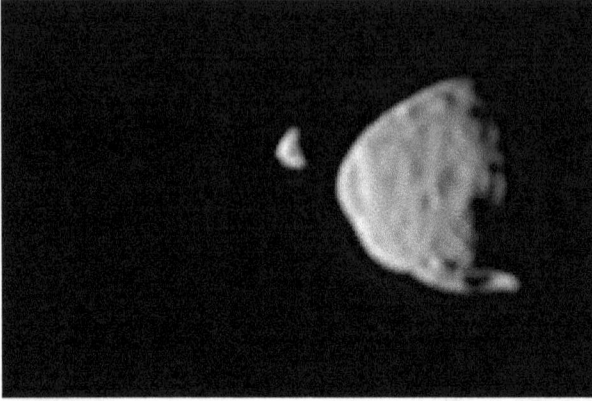

Figure 216: *Curiosity's view of the Martian moons: Phobos passing in front of Deimos – in real-time (video-gif, 1 August 2013)*

History

Early speculation

The existence of the moons of Mars had been speculated about since the moons of Jupiter were discovered. It is known that when Galileo, as a hidden report about him having observed two bumps on the sides of Saturn (later discovered to be its rings), used the anagram *smaismrmilmepoetaleumibunenugttauiras* for *Altissimum planetam tergeminum observavi* ("I have observed the most distant planet to have a triple form"), Johannes Kepler had misinterpreted it to mean *Salve umbistineum geminatum Martia proles* (Hello, furious twins, sons of Mars).[307]

Perhaps inspired by Kepler (and quoting Kepler's third law of planetary motion), Jonathan Swift's satire *Gulliver's Travels* (1726) refers to two moons in Part 3, Chapter 3 (the "Voyage to Laputa"), in which Laputa's astronomers are described as having discovered two satellites of Mars orbiting at distances of 3 and 5 Martian diameters with periods of 10 and 21.5 hours. Phobos and Deimos (both found in 1877, more than a century after Swift's novel) have actual orbital distances of 1.4 and 3.5 Martian diameters, and their respective orbital periods are 7.66 and 30.35 hours.[308] In the 20th century, V. G. Perminov, a spacecraft designer of early Soviet Mars and Venus spacecraft, speculated Swift found and deciphered records that Martians left on Earth.[309] However, the view of most astronomers is that Swift was simply employing a common argument of the time, that as the inner planets Venus and Mercury had no satellites, Earth had one and Jupiter had four (known at the time), that Mars by analogy must have two. Furthermore, as they had not yet been discovered, it was reasoned that they must be small and close to Mars. This would lead Swift to making a roughly accurate estimate of their orbital distances and revolution periods. In addition Swift could have been helped in his calculations by his friend, the mathematician John Arbuthnot.[310]

Voltaire's 1752 short story "Micromégas", about an alien visitor to Earth, also refers to two moons of Mars. Voltaire was presumably influenced by Swift.[311,312] In recognition of these 'predictions', two craters on Deimos are named Swift and Voltaire, while on Phobos there is one named regio, *Laputa Regio*, and one named planitia, *Lagado Planitia*, both of which are named after places in *Gulliver's Travels* (the fictional Laputa, a flying island, and Lagado, imaginary capital of the fictional nation Balnibarbi).[313] Many of the craters on Phobos are also named after characters in *Gulliver's Travels*.

Discovery

Asaph Hall discovered Deimos on 12 August 1877 at about 07:48 UTC and Phobos on 18 August 1877, at the US Naval Observatory in Washington, D.C., at about 09:14 GMT (contemporary sources, using the pre-1925 astronomical convention that began the day at noon, give the time of discovery as 11 August 14:40 and 17 August 16:06 Washington mean time respectively).[314] At the time, he was deliberately searching for Martian moons. Hall had previously seen what appeared to be a Martian moon on 10 August, but due to bad weather, he could not definitively identify them until later.

Hall recorded his discovery of Phobos in his notebook as follows:

<templatestyles src="Template:Quote/styles.css"/>

> *"I repeated the examination in the early part of the night of 11th [August 1877], and again found nothing, but trying again some hours later I found a faint object on the following side and a little north of the planet. I had barely time to secure an observation of its position when fog from the River stopped the work. This was at half past two o'clock on the night of the 11th. Cloudy weather intervened for several days.*
>
> *"On 15 August the weather looking more promising, I slept at the Observatory. The sky cleared off with a thunderstorm at 11 o'clock and the search was resumed. The atmosphere however was in a very bad condition and Mars was so blazing and unsteady that nothing could be seen of the object, which we now know was at that time so near the planet as to be invisible.*

Figure 217: *The telescope used to discover the Martian moons*

"On 16 August the object was found again on the following side of the planet, and the observations of that night showed that it was moving with the planet, and if a satellite, was near one of its elongations. Until this time I had said nothing to anyone at the Observatory of my search for a satellite of Mars, but on leaving the observatory after these observations of the 16th, at about three o'clock in the morning, I told my assistant, George Anderson, to whom I had shown the object, that I thought I had discovered a satellite of Mars. I told him also to keep quiet as I did not wish anything said until the matter was beyond doubt. He said nothing, but the thing was too good to keep and I let it out myself. On 17 August between one and two o'clock, while I was reducing my observations, Professor Newcomb came into my room to eat his lunch and I showed him my measures of the faint object near Mars which proved that it was moving with the planet.

"On 17 August while waiting and watching for the outer moon, the inner one was discovered. The observations of the 17th and 18th put beyond doubt the character of these objects and the discovery was publicly announced by Admiral Rodgers."

The telescope used for the discovery was the 26-inch (66 cm) refractor (telescope with a lens) then located at Foggy Bottom.[315] In 1893 the lens was remounted and put in a new dome, where it remains into the 21st century.[316]

The names, originally spelled *Phobus* and *Deimus*, respectively, were suggested by Henry Madan (1838–1901), Science Master of Eton, from Book XV of the Iliad, where Ares summons Fear and Fright.

Mars moon hoax

In 1959, Walter Scott Houston perpetrated a celebrated April Fool's hoax in the April edition of the *Great Plains Observer*, claiming that "Dr. Arthur Hayall of the University of the Sierras reports that the moons of Mars are actually artificial satellites". Both Dr. Hayall and the University of the Sierras were fictitious. The hoax gained worldwide attention when Houston's claim was repeated in earnest by a Soviet scientist, Iosif Shklovsky,[317] who, based on a later-disproven density estimate, suggested Phobos was a hollow metal shell.

Recent surveys

Searches have been conducted for additional satellites. In 2003, Scott S. Sheppard and David C. Jewitt surveyed early the entire Hill sphere of Mars for irregular satellites. However scattered light from Mars prevented them from searching the inner few arcminutes where the satellites Phobos and Deimos reside. No new satellites were found to an apparent limiting red magnitude of 23.5, which corresponds to radii of about 0.09 km using an albedo of 0.07.[318]

Characteristics

If viewed from Mars's surface near its equator, full Phobos looks about one third as big as a full moon on Earth. It has an angular diameter of between 8' (rising) and 12' (overhead). It would look smaller when the observer is further away from the Martian equator, and is completely invisible from Mars's polar ice caps. Deimos looks more like a bright star or planet for an observer on Mars, only slightly bigger than Venus looks from Earth; it has an angular diameter of about 2'. The Sun's angular diameter as seen from Mars, by contrast, is about 21'. Thus there are no total solar eclipses on Mars, as the moons are far too small to completely cover the Sun. On the other hand, total lunar eclipses of Phobos are very common, happening almost every night.[319]

The motions of Phobos and Deimos would appear very different from that of our own Moon. Speedy Phobos rises in the west, sets in the east, and rises again in just eleven hours, while Deimos, being only just outside synchronous orbit, rises as expected in the east but very slowly. Despite its 30-hour orbit, it takes 2.7 days to set in the west as it slowly falls behind the rotation of Mars.

Both moons are tidally locked, always presenting the same face towards Mars. Since Phobos orbits Mars faster than the planet itself rotates, tidal forces are

Figure 218: *Apparent sizes of the moons of Mars, Deimos and Phobos, and the Moon as viewed from the surface of their respective planets (Curiosity rover, 1 August 2013)*

slowly but steadily decreasing its orbital radius. At some point in the future, when it approaches Mars closely enough (see Roche limit), Phobos will be broken up by these tidal forces. Several strings of craters on the Martian surface, inclined further from the equator the older they are, suggest that there may have been other small moons that suffered the fate expected of Phobos, and that the Martian crust as a whole shifted between these events.[320] Deimos, on the other hand, is far enough away that its orbit is being slowly boosted instead,[321] as in the case of our own Moon.

Orbital details

Name and pronunciation		Image	Diameter (km)	Surface Area (km)	Mass (kg)	Semi-major axis (km)	Orbital period (h)	Average moonrise period (h, d)
Mars I	Phobos /ˈfoʊbəs/ *FOH-bes*		22.2 km (13.8 mi) (27×21.6×18.8 km)	1,548 km	1.08×10^{16}	9,377 km (5,827 mi)	7.66	11.12 h (0.463 d)
Mars II	Deimos /ˈdaɪmɒs/ *DY-moss*		12.6 km (7.8 mi) (10×12×16 km)	483 km	2×10^{15}	23,460 km (14,580 mi)	30.35	131 h (5.44 d)

The relative sizes of and distance between Mars, Phobos, and Deimos, to scale (*Load the image in full size to see both Moons of Mars.*)

Origin

The origin of the Martian moons is still controversial.[322] Phobos and Deimos both have much in common with carbonaceous C-type asteroids, with spectra, albedo, and density very similar to those of C- or D-type asteroids. Based on their similarity, one hypothesis is that both moons may be captured main-belt asteroids.[323] Both moons have very circular orbits which lie almost exactly in Mars's equatorial plane, and hence a capture origin requires a mechanism for circularizing the initially highly eccentric orbit, and adjusting its inclination into the equatorial plane, most probably by a combination of atmospheric drag and tidal forces, although it is not clear that sufficient time is available for this to occur for Deimos. Capture also requires dissipation of energy. The current atmosphere of Mars is too thin to capture a Phobos-sized object by atmospheric braking. Geoffrey Landis has pointed out that the capture could have occurred if the original body was a binary asteroid that separated under tidal forces.

Phobos could be a second-generation Solar System object that coalesced in orbit after Mars formed, rather than forming concurrently out of the same birth cloud as Mars.

Another hypothesis is that Mars was once surrounded by many Phobos- and Deimos-sized bodies, perhaps ejected into orbit around it by a collision with a large planetesimal.[324] The high porosity of the interior of Phobos (based on the density of 1.88 g/cm^3, voids are estimated to comprise 25 to 35 percent of Phobos' volume) is inconsistent with an asteroidal origin. Observations of Phobos in the thermal infrared suggest a composition containing mainly phyllosilicates, which are well known from the surface of Mars. The spectra are distinct from those of all classes of chondrite meteorites, again pointing away from an asteroidal origin. Both sets of findings support an origin of Phobos from material ejected by an impact on Mars that reaccreted in Martian orbit, similar to the prevailing theory for the origin of Earth's moon.

The moons of Mars may have started with a huge collision with a protoplanet one third the mass of Mars that formed a ring around Mars. The inner part of the ring formed a large moon. Gravitational interactions between this moon and the outer ring formed Phobos and Deimos. Later, the large moon crashed into Mars, but the two small moons remained in orbit. This theory agrees with the fine-grained surface of the moons and their high porosity. The outer disk would create fine-grained material.[325,326]

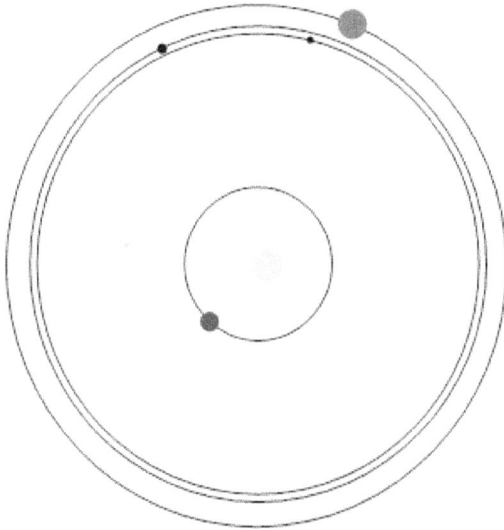

Figure 219: *Animation illustrating the asteroid-belt origin for the moons*

Exploration

While many Martian probes provided images and other data about Phobos and Deimos, only few were dedicated to these satellites and intended to perform a flyby or landing on the surface.

Two probes under the Soviet Phobos program were successfully launched in 1988, but neither conducted the intended jumping landings on Phobos and Deimos due to failures (although *Phobos 2* successfully photographed Phobos). The post-Soviet Russian *Fobos-Grunt* probe was intended to be the first sample return mission from Phobos, but a rocket failure left it stranded in Earth orbit in 2011.

In 1997 and 1998, the Aladdin mission was selected as a finalist in the NASA Discovery Program. The plan was to visit both Phobos and Deimos, and launch projectiles at the satellites. The probe would collect the ejecta as it performed a slow flyby. These samples would be returned to Earth for study three years later. Ultimately, NASA rejected this proposal in favor of MESSENGER, a probe to Mercury.

In 2007, the European Space Agency and EADS Astrium proposed and developed a mission to Phobos in 2016 with a lander and sample return, but this mission was never flown. Since 2007 the Canadian Space Agency has been considering the PRIME mission to Phobos with orbiter and lander. Since 2008 NASA Glenn Research Center began studying a Phobos and Deimos sample return mission. Since 2013 NASA developed the Phobos Surveyor mission concept with an orbiter and a small rover that is proposed to launch sometime after 2023.Wikipedia:Citation needed Proposed NASA's PADME mission proposes to launch in 2020 and reach Mars orbit in 2021 to conduct multiple flybys of the Martian moons.Wikipedia:Citation needed Also, NASA is assessing the OSIRIS-REx II, concept mission for a sample return from Phobos.[327] Another sample return mission from Deimos, called Gulliver. has been conceptualized.Wikipedia:Citation needed Russia plans to repeat Fobos-Grunt mission around 2024.Wikipedia:Citation needed

Gallery

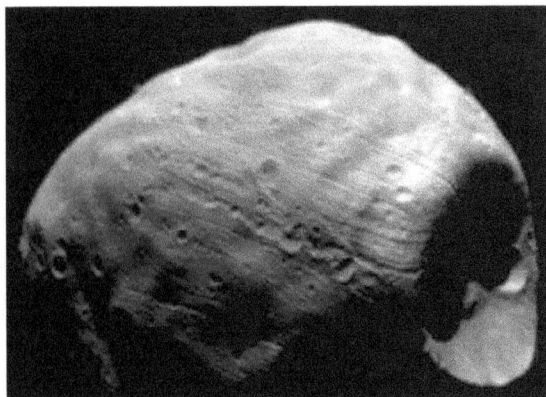

Figure 220: *Phobos, with Stickney Crater on the right (2003).*

Figure 221: *Phobos (1998). UNIQ-ref-0-086cfa95efb485b4-QINU*

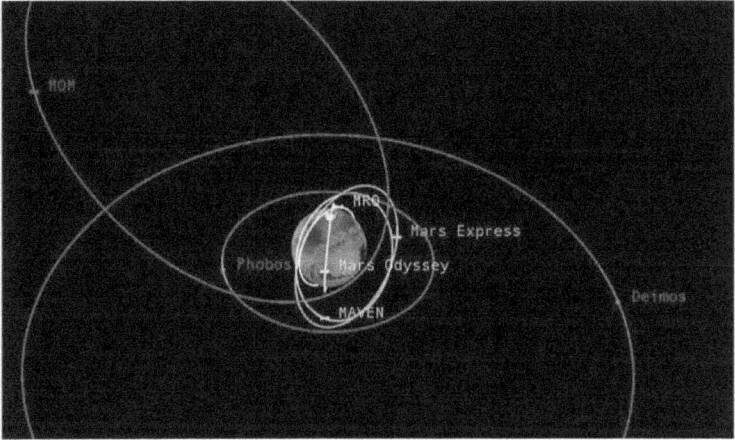

Figure 223:
Orbits of moons and spacecraft orbiting Mars.

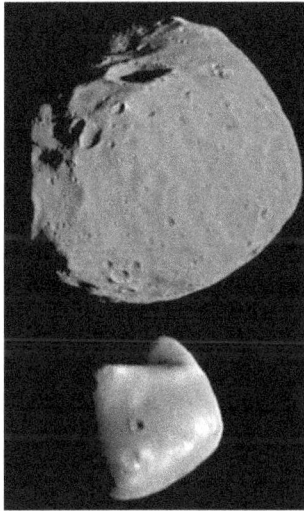

Figure 222: *Comparison - Phobos (top) and Deimos (bottom) (2005).*

Further reading

- First International Conference On The Exploration Of Phobos And Deimos (2007)[328]

- Pascal Lee – *A Case for the Human Exploration of the Moons of Mars* (2007)[329] (pdf)

External links

- Gazetteer of Planetary Nomenclature – Mars (USGS)[330]

Exploration of Mars

Exploration of Mars

Active missions at Mars from 2001 to present

Year	Missions
2018	8
2017	8
2016	8
2015	7
2014	7
2013	5
2012	5
2011	4
2010	5
2009	5
2008	6
2007	5
2006	6
2005	5

2004 5	
2003 3	
2002 2	
2001 2	

Active spacecraft at Mars 1971–2000

Year	Spacecraft	
2000	1	
1999	1	
1998	1	
1997	2	
1990–1996	0	
1989	1	
1983–1988	0	
1982	1	
1981	1	
1980	3	
1979	3	
1978	4	
1976	4	
1975	4	
1974	3	
1973	0	
1972	3	
1971	5	

The planet Mars has been explored remotely by spacecraft. Probes sent from Earth, beginning in the late 20th century, have yielded a dramatic increase in knowledge about the Martian system, focused primarily on understanding its geology and habitability potential.

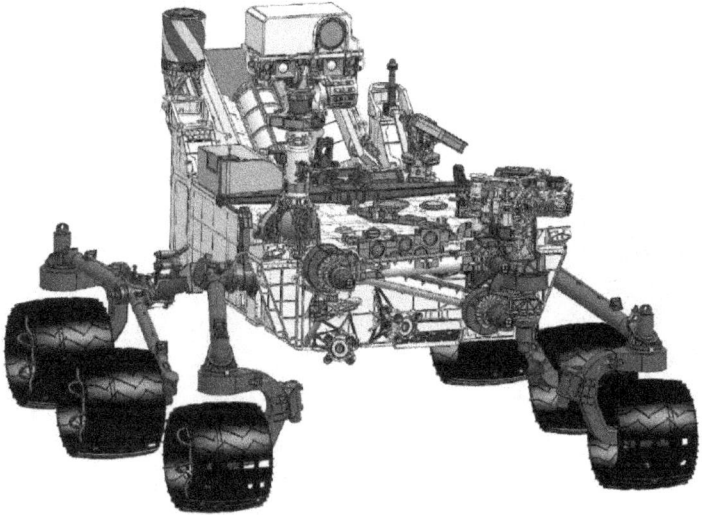

Figure 224: *A diagram of the Curiosity rover, landed on Mars in 2012.*

Current status

Engineering interplanetary journeys is complicated and the exploration of Mars has experienced a high failure rate, especially the early attempts. Roughly two-thirds of all spacecraft destined for Mars failed before completing their missions and some failed before their observations could begin. Some missions have met with unexpected success, such as the twin Mars Exploration Rovers, which operated for years beyond their specification. As of 16 October 2017, two scientific rovers were on the surface of Mars beaming signals back to Earth (*Opportunity* of the *Mars Exploration Rover* mission and *Curiosity* of the *Mars Science Laboratory* mission), with six orbiters surveying the planet: *Mars Odyssey*, *Mars Express*, *Mars Reconnaissance Orbiter*, *Mars Orbiter Mission*, MAVEN, and the Trace Gas Orbiter, which have contributed massive amounts of information about Mars. No sample return missions have been attempted for Mars and an attempted return mission for Mars' moon Phobos (*Fobos-Grunt*) failed.

On 24 January 2014, NASA reported that current studies on the planet Mars by the *Curiosity* and *Opportunity* rovers will search for evidence of ancient life, including a biosphere based on autotrophic, chemotrophic and/or chemolithoautotrophic microorganisms, as well as ancient water, including fluvio-lacustrine environments (plains related to ancient rivers or lakes) that may have been habitable. The search for evidence of habitability, taphonomy

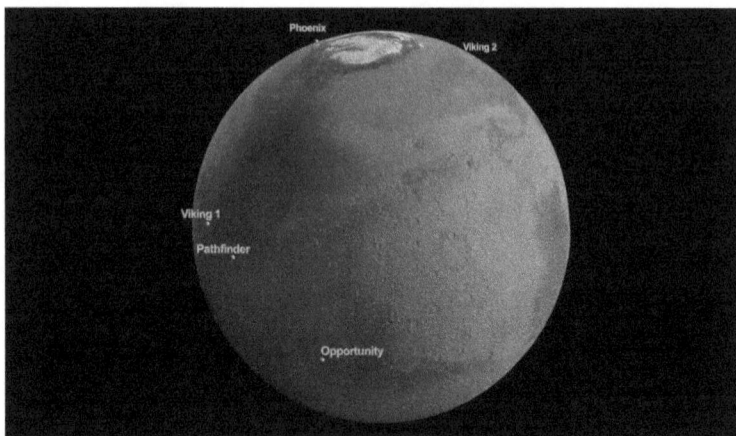

Figure 225: *The landing site of each Mars mission can be seen on this rotating globe.*

(related to fossils), and organic carbon on the planet Mars is now a primary NASA objective.

Martian system

Mars has long been the subject of human interest. Early telescopic observations revealed color changes on the surface that were attributed to seasonal vegetation and apparent linear features were ascribed to intelligent design. Further telescopic observations found two moons, Phobos and Deimos, polar ice caps and the feature now known as Olympus Mons, the solar system's tallest mountain. The discoveries piqued further interest in the study and exploration of the red planet. Mars is a rocky planet, like Earth, that formed around the same time, yet with only half the diameter of Earth, and a far thinner atmosphere; it has a cold and desert-like surface.

Launch windows

Opportunities 2013–2020

Year	Launch	Spacecraft (launched or planned)
2013	Nov 2013	MAVEN, Mars Orbiter Mission
2016	Jan 2016 – Apr 2016	ExoMars TGO
2018	May 5, 2018 – June 8, 2018	InSight
2020	Jul 2020 – Sep 2020	ExoMars rover, Mars 2020, Mars Hope, 2020 Chinese Mars Mission, Mars Orbiter Mission 2 (MOM-2)

The minimum-energy launch windows for a Martian expedition occur at intervals of approximately two years and two months (specifically 780 days, the planet's synodic period with respect to Earth).[331] In addition, the lowest available transfer energy varies on a roughly 16-year cycle. For example, a minimum occurred in the 1969 and 1971 launch windows, rising to a peak in the late 1970s, and hitting another low in 1986 and 1988.

Past and current missions

Launches to Mars

Decade	
1960s	13
1970s	11
1980s	2
1990s	8
2000s	8
2010s	5

<templatestyles src="Multiple_image/styles.css" />

Martian sunset by *Spirit* rover, 2005.

North polar view by *Phoenix* lander, 2008.

Starting in 1960, the Soviets launched a series of probes to Mars including the first intended flybys and hard (impact) landing (Mars 1962B). The first successful fly-by of Mars was on 14–15 July 1965, by NASA's Mariner 4. On November 14, 1971 Mariner 9 became the first space probe to orbit another planet when it entered into orbit around Mars. The amount of data returned by probes increased dramatically as technology improved.[332]

The first to contact the surface were two Soviet probes: Mars 2 lander on November 27 and Mars 3 lander on December 2, 1971—Mars 2 failed during descent and Mars 3 about twenty seconds after the first Martian soft landing.[333] Mars 6 failed during descent but did return some corrupted atmospheric data in 1974.[334] The 1975 NASA launches of the Viking program consisted of two orbiters, each with a lander that successfully soft landed in 1976. Viking 1 remained operational for six years, Viking 2 for three. The Viking landers relayed the first color panoramas of Mars.

The Soviet probes Phobos 1 and 2 were sent to Mars in 1988 to study Mars and its two moons, with a focus on Phobos. Phobos 1 lost contact on the way to Mars. Phobos 2, while successfully photographing Mars and Phobos, failed before it was set to release two landers to the surface of Phobos.

Roughly two-thirds of all spacecraft destined for Mars have failed without completing their missions, and it has a reputation as a difficult space exploration target.

Missions that ended prematurely after Phobos 1 & 2 (1988) include (see Probing difficulties section for more details):

- Mars Observer (launched in 1992)
- Mars 96 (1996)
- Mars Climate Orbiter (1999)
- Mars Polar Lander with Deep Space 2 (1999)
- Nozomi (2003)
- Beagle 2 (2003)
- Fobos-Grunt with Yinghuo-1 (2011)
- *Schiaparelli* lander (2016)

Following the 1993 failure of the Mars Observer orbiter, the NASA Mars Global Surveyor achieved Mars orbit in 1997. This mission was a complete success, having finished its primary mapping mission in early 2001. Contact was lost with the probe in November 2006 during its third extended program, spending exactly 10 operational years in space. The NASA Mars Pathfinder, carrying a robotic exploration vehicle Sojourner, landed in the Ares Vallis on Mars in the summer of 1997, returning many images.

Phoenix landed on the north polar region of Mars on May 25, 2008. Its robotic arm dug into the Martian soil and the presence of water ice was confirmed on June 20, 2008. The mission concluded on November 10, 2008 after contact was lost. In 2008, the price of transporting material from the surface of Earth to the surface of Mars was approximately US\$309,000 per kilogram.

Rosetta came within 250 km of Mars during its 2007 flyby.[335] Dawn flew by Mars in February 2009 for a gravity assist on its way to investigate Vesta and Ceres.

Figure 226: *Curiosity's self-portrait on the planet Mars at "Rocknest" (October 31, 2012).*

Recent missions

NASA's Mars Odyssey orbiter entered Mars orbit in 2001. Odyssey's Gamma Ray Spectrometer detected significant amounts of hydrogen in the upper metre or so of regolith on Mars. This hydrogen is thought to be contained in large deposits of water ice.

The Mars Express mission of the European Space Agency (ESA) reached Mars in 2003. It carried the Beagle 2 lander, which was not heard from after being released and was declared lost in February 2004. Beagle 2 was located in January 2015 by HiRise camera on NASA's Mars Reconnaissance Orbiter (MRO) having landed safely but failed to fully deploy its solar panels and antenna. In early 2004 the Mars Express Planetary Fourier Spectrometer team announced the orbiter had detected methane in the Martian atmosphere. ESA announced in June 2006 the discovery of aurorae on Mars.

In January 2004, the NASA twin Mars Exploration Rovers named *Spirit* (MER-A) and *Opportunity* (MER-B) landed on the surface of Mars. Both have met or exceeded all their targets. Among the most significant scientific returns has been conclusive evidence that liquid water existed at some time in the past at both landing sites. Martian dust devils and windstorms have occasionally cleaned both rovers' solar panels, and thus increased their lifespan.

Figure 227: *The Electra radio of the MAVEN orbiter*

Spirit Rover (MER-A) was active until 2010, when it stopped sending data because it had fallen into a sand dune.

On 10 March 2006, the NASA Mars Reconnaissance Orbiter (MRO) probe arrived in orbit to conduct a two-year science survey. The orbiter began mapping the Martian terrain and weather to find suitable landing sites for upcoming lander missions. The MRO snapped the first image of a series of active avalanches near the planet's north pole, scientists said March 3, 2008.

The Mars Science Laboratory mission was launched on November 26, 2011 and it delivered the *Curiosity* rover, on the surface of Mars on August 6, 2012 UTC. It is larger and more advanced than the Mars Exploration Rovers, with a velocity of up to 90 meters per hour (295 feet per hour). Experiments include a laser chemical sampler that can deduce the make-up of rocks at a distance of 7 meters.

MAVEN was successfully launched aboard an Atlas V launch vehicle at the beginning of the first launch window on November 18, 2013. Following the first engine burn of the Centaur second stage, the vehicle coasted in low Earth orbit for 27 minutes before a second Centaur burn of five minutes to insert it into a heliocentric Mars transit orbit.

On September 22, 2014, MAVEN reached Mars and was inserted into an areocentric elliptic orbit 6,200 km (3,900 mi) by 150 km (93 mi) above the planet's surface.

Figure 228: *Mars 1M spacecraft.*

The Indian Space Research Organisation (ISRO) launched its Mars Orbiter Mission (MOM) on November 5, 2013. It was successfully inserted into Mars orbit on 24 September 2014. India's ISRO is the fourth space agency to reach Mars, after the Soviet space program, NASA and ESA. India became the first country to successfully get a spacecraft into Mars orbit on its maiden attempt.

The ExoMars Trace Gas Orbiter arrived at Mars in 2016 and deployed the Schiaparelli EDM lander, a test lander that had partial success. Crash-landed on surface, but transmitted data during descent.

Overview of missions

The following entails a brief overview of Mars exploration, oriented towards orbiters and flybys; see also Mars landing and Mars rover.

Early Soviet missions

1960s

Between 1960 and 1969, the Soviet Union launched nine probes intended to reach Mars. They all failed: three at launch; three failed to reach near-Earth orbit; one during the burn to put the spacecraft into trans-Mars trajectory; and two during the interplanetary orbit.

The Mars 1M programs (sometimes dubbed Marsnik in Western media) was the first Soviet unmanned spacecraft interplanetary exploration program, which consisted of two flyby probes launched towards Mars in October 1960, Mars 1960A and Mars 1960B (also known as *Korabl 4* and *Korabl 5* respectively). After launch, the third stage pumps on both launchers were unable to develop enough pressure to commence ignition, so Earth parking orbit was not achieved. The spacecraft reached an altitude of 120 km before reentry.

Mars 1962A was a Mars fly-by mission, launched on October 24, 1962 and Mars 1962B an intended first Mars lander mission, launched in late December of the same year (1962). Both failed from either breaking up as they were going into Earth orbit or having the upper stage explode in orbit during the burn to put the spacecraft into trans-Mars trajectory.

The first success

Selected Soviet Mars probes

Spacecraft	Orbiter or flyby outcome	Lander outcome
Mars 1	Failure	Failure
Mars 2	Success	Failure
Mars 3	Partial success	Partial success
Mars 4	Failure	N/A
Mars 5	Partial success	N/A
Mars 6	Success	Failure
Mars 7	Success	Failure
Phobos 1	Failure	Not deployed
Phobos 2	Partial success	Not deployed

Mars 1 (1962 Beta Nu 1), an automatic interplanetary spacecraft launched to Mars on November 1, 1962, was the first probe of the Soviet Mars probe program to achieve interplanetary orbit. Mars 1 was intended to fly by the planet at a distance of about 11,000 km and take images of the surface as well as send back data on cosmic radiation, micrometeoroid impacts and Mars' magnetic field, radiation environment, atmospheric structure, and possible organic compounds. Sixty-one radio transmissions were held, initially at two-day intervals and later at 5 day intervals, from which a large amount of interplanetary data was collected. On 21 March 1963, when the spacecraft was at a distance of 106,760,000 km from Earth, on its way to Mars, communications ceased due to failure of its antenna orientation system.

In 1964, both Soviet probe launches, of Zond 1964A on June 4, and Zond 2 on November 30, (part of the Zond program), resulted in failures. Zond 1964A had a failure at launch, while communication was lost with Zond 2 en route to Mars after a mid-course maneuver, in early May 1965.

In 1969, and as part of the Mars probe program, the Soviet Union prepared two identical 5-ton orbiters called M-69, dubbed by NASA as Mars 1969A and Mars 1969B. Both probes were lost in launch-related complications with the newly developed Proton rocket.

1970s

The USSR intended to have the first artificial satellite of Mars beating the planned American Mariner 8 and Mariner 9 Mars orbiters. In May 1971, one day after Mariner 8 malfunctioned at launch and failed to reach orbit, Cosmos 419 (Mars 1971C), a heavy probe of the Soviet Mars program M-71, also failed to launch. This spacecraft was designed as an orbiter only, while the next two probes of project M-71, Mars 2 and Mars 3, were multipurpose combinations of an orbiter and a lander with small skis-walking rovers that would be the first planet rovers outside the Moon. They were successfully launched in mid-May 1971 and reached Mars about seven months later. On November 27, 1971 the lander of Mars 2 crash-landed due to an on-board computer malfunction and became the first man-made object to reach the surface of Mars. On 2 December 1971, the Mars 3 lander became the first spacecraft to achieve a soft landing, but its transmission was interrupted after 14.5 seconds.

The Mars 2 and 3 orbiters sent back a relatively large volume of data covering the period from December 1971 to March 1972, although transmissions continued through to August. By 22 August 1972, after sending back data and a total of 60 pictures, Mars 2 and 3 concluded their missions. The images and data enabled creation of surface relief maps, and gave information on the Martian gravity and magnetic fields.

In 1973, the Soviet Union sent four more probes to Mars: the Mars 4 and Mars 5 orbiters and the Mars 6 and Mars 7 fly-by/lander combinations. All missions except Mars 7 sent back data, with Mars 5 being most successful. Mars 5 transmitted just 60 images before a loss of pressurization in the transmitter housing ended the mission. Mars 6 lander transmitted data during descent, but failed upon impact. Mars 4 flew by the planet at a range of 2200 km returning one swath of pictures and radio occultation data, which constituted the first detection of the nightside ionosphere on Mars. Mars 7 probe separated prematurely from the carrying vehicle due to a problem in the operation of one of the onboard systems (attitude control or retro-rockets) and missed the planet by 1,300 kilometres (8.7×10^{-6} au).Wikipedia:Citation needed

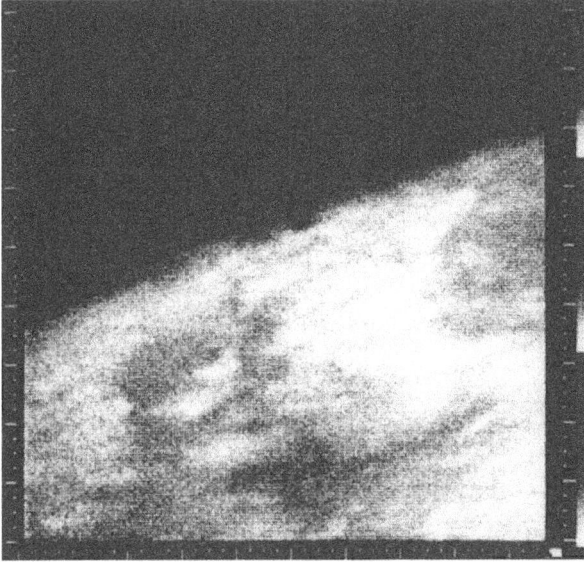

Figure 229: *The first close-up images taken of Mars in 1965 from Mariner 4 show an area about 330 km across by 1200 km from limb to bottom of frame.*

Mariner program

In 1964, NASA's Jet Propulsion Laboratory made two attempts at reaching Mars. Mariner 3 and Mariner 4 were identical spacecraft designed to carry out the first flybys of Mars. Mariner 3 was launched on November 5, 1964, but the shroud encasing the spacecraft atop its rocket failed to open properly, dooming the mission. Three weeks later, on November 28, 1964, Mariner 4 was launched successfully on a 7½-month voyage to Mars..Wikipedia:Citation needed

Mariner 4 flew past Mars on July 14, 1965, providing the first close-up photographs of another planet. The pictures, gradually played back to Earth from a small tape recorder on the probe, showed impact craters. It provided radically more accurate data about the planet; a surface atmospheric pressure of about 1% of Earth's and daytime temperatures of –100 °C (–148 °F) were estimated. No magnetic field or Martian radiation belts were detected. The new data meant redesigns for then planned Martian landers, and showed life would have a more difficult time surviving there than previously anticipated.

NASA continued the Mariner program with another pair of Mars flyby probes, Mariner 6 and 7. They were sent at the next launch window, and reached the planet in 1969. During the following launch window the Mariner program

Figure 230: *Mariner Crater, as seen by Mariner 4. The location is Phaethontis quadrangle.*

again suffered the loss of one of a pair of probes. Mariner 9 successfully entered orbit about Mars, the first spacecraft ever to do so, after the launch time failure of its sister ship, Mariner 8. When Mariner 9 reached Mars in 1971, it and two Soviet orbiters (Mars 2 and Mars 3, see *Mars probe program* below) found that a planet-wide dust storm was in progress. The mission controllers used the time spent waiting for the storm to clear to have the probe rendezvous with, and photograph, Phobos. When the storm cleared sufficiently for Mars' surface to be photographed by Mariner 9, the pictures returned represented a substantial advance over previous missions. These pictures were the first to offer more detailed evidence that liquid water might at one time have flowed on the planetary surface. They also finally discerned the true nature of many Martian albedo features. For example, Nix Olympica was one of only a few features that could be seen during the planetary duststorm, revealing it to be the highest mountain (volcano, to be exact) on any planet in the entire Solar System, and leading to its reclassification as Olympus Mons.Wikipedia:Citation needed

Viking program

The Viking program launched Viking 1 and 2 spacecraft to Mars in 1975; The program consisted of two orbiters and two landers – these were the first two spacecraft to successfully land and operate on Mars.

<templatestyles src=">Multiple_image/styles.css" />

Viking 1 lander site (1st color, July 21, 1976).

Viking 2 lander site (1st color, September 5, 1976).

Viking 2 lander site (September 25, 1977).

(False color image) Frost at Viking 2 site (May 18, 1979).

Martian sunset over Chryse Planitia at Viking 1 site (August 20, 1976).

The primary scientific objectives of the lander mission were to search for biosignatures and observe meteorologic, seismic and magnetic properties of Mars. The results of the biological experiments on board the Viking landers remain inconclusive, with a reanalysis of the Viking data published in 2012 suggesting signs of microbial life on Mars.

<templatestyles src="Multiple_image/styles.css" />

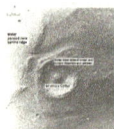

Flood erosion at Dromore crater.

Tear-drop shaped islands at Oxia Palus.

Streamlined islands in Lunae Palus.

Scour patterns located in Lunae Palus.

The Viking orbiters revealed that large floods of water carved deep valleys, eroded grooves into bedrock, and traveled thousands of kilometers. Areas of branched streams, in the southern hemisphere, suggest that rain once fell.[336]

Mars Pathfinder

Mars Pathfinder was a U.S. spacecraft that landed a base station with a roving probe on Mars on July 4, 1997. It consisted of a lander and a small 10.6 kilograms (23 lb) wheeled robotic rover named Sojourner, which was the first rover to operate on the surface of Mars.[337] In addition to scientific objectives, the Mars Pathfinder mission was also a "proof-of-concept" for various technologies, such as an airbag landing system and automated obstacle avoidance, both later exploited by the Mars Exploration Rovers.

Figure 231: *Sojourner takes Alpha Proton X-ray Spectrometer measurements of the Yogi Rock.*

Mars Global Surveyor

After the 1992 failure of NASA's Mars Observer orbiter, NASA retooled and launched Mars Global Surveyor (MGS). Mars Global Surveyor launched on November 7, 1996, and entered orbit on September 12, 1997. After a year and a half trimming its orbit from a looping ellipse to a circular track around the planet, the spacecraft began its primary mapping mission in March 1999. It observed the planet from a low-altitude, nearly polar orbit over the course of one complete Martian year, the equivalent of nearly two Earth years. Mars Global Surveyor completed its primary mission on January 31, 2001, and completed several extended mission phases.Wikipedia:Citation needed

The mission studied the entire Martian surface, atmosphere, and interior, and returned more data about the red planet than all previous Mars missions combined. The data has been archived and remains available publicly.

Among key scientific findings, Global Surveyor took pictures of gullies and debris flow features that suggest there may be current sources of liquid water, similar to an aquifer, at or near the surface of the planet. Similar channels on Earth are formed by flowing water, but on Mars the temperature is normally too cold and the atmosphere too thin to sustain liquid water. Nevertheless, many scientists hypothesize that liquid groundwater can sometimes surface on Mars, erode gullies and channels, and pool at the bottom before freezing and evaporating.Wikipedia:Citation needed

Magnetometer readings showed that the planet's magnetic field is not globally generated in the planet's core, but is localized in particular areas of the

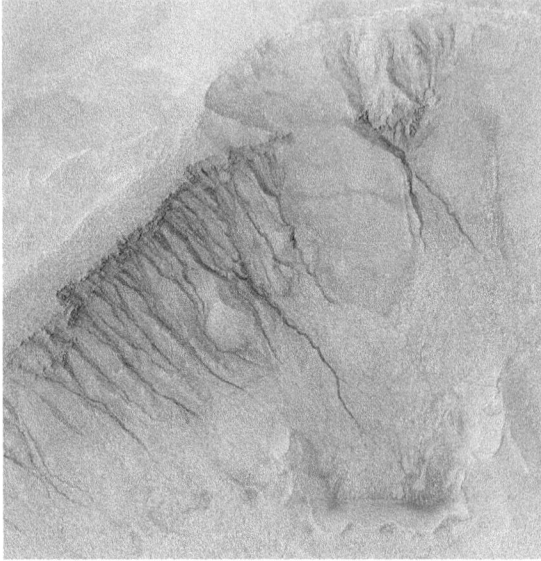

Figure 232: *Gullies, similar to those formed on Earth, are visible on this image from Mars Global Surveyor.*

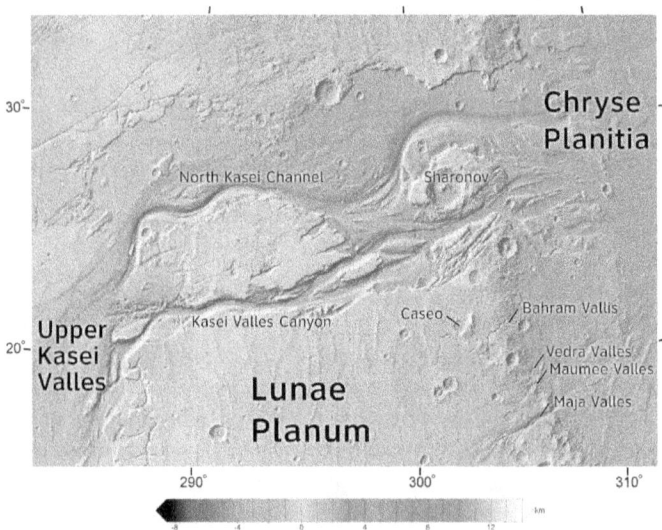

Figure 233: *A color-coded elevation map produced from data collected by Mars Global Surveyor indicating the result of floods on Mars.*

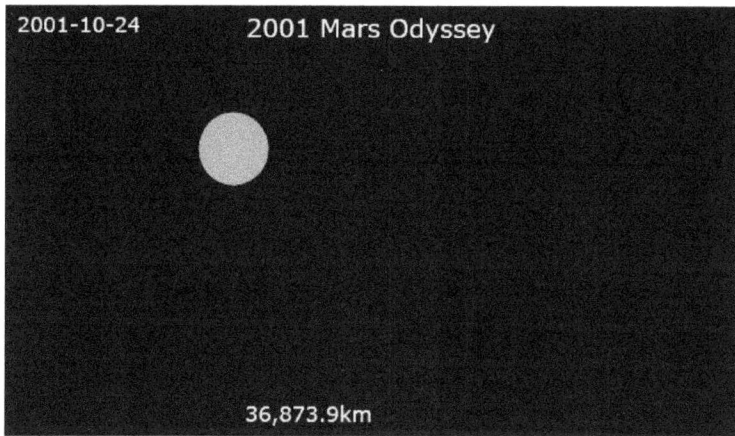

Figure 234: *Animation of 2001 Mars Odyssey's trajectory around Mars from 24 October 2001 to 24 October 2002*
2001 Mars Odyssey · Mars

crust. New temperature data and closeup images of the Martian moon Phobos showed that its surface is composed of powdery material at least 1 metre (3 feet) thick, caused by millions of years of meteoroid impacts. Data from the spacecraft's laser altimeter gave scientists their first 3-D views of Mars' north polar ice cap.Wikipedia:Citation needed On November 5, 2006 MGS lost contact with Earth. NASA ended efforts to restore communication on January 28, 2007.

Mars Odyssey and Mars Express

In 2001, NASA's Mars Odyssey orbiter arrived at Mars. Its mission is to use spectrometers and imagers to hunt for evidence of past or present water and volcanic activity on Mars. In 2002, it was announced that the probe's gamma-ray spectrometer and neutron spectrometer had detected large amounts of hydrogen, indicating that there are vast deposits of water ice in the upper three meters of Mars' soil within 60° latitude of the south pole.Wikipedia:Citation needed

On June 2, 2003, the European Space Agency's Mars Express set off from Baikonur Cosmodrome to Mars. The Mars Express craft consists of the Mars Express Orbiter and the stationary lander Beagle 2. The lander carried a digging device and the smallest mass spectrometer created to date, as well as a range of other devices, on a robotic arm in order to accurately analyze soil beneath the dusty surface to look for biosignatures and biomolecules.Wikipedia:Citation needed

Figure 235: *Animation of Mars Express's trajectory
around Mars from 5 December 2003 to 1 January 2010*
Mars Express · Mars

The orbiter entered Mars orbit on December 25, 2003, and Beagle 2 entered
Mars' atmosphere the same day. However, attempts to contact the lander
failed. Communications attempts continued throughout January, but Beagle
2 was declared lost in mid-February, and a joint inquiry was launched by the
UK and ESA. The Mars Express Orbiter confirmed the presence of water ice
and carbon dioxide ice at the planet's south pole, while NASA had previously
confirmed their presence at the north pole of Mars.Wikipedia:Citation needed

The lander's fate remained a mystery until it was located intact on the surface
of Mars in a series of images from the Mars Reconnaissance Orbiter. The
images suggest that two of the spacecraft's four solar panels failed to deploy,
blocking the spacecraft's communications antenna. *Beagle 2* is the first British
and first European probe to achieve a soft landing on Mars.Wikipedia:Citation
needed

MER and Phoenix

NASA's Mars Exploration Rover Mission (MER), started in 2003, is an ongoing robotic space mission involving two rovers, Spirit (MER-A) and Opportunity, (MER-B) exploring the Martian surface geology.Wikipedia:Citation
needed The mission's scientific objective is to search for and characterize a
wide range of rocks and soils that hold clues to past water activity on Mars.
The mission is part of NASA's Mars Exploration Program, which includes
three previous successful landers: the two Viking program landers in 1976;
and Mars Pathfinder probe in 1997.Wikipedia:Citation needed

Figure 236: *Polar surface as seen by the Phoenix lander.*

Mars Reconnaissance Orbiter

The *Mars Reconnaissance Orbiter* (MRO) is a multipurpose spacecraft designed to conduct reconnaissance and exploration of Mars from orbit. The $720 million USD spacecraft was built by Lockheed Martin under the supervision of the Jet Propulsion Laboratory, launched August 12, 2005, and entered Mars orbit on March 10, 2006.

The MRO contains a host of scientific instruments such as the HiRISE camera, CTX camera, CRISM, and SHARAD. The HiRISE camera is used to analyze Martian landforms, whereas CRISM and SHARAD can detect water, ice, and minerals on and below the surface. Additionally, MRO is paving the way for upcoming generations of spacecraft through daily monitoring of Martian weather and surface conditions, searching for future landing sites, and testing a new telecommunications system that enable it to send and receive information at an unprecedented bitrate, compared to previous Mars spacecraft. Data transfer to and from the spacecraft occurs faster than all previous interplanetary missions combined and allows it to serve as an important relay satellite for other missions.Wikipedia:Citation needed

Figure 237: *Slope streaks as seen by HiRise*

Rosetta and *Dawn* swingbys

The ESA *Rosetta* space probe mission to the comet 67P/Churyumov-Gerasimenko flew within 250 km of Mars on February 25, 2007, in a gravitational slingshot designed to slow and redirect the spacecraft.

The NASA *Dawn* spacecraft used the gravity of Mars in 2009 to change direction and velocity on its way to Vesta, and tested out *Dawn*'s cameras and other instruments on Mars.

Fobos-Grunt

On November 8, 2011, Russia's Roscosmos launched an ambitious mission called Fobos-Grunt. It consisted of a lander aimed to retrieve a sample back to Earth from Mars' moon Phobos, and place the Chinese Yinghuo-1 probe in Mars' orbit. The Fobos-Grunt mission suffered a complete control and communications failure shortly after launch and was left stranded in low Earth orbit, later falling back to Earth.[338] The Yinghuo-1 satellite and Fobos-Grunt underwent destructive re-entry on January 15, 2012, finally disintegrating over the Pacific Ocean.[339,340,341]

Figure 238: *Curiosity's view of Aeolis Mons ("Mount Sharp")*
foothills on August 9, 2012 EDT (white balanced image).

Curiosity rover

The NASA Mars Science Laboratory mission with its rover named *Curiosity*, was launched on November 26, 2011, and landed on Mars on August 6, 2012 on Aeolis Palus in Gale Crater. The rover carries instruments designed to look for past or present conditions relevant to the past or present habitability of Mars.

MAVEN

NASA's MAVEN is an orbiter mission to study the upper atmosphere of Mars. It will also serve as a communications relay satellite for robotic landers and rovers on the surface of Mars. MAVEN was launched 18 November 2013 and reached Mars on 22 September 2014.Wikipedia:Citation needed

Mars Orbiter Mission

The Mars Orbiter Mission, also called *Mangalyaan*, was launched on 5 November 2013 by the Indian Space Research Organisation (ISRO). It was successfully inserted into Martian orbit on 24 September 2014. The mission is a technology demonstrator, and as secondary objective, it will also study the Martian atmosphere. This is India's first mission to Mars, and with it, ISRO became the fourth space agency to successfully reach Mars after the Soviet Union, NASA (USA) and ESA (Europe). It also made ISRO the second space agency to reach Mars orbit on its first attempt (the first national one, after the international ESA), and also the first Asian country to successfully send an orbiter to Mars. It was completed in a record low budget of $71 million, making it the least-expensive Mars mission to date.

Figure 239: *Computer-design drawing for NASA's 2020 Mars Rover.*

Trace Gas Orbiter and EDM

The ExoMars Trace Gas Orbiter is an atmospheric research orbiter built in collaboration between ESA and Roscosmos. It was injected into Mars orbit on 19 October 2016 to gain a better understanding of methane (CH_4) and other trace gases present in the Martian atmosphere that could be evidence for possible biological or geological activity.

InSight

In August 2012, NASA selected *InSight*, a $425 million lander mission, with a drill and seismometer to determine the interior structure of Mars.[342,343] Two flyby CubeSats called MarCO were launched with *InSight* to provide real-time telemetry during the entry and landing of *InSight*. The CubeSats separated from the Atlas V booster 1.5 hours after launch and are traveling their own trajectories to Mars. The mission was launched on 5 May 2018 and is expected to reach Mars on 26 November 2018.

Future missions

- As part of the ExoMars program, ESA and the Russian Federal Space Agency plan to send the ExoMars rover in 2020 to search for past or present microscopic life on Mars.

- The Mars 2020 rover mission by NASA will be launched in 2020, and it is based on the Mars Science Laboratory design. The scientific payload will be focused on astrobiology.[344] It will include a miniature robotic Mars Helicopter Scout.
- The 2020 Chinese Mars Mission is planned to comprise an orbiter, a lander and a small rover.
- The United Arab Emirates will send an orbiter to Mars, the Hope Mars Mission, in 2020.[345,346]
- The ISRO plans to send a follow up mission to its Mars Orbiter Mission in the 2021–2022 timeframe; it is called Mars Orbiter Mission 2 (MOM-2). This mission will consist of an orbiter, and probably a rover.

Proposals

- The Finnish-Russian Mars MetNet concept would use multiple small meteorological stations on Mars to establish a widespread observation network to investigate the planet's atmospheric structure, physics and meteorology. The MetNet precursor or demonstrator was considered for a piggyback launch on Fobos-Grunt, and on the two proposed to fly on the 2016 and 2020 ExoMars spacecraft.
- The Mars-Grunt is a Russian mission concept to bring a sample of Martian soil to Earth.
- A ESA-NASA team produced a three-launch architecture concept for a Mars sample return, which uses a rover to cache small samples, a Mars ascent stage to send it into orbit, and an orbiter to rendezvous with it above Mars and take it to Earth.[347] Solar-electric propulsion could allow a one launch sample return instead of three.[348]
- The Mars Scout Program's SCIM would involve a probe grazing the upper atmosphere of Mars to collect dust and air for return to Earth.[349]
- On 10 Nov 2014, China unveiled a prototype model of a Mars rover based on its lunar rover Yutu at an annual air show at Zhuhai. The CASC also said that a mission including an orbiter, lander and the rover (2020 Chinese Mars Mission) will be sent in 2020.
- Japan is working on a mission concept called MELOS rover that would look for biosignatures of extant life on Mars.

Other future mission concepts include polar probes, Martian aircraft and a network of small meteorological stations. Longterm areas of study may include Martian lava tubes, resource utilization, and electronic charge carriers in rocks.[350,351] Micromissions are another possibility, such as piggybacking a small spacecraft on an Ariane 5 rocket and using a lunar gravity assist to get to Mars.[352]

Figure 240: *Concept for NASA Design Reference Mission Architecture 5.0 (2009).*

Human mission proposals

Many people have long advocated a human mission to Mars, perhaps eventually leading to the permanent colonization of Mars, as the next logical step for a human space program after lunar exploration. Aside from the prestige such a mission would bring, advocates argue that humans would easily be able to outperform robotic explorers, justifying the expense. Aerospace engineer Bob Zubrin is one of the proponents of such missions. Some critics contend unmanned robots can perform better than humans at a fraction of the expense. If life exists on Mars, a human mission could contaminate it by introducing earthly microbes, so robotic exploration would be preferable.[353]

NASA

Human exploration by the United States was identified as a long-term goal in the Vision for Space Exploration announced in 2004 by then US President George W. Bush. The planned *Orion* spacecraft would be used to send a human expedition to Earth's moon by 2020 as a stepping stone to a Mars expedition. On September 28, 2007, NASA administrator Michael D. Griffin stated that NASA aims to put a person on Mars by 2037.

On December 2, 2014, NASA's Advanced Human Exploration Systems and Operations Mission Director Jason Crusan and Deputy Associate Administrator for Programs James Reuthner announced tentative support for the Boeing "Affordable Mars Mission Design" including radiation shielding, centrifugal artificial gravity, in-transit consumable resupply, and a lander which can return.[354,355] Reuthner suggested that if adequate funding was forthcoming, the proposed mission would be expected in the early 2030s.[356]

Figure 241: *Artistic simulated photo looking out a portal spacecraft coming for a Mars landing.*

On October 8, 2015, NASA published its official plan for human exploration and colonization of Mars. They called it "Journey to Mars". The plan operates through three distinct phases leading up to fully sustained colonization.

- The first stage, already underway, is the "Earth Reliant" phase. This phase continues utilizing the International Space Station until 2024; validating deep space technologies and studying the effects of long duration space missions on the human body.
- The second stage, "Proving Ground," moves away from Earth reliance and ventures into cislunar space for most of its tasks. This is when NASA plans to capture an asteroid (planned for 2020), test deep space habitation facilities, and validate capabilities required for human exploration of Mars. Finally, phase three is the transition to independence from Earth resources.
- The last stage, the "Earth Independent" phase, includes long term missions on the lunar surface which leverage surface habitats that only require routine maintenance, and the harvesting of Martian resources for fuel, water, and building materials. NASA is still aiming for human missions to Mars in the 2030s, though Earth independence could take decades longer.

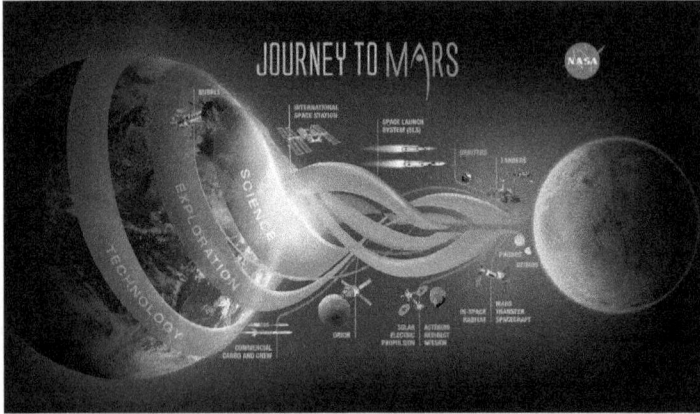

File:NASA-JourneyToMars-ScienceExplorationTechnology-20141202.jpg

Journey to Mars – Science, Exploration, Technology.

On August 28, 2015, NASA funded a year long simulation to study the effects of a year long Mars mission on six scientists. The scientists lived in a bio dome on a Mauna Loa mountain in Hawaii with limited connection to the outside world and were only allowed outside if they were wearing spacesuits.

NASAs human Mars exploration plans have evolved through the NASA Mars Design Reference Missions, a series of design studies for human exploration of Mars.

Zubrin

Mars Direct, a low-cost human mission proposed by Robert Zubrin, founder of the Mars Society, would use heavy-lift Saturn V class rockets, such as the Ares V, to skip orbital construction, LEO rendezvous, and lunar fuel depots. A modified proposal, called "Mars to Stay", involves not returning the first immigrant explorers immediately, if ever (see Colonization of Mars).

Probing difficulties

Mars Spacecraft 1988–1999

Spacecraft	Outcome
Phobos 1	Failure
Phobos 2	Failure
Mars Observer	Failure
Mars 96	Failure

Mars Pathfinder	Success
Mars Global Surveyor	Success
Mars Climate Orbiter	Failure
Mars Polar Lander	Failure
Deep Space 2	Failure
Nozomi	Failure

The challenge, complexity and length of Mars missions have led to many mission failures.[357] The high failure rate of missions launched from Earth attempting to explore Mars is informally called the "Mars Curse" or "Martian Curse". The phrase "Galactic Ghoul" or "Great Galactic Ghoul", referring to a fictitious space monster that subsists on a diet of Mars probes, was coined in 1997 by Time Magazine journalist Donald Neff, and is sometimes facetiously used to "explain" the recurring difficulties.[358,359,360]

Two Soviet probes were sent to Mars in 1988 as part of the Phobos program. Phobos 1 operated normally until an expected communications session on 2 September 1988 failed to occur. The problem was traced to a software error, which deactivated attitude thrusters causing the spacecrafts' solar arrays to no longer point at the Sun, depleting Phobos 1 batteries. Phobos 2 operated normally throughout its cruise and Mars orbital insertion phases on January 29, 1989, gathering data on the Sun, interplanetary medium, Mars, and Phobos. Shortly before the final phase of the mission, during which the spacecraft was to approach within 50 m of Phobos' surface and release two landers, one a mobile 'hopper', the other a stationary platform, contact with Phobos 2 was lost. The mission ended when the spacecraft signal failed to be successfully reacquired on March 27, 1989. The cause of the failure was determined to be a malfunction of the on-board computer.Wikipedia:Citation needed

Just a few years later in 1992 Mars Observer, launched by NASA, failed as it approached Mars. Mars 96, an orbiter launched on November 16, 1996 by Russia failed, when the planned second burn of the Block D-2 fourth stage did not occur.

Following the success of Global Surveyor and Pathfinder, another spate of failures occurred in 1998 and 1999, with the Japanese Nozomi orbiter and NASA's Mars Climate Orbiter, Mars Polar Lander, and Deep Space 2 penetrators all suffering various fatal errors. The Mars Climate Orbiter was noted for mixing up U.S. customary units with metric units, causing the orbiter to burn up while entering Mars' atmosphere.

The European Space Agency has also attempted to land two probes on the Martian surface; Beagle 2, a British-built lander that failed to deploy its solar arrays properly after touchdown in December 2003, and *Schiaparelli*, which

Figure 242: *Deep Space 2 technology*

was flown along the ExoMars Trace Gas Orbiter. Contact with the *Schiaparelli* EDM lander was lost 50 seconds before touchdown. It was later confirmed that the lander struck the surface at a high velocity, possibly exploding.

Timeline of Mars exploration

Overview of missions to Mars[361]

Mission type	Success rate	Total attempts	Suc- cess	Partial success	Launch failure	Failed en route	Failed to orbit/land
Flyby	45%	11	5	0	4	2	0
Orbiter	50%	23	10	2	5	3	3
Lander	53%	15	7	1	0	3	4
Rover	66%	6	4	0	0	0	2
Total	53%	55	26	3	9	8	9

Yearly statistics

This chart lists launches to Mars, but it ignores how long the mission lasted. For example, few missions were launched in the late 1970s, but the Viking program had two orbiters and two landers active at Mars at this time, and one lander remained active until 1982.

- failure
- partial success
- success
- in transit
- scheduled

Timeline

Mission (1960–1969)	Launch	Mars arrival	Termination	Elements	Outcome	Mission budget, bn USD	Launch mass, t	Mass of orbiter / lander / rover, t
Mars 1M No. 1	10 October 1960		10 October 1960	Flyby	Failure (at launch)		0.66	
Mars 1M No. 2	14 October 1960		14 October 1960	Flyby	Failure (at launch)		0.66	
Mars 2MV-4 No.1	24 October 1962		24 October 1962	Flyby	Failure (broke up during TMIburn)		0.88	
Mars 1	1 November 1962	19 June 1963	21 March 1963	Flyby	Partial success: some data collected, but lost contact before reaching Mars, flyby at approx. 193,000 km		0.89	
Mars 2MV-3 No.1	4 November 1962		19 January 1963	Lander	Failure (before leaving Earth's orbit)		0.90	0.26
Mariner 3	5 November 1964		5 November 1964	Flyby	Failure (fairing separation ruined trajectory)		0.26	
Mariner 4	28 November 1964	14 July 1965	21 December 1967	Flyby	Success (21 images returned)	0.08	0.26	
Zond 2	30 November 1964		May 1965	Flyby (intended lander)	Failure (communication lost three months before reaching Mars)		0.98	
Mariner 6	25 February 1969	31 July 1969	August 1969	Flyby	Success		0.41	
Mariner 7	27 March 1969	5 August 1969	August 1969	Flyby	Success		0.41	

Mission (1970–1989)	Launch	Arrival at Mars	Termination	Elements	Outcome	Mission budget, bn USD	Launch mass, t	Mass of orbiter / lander / rover, t
Mars 2M No. 521	27 March 1969		27 March 1969	Orbiter	Failure (at launch)		3.55	2.10
Mars 2M No. 522	2 April 1969		2 April 1969	Orbiter	Failure (at launch)		3.55	2.10
Mariner 8	8 May 1971		8 May 1971	Orbiter	Failure (at launch)		1.00	0.60
Kosmos 419	10 May 1971		12 May 1971	Orbiter	Failure (at launch)		3.80	2.50
Mariner 9	30 May 1971	14 November 1971	27 October 1972	Orbiter	Success (first successful orbit)		1.00	0.52
Mars 2	19 May 1971	27 November 1971	22 August 1972	Orbiter	Success		4.65	2.50
			27 November 1971	Lander, rover	Failure. Crashed on surface of Mars			0.85
Mars 3	28 May 1971	2 December 1971	22 August 1972	Orbiter	Success		4.65	2.50
			2 December 1971	Lander, rover	Partial success. First successful landing; landed softly but ceased transmission within 15 seconds			0.85
Mars 4	21 July 1973	10 February 1974	10 February 1974	Orbiter	Partial success (could not enter orbit, made a close flyby)		3.55	2.40

Mission	Launch		Arrival	Type	Notes			
Mars 5	25 July 1973	2 February 1974	21 February 1974	Orbiter	Partial success. Entered orbit and returned data, but failed within 9 days		3.55	2.40
Mars 6	5 August 1973	12 March 1974	12 March 1974	Lander	Partial success. Data returned during descent but not after landing on Mars		4.55	0.85
Mars 7	9 August 1973	9 March 1974	9 March 1974	Lander	Failure. Landing probe separated prematurely; entered heliocentric orbit		4.55	0.85
Viking 1	20 August 1975	20 July 1976	17 August 1980	Orbiter	Success	0.5	3.53	2.33
			13 November 1982	Lander	Success			0.61
Viking 2	9 September 1975	3 September 1976	25 July 1978	Orbiter	Success	0.5	3.53	2.33
			11 April 1980	Lander	Success			0.61
Phobos 1	7 July 1988		2 September 1988	Orbiter	Partial success. Returned some data. Contact lost while en route to Mars		6.22	
				Lander	Failure. Not deployed			0.09
Phobos 2	12 July 1988	29 January 1989	27 March 1989	Orbiter	Partial success: entered orbit and returned some data. Contact lost just before deployment of landers		6.22	
				Landers	Failure. Not deployed			0.07

Mission (1990–1999)	Launch	Arrival at Mars	Termination	Elements	Outcome	Mission budget, bn USD	Launch mass, t	Mass of orbiter / lander / rover, t
Mars Observer	25 September 1992	24 August 1993	21 August 1993	Orbiter	Failure. Lost contact just before arrival	0.8	2.5	
Mars Global Surveyor	7 November 1996	11 September 1997	5 November 2006	Orbiter	Success		1.1	0.74
Mars 96	16 November 1996		17 November 1996	Orbiter, lander, penetrator	Failure (at launch)		6.83	2.59
Mars Pathfinder	4 December 1996	4 July 1997	27 September 1997	Lander	Success		0.89	0.36
Sojourner				Rover	Success			
Nozomi (Planet-B)	3 July 1998		9 December 2003	Orbiter	Failure. Complications while en route; Never entered orbit		0.54	0.26
Mars Climate Orbiter	11 December 1998	23 September 1999	23 September 1999	Orbiter	Failure. Crashed on surface due to metric-imperial mix-up		0.63	0.54
Mars Polar Lander	3 January 1999	3 December 1999	3 December 1999	Lander	Failure. Crash-landed on surface due to improper hardware testing		0.58	0.29
Deep Space 2 (DS2)				Hard landers				

Mission (2000–2009)	Launch	Arrival at Mars	Termination	Elements	Outcome	Mission budget, bn USD	Launch mass, t	Mass of orbiter / lander / rover, t
2001 Mars Odyssey	7 April 2001	24 October 2001	Operational	Orbiter	Success	0.3^{362}	0.73	0.33
Mars Express	2 June 2003	25 December 2003	Operational	Orbiter	Success	0.3^{363}	1.12	0.60
Beagle 2			6 February 2004	Lander	Failure. Landed safely but failed to fully deploy. Could not return any data.			0.06
MER-A Spirit	10 June 2003	4 January 2004	22 March 2011	Rover	Success			
MER-B Opportunity	7 July 2003	25 January 2004	Operational	Rover	Success	0.4		
Rosetta	2 March 2004	25 February 2007	30 September 2016	Flyby/Gravity assist en route to comet 67P/-Churyumov-Gerasimenko	Success (successful Mars flyby).	1.8		
Mars Reconnaissance Orbiter	12 August 2005	10 March 2006	Operational	Orbiter	Success	0.7		
Phoenix	4 August 2007	25 May 2008	10 November 2008	Lander	Success	0.4		
Dawn	27 September 2007	17 February 2009	Operational	Flyby – gravity assist to Vesta	Success (successful Mars flyby).	0.4		

Mission (2010–2019)	Launch	Arrival at Mars	Termination	Elements	Outcome	Mission budget, bn USD	Launch mass, t	Mass of orbiter / lander / rover, t
Fobos-Grunt	8 November 2011		8 November 2011	Phobos lander, sample return	Failure. Failed to leave Earth orbit. Fell back to Earth.	0.2	13.5	2.30
Yinghuo-1			8 November 2011	Orbiter				0.12
MSL *Curiosity*	26 November 2011	6 August 2012	Operational	Rover	Success	2.5	3.89	2.91
Mars Orbiter Mission	5 November 2013	24 September 2014	Operational	Orbiter	Success	0.07	1.34	0.50
MAVEN	18 November 2013	22 September 2014	Operational	Orbiter	Success	0.7	2.45	0.81
ExoMars TGO	March 14, 2016	October 19, 2016	Operational	Orbiter	Success	1.2	4.33	1.43
Schiaparelli			October 19, 2016	Lander	Partial success. Crash-landed on surface, but transmitted data during descent.			0.60
InSight	5 May 2018	November 2018	On route	Lander	Study interior structure of Mars.			358 kg (789 lb)[364]
Mars Cube One		November 2018	On route	2 probes, flyby	To provide telemetry during atmospheric entry and landing of InSight.			13.5 kg (30 lb) each

Planned missions

Name	Estimated launch	Elements	Notes
ExoMars	2020	Surface platform	Meteorological tests, and deployment of rover.
		Rover	Search for the existence of past or present life on Mars.
Mars 2020	2020	Rover	Astrobiology objectives; rover is based on the *Curiosity* rover.
Mars Hope	2020	Orbiter	Atmospheric studies; would become the first Arab probe to Mars.
Mars Global Remote Sensing Orbiter and Small Rover	2020	Orbiter, rover	Technology demonstration; science[365,366]
Mars Orbiter Mission 2	2022	Orbiter	To be launched launched with a GSLV III.

Proposals under study

Name	Proposed launch	Elements	Notes
Mars MetNet precursor	2018 or later	Single impact lander test	Precursor for multi-lander network.
Mars MetNet	after precursor	Multi-lander network	Simultaneous meteorological measurements at multiple locations.
Mars Geyser Hopper	2018	Lander	Would have the ability to fly or "hop" at least twice from its landed location to reposition itself close to a CO_2 geyser site.
Northern Light	2018	Lander, rover	Mission designed by Canadian organisations and Thoth Technology Inc.[367]
Icebreaker Life	2018 or 2020	Stationary lander	Based on the 2008 *Phoenix* lander, would perform astrobiology tests on sub-surface ice.
PADME	2020	Orbiter	Would study Phobos and Deimos
Inspiration Mars Foundation	2021	Manned flyby	Private mission to send two humans around Mars on a free return trajectory, without landing.
Martian Moons Exploration	2022	Lander, sample return	Sample return from Phobos and remote sensing of Deimos; will also observe the atmosphere of Mars
Mars 2022	2022	Orbiter	Communications relay, mapping

Phootprint	2024	Lander and ascent stage	Mars moon sample return mission.
Fobos-Grunt (repeat mission)	2024	Lander, ascent stage	Phobos sample return.
Mars Orbiter	2027	Orbiter	First South Korean Mars exploration.
MELOS	2020s	Rover	A rover; may include a small aircraft[368]
Mars-Grunt	2020s	Orbiter, lander, ascent stage	Single launch Mars sample return.
BOLD	2020s	6 impact landers	The Biological Oxidant and Life Detection would perform astrobiology tests on sub-surface soil.[369]
Mars Lander	2030	Lander	First South Korean Mars landing mission.[370,371]

Undeveloped concepts

1970s

- *Mars 4NM* and *Mars 5NM* – projects intended by the Soviet Union for heavy Marsokhod (in 1973 according to initial plan of 1970) and Mars sample return (planned for 1975) missions by launching on N1 (rocket) that has never flown successfully.[372]
- *Mars 5M (Mars-79)* – double-launching Soviet sample return mission planned to 1979 but cancelled due to complexity and technical problems
- *Voyager-Mars* – USA, 1970s – Two orbiters and two landers, launched by a single Saturn V rocket.

1990s

- *Vesta* – the multiaimed Soviet mission, developed in cooperation with European countries for realisation in 1991–1994 but canceled due to the Soviet Union disbanding, included the flyby of Mars with delivering the aerostat and small landers or penetrators followed by flybys of 1 Ceres or 4 Vesta and some other asteroids with impact of penetrator on the one of them.
- *Mars Aerostat* – Russian/French balloon part for cancelled Vesta mission and then for failed Mars 96 mission,[373] originally planned for the 1992 launch window, postponed to 1994 and then to 1996 before being cancelled.[374]
- Mars Together, combined U.S. and Russian mission study in the 1990s. To be launched by a Molinya with possible U.S. orbiter or lander.
- *Mars Environmental Survey* – set of 16 landers planned for 1999–2009

- *Mars-98* – Russian mission including an orbiter, lander, and rover, planned for 1998 launch opportunity as repeat of failed Mars 96 mission and cancelled due to lack of funding

2000s

- *Mars Surveyor 2001 Lander* – October 2001 – Mars lander (refurbished, became Phoenix lander)
- *Kitty Hawk* – Mars airplane micromission, proposed for December 17, 2003, the centennial of the Wright brothers' first flight.[375] Its funding was eventually given to the 2003 Mars Network project.[376]
- *NetLander – 2007 or 2009 – Mars netlanders*
- *Beagle 3* – 2009 British lander mission meant to search for life, past or present.
- *Mars Telecommunications Orbiter* – September 2009 – Mars orbiter for telecommunications

2010s

- *Sky-Sailor* – 2014 – Plane developed by Switzerland to take detailed pictures of Mars surface
- *Mars Astrobiology Explorer-Cacher* – 2018 rover
- *Red Dragon* – Derivative of a Dragon 2 capsule by SpaceX, designed to land by aerobraking and retropropulsion. Planned for 2018, then 2020. Cancelled in favor of an unspecified new landing method on future larger spacecraft.
- Tumbleweed rover.[377]
- Mars One, orbiters, lander, rover.

Bibliography

- *Mars – A Warmer, Wetter Planet* by Jeffrey S. Kargel (published July 2004; ISBN 978-1-85233-568-7)
- *The Compact NASA Atlas of the Solar System* by Ronald Greeley and Raymond Batson (published January 2002; ISBN 0-521-80633-X)
- *Mars: The NASA Mission Reports /* edited by Robert Godwin (2000) ISBN 1-896522-62-9

External links

Wikimedia Commons has media related to *Mars exploration*.

- NASA Mars exploration website[378]
- Mars Exploration[379] Scientific American Maps and Articles
- Next on Mars[380] (Bruce Moomaw, Space Daily, 9 March 2005): An extensive overview of NASA's Mars exploration plans
- Catalog of Soviet Mars images[381] Collection of Russian Mars probes' images.
- Simplified study of orbits to land on Mars and return to Earth[382] (High School level)
- Planetary Society Mars page[383]

Astronomy on Mars

Astronomy on Mars

In many cases astronomical phenomena viewed from the planet Mars are the same or similar to those seen from Earth but sometimes (as with the view of Earth as an evening/morning star) they can be quite different. For example, because the atmosphere of Mars does not contain an ozone layer, it is also possible to make UV observations from the surface of Mars.

Seasons

Mars has an axial tilt of 25.19°, quite close to the value of 23.44° for Earth, and thus Mars has seasons of spring, summer, autumn, winter as Earth does. As on Earth, the southern and northern hemispheres have summer and winter at opposing times.

However, the orbit of Mars has significantly greater eccentricity than that of Earth. Therefore, the seasons are of unequal length, much more so than on Earth:

Season	Sols (on Mars)	Days (on Earth)
Northern spring, southern autumn:	193.30	92.764
Northern summer, southern winter:	178.64	93.647
Northern autumn, southern spring:	142.70	89.836
Northern winter, southern summer:	153.95	88.997

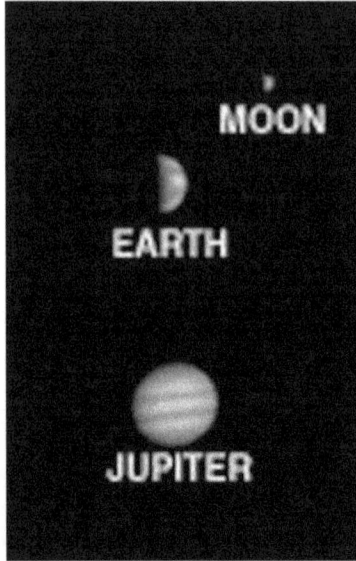

Figure 243: *Mosaic of two different Mars Global Surveyor Mars Orbiter Camera (MOC) exposures of Earth, the Moon, and Jupiter from 2003*

Figure 244: *Mars sky turned violet by water ice clouds*

Figure 245: *Close-up of Mars sky at sunset, showing more color variation, as imaged by Mars Pathfinder*

Figure 246: *Mars sky at noon, as imaged by Mars Pathfinder (June, 1999).*

In practical terms, this means that summers and winters have different lengths and intensities in the northern and southern hemispheres. Winters in the north are warm and short (because Mars is moving fast near its perihelion), while winters in the south are long and cold (Mars is moving slowly near aphelion). Similarly, summers in the north are long and cool, while summers in the south are short and hot. Therefore, extremes of temperature are considerably wider in the southern hemisphere than in the north.

The seasonal lag on Mars is no more than a couple of days,[384] due to its lack of large bodies of water and similar factors that would provide a buffering effect. Thus, for temperatures on Mars, "spring" is approximately the mirror image

Figure 247: *Mars sky at sunset, as imaged by Mars Pathfinder (June, 1999).*

Figure 248: *Mars sky at sunset, as imaged by the Spirit rover (May, 2005).*

of "summer" and "autumn" is approximately the mirror image of "winter" (if you consider the solstices and equinoxes to be the beginnings of their respective seasons), and if Mars had a circular orbit the maximum and minimum temperatures would occur a couple of days after the summer and winter solstices rather than about one month after as on Earth. The only difference between spring temperatures and summer temperatures is due to the relatively high eccentricity of Mars's orbit: in northern spring Mars is farther from the Sun than during northern summer, and therefore by coincidence spring is slightly cooler than summer and autumn is slightly warmer than winter. However, in the southern hemisphere the opposite is true.

The temperature variations between spring and summer are much less than

Figure 249: *Mars sky at sunset, as imaged by the Curiosity rover (February 2013; sun simulated by artist).*

the very sharp variations that occur within a single Martian sol (solar day). On a daily basis, temperatures peak at local solar noon and reach a minimum at local midnight. This is similar to the effect in Earth's deserts, only much more pronounced.

The axial tilt and eccentricity of Earth (or Mars) are by no means fixed, but rather vary due to gravitational perturbations from other planets in the solar system on a timescale of tens of thousands or hundreds of thousands of years. Thus, for example Earth's eccentricity of about 1% regularly fluctuates and can increase up to 6%, and at some point in the distant future the Earth will also have to deal with the calendrical implications of seasons of widely differing length and the major climate disruptions that go along with it.

Aside from the eccentricity, the Earth's axial tilt can also vary from 21.5° to 24.5°, and the length of this "obliquity cycle" is 41,000 years. These and other similar cyclical changes are thought to be responsible for ice ages (see Milankovitch cycles). By contrast, the obliquity cycle for Mars is much more extreme: from 15° to 35° over a 124,000-year cycle. Some recent studies even suggest that over tens of millions of years, the swing may be as much as 0° to 60°.[385] Earth's large Moon apparently plays an important role in keeping Earth's axial tilt within reasonable bounds; Mars has no such stabilizing influence and its axial tilt can vary more chaotically.

The color of the sky

The normal hue of the sky during the daytime is a pinkish-red, however in the vicinity of the setting or rising sun it is blue. This is the exact opposite of the situation on Earth. However, during the day the sky is a yellow-brown "butterscotch" color. On Mars, Rayleigh scattering is usually a very small effect. It is believed that the color of the sky is caused by the presence of 1% by volume of magnetite in the dust particles. Twilight lasts a long time after the Sun has set and before it rises, because of all the dust in Mars's atmosphere. At times, the Martian sky takes on a violet color, due to scattering of light by very small water ice particles in clouds.[386]

Generating accurate true-color images of Mars's surface is surprisingly complicated.[387] There is much variation in the color of the sky as reproduced in published images; many of those images, however, are using filters to maximize the science value and are not trying to show true color. Nevertheless, for many years, the sky on Mars was thought to be more pinkish than it now is believed to be.

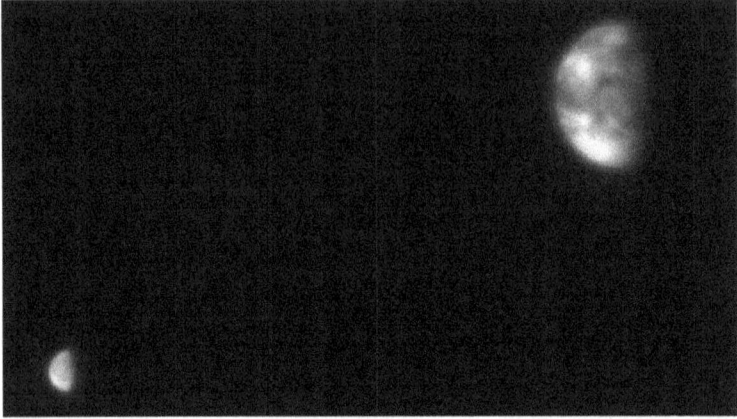

Figure 250:
Earth and the Moon as viewed from Mars
(MRO; HiRISE; November 20, 2016)

Astronomical phenomena

Earth and Moon

As seen from Mars, the Earth is an inner planet like Venus (a "morning star" or "evening star"). The Earth and Moon appear starlike to the naked eye, but observers with telescopes would see them as crescents, with some detail visible.

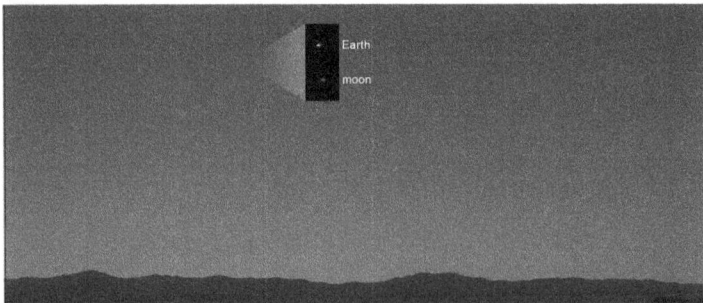

File:PIA17936-f2-MarsCuriosityRover-EarthMoon-20140131.jpg

Curiosity's first view of the Earth and the Moon
from the surface of Mars (January 31, 2014).

An observer on Mars would be able to see the Moon orbiting around the Earth, and this would easily be visible to the naked eye. By contrast, observers on

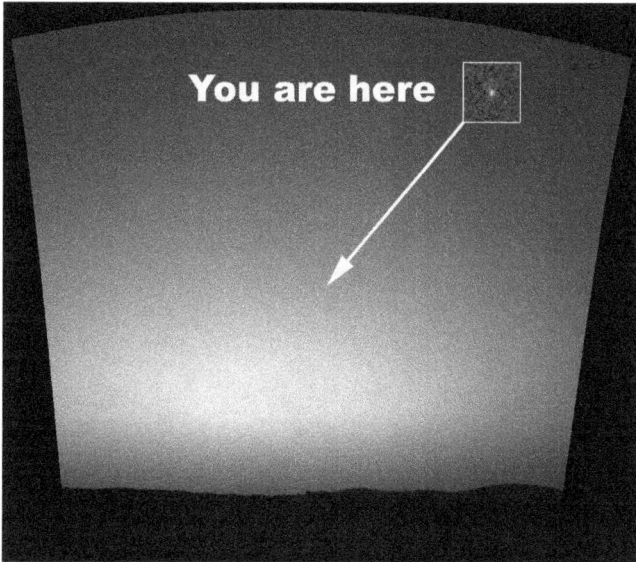

Figure 251: *Earth as morning star, imaged by MER Spirit on March 7, 2004*

Earth cannot see any other planet's satellites with the naked eye, and it was not until soon after the invention of the telescope that the first such satellites were discovered (Jupiter's Galilean moons).

At maximum angular separation, the Earth and Moon would be easily distinguished as a double planet, but about one week later they would merge into a single point of light (to the naked eye), and then about a week after that, the Moon would reach maximum angular separation on the opposite side. The maximum angular separation of the Earth and Moon varies considerably according to the relative distance between the Earth and Mars: it is about 17′ when Earth is closest to Mars (near inferior conjunction) but only about 3.5′ when the Earth is farthest from Mars (near superior conjunction). For comparison, the apparent diameter of the Moon from Earth is 31′.

The minimum angular separation would be less than 1′, and occasionally the Moon would be seen to transit in front of or pass behind (be occulted by) the Earth. The former case would correspond to a lunar occultation of Mars as seen from Earth, and because the Moon's albedo is considerably less than that of the Earth, a dip in overall brightness would occur, although this would be too small to be noticeable by casual naked eye observers because the size of the Moon is much smaller than that of the Earth and it would cover only a small fraction of the Earth's disk.

Figure 252: *Earth and Moon from Mars, imaged by Mars Global Surveyor on May 8, 2003, 13:00 UTC. South America is visible.*

Mars Global Surveyor imaged the Earth and Moon on May 8, 2003 13:00 UTC, very close to maximum angular elongation from the Sun and at a distance of 0.930 AU from Mars. The apparent magnitudes were given as –2.5 and +0.9.[388] At different times the actual magnitudes will vary considerably depending on distance and the phases of the Earth and Moon.

From one day to the next, the view of the Moon would change very differently for an observer on Mars than for an observer on Earth. The phase of the Moon as seen from Mars would not change much from day to day; it would match the phase of the Earth, and would only gradually change as both Earth and Moon move in their orbits around the Sun. On the other hand, an observer on Mars would see the Moon rotate, with the same period as its orbital period, and would see far side features that can never be seen from Earth.

Since Earth is an inner planet, observers on Mars can occasionally view transits of Earth across the Sun. The next one will take place in 2084. They can also view transits of Mercury and transits of Venus.

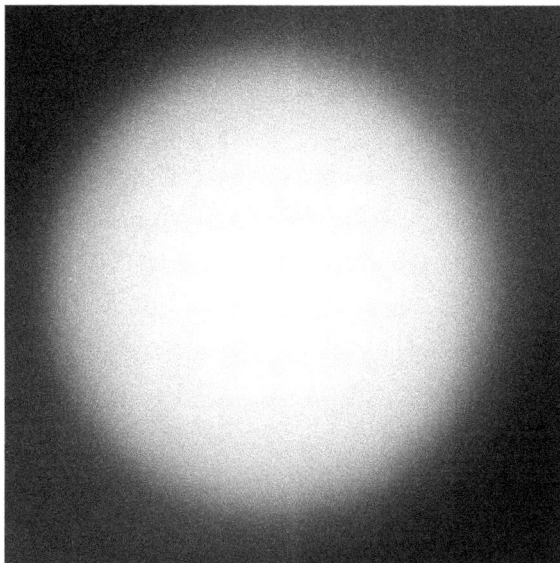

Figure 253: *Phobos eclipses the Sun, imaged by MER*

Phobos and Deimos

The moon Phobos appears about one third the angular diameter that the full Moon appears from Earth; on the other hand, Deimos appears more or less starlike with a disk barely discernible if at all. Phobos orbits so fast (with a period of just under one third of a sol) that it rises in the west and sets in the east, and does so twice per sol; Deimos on the other hand rises in the east and sets in the west, but orbits only a few hours slower than a Martian sol, so it spends about two and a half sols above the horizon at a time.

The maximum brightness of Phobos at "full moon" is about magnitude –9 or –10, while for Deimos it is about –5.[389] By comparison, the full Moon as seen from Earth is considerably brighter at magnitude –12.7. Phobos is still bright enough to cast shadows; Deimos is only slightly brighter than Venus is from Earth. Just like Earth's Moon, both Phobos and Deimos are considerably fainter at non-full phases. Unlike Earth's Moon, Phobos's phases and angular diameter visibly change from hour to hour; Deimos is too small for its phases to be visible with the naked eye.

Both Phobos and Deimos have low-inclination equatorial orbits and orbit fairly close to Mars. As a result, Phobos is not visible from latitudes north of 70.4°N or south of 70.4°S; Deimos is not visible from latitudes north of 82.7°N or

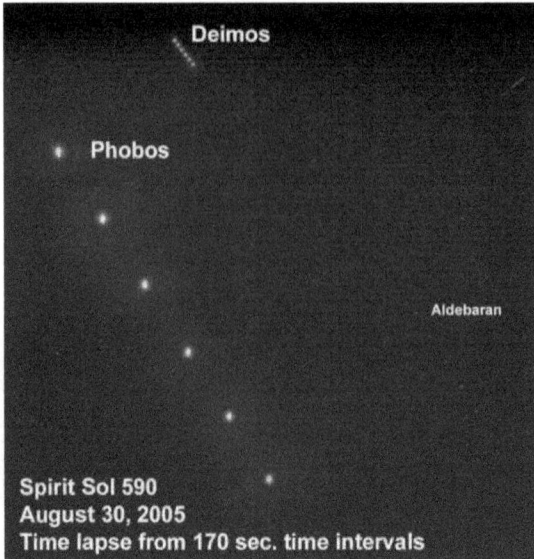

Figure 254: *Phobos and Deimos from the
Spirit rover. Courtesy NASA/JPL-Caltech*

south of 82.7°S. Observers at high latitudes (less than 70.4°) would see a noticeably smaller angular diameter for Phobos because they are farther away
from it. Similarly, equatorial observers of Phobos would see a noticeably
smaller angular diameter for Phobos when it is rising and setting, compared
to when it is overhead.

Observers on Mars can view transits of Phobos and transits of Deimos across
the Sun. The transits of Phobos could also be called partial eclipses of the
Sun by Phobos, since the angular diameter of Phobos is up to half the angular
diameter of the Sun. However, in the case of Deimos the term "transit" is
appropriate, since it appears as a small dot on the Sun's disk.

Since Phobos orbits in a low-inclination equatorial orbit, there is a seasonal
variation in the latitude of the position of Phobos's shadow projected onto the
Martian surface, cycling from far north to far south and back again. At any
given fixed geographical location on Mars, there are two intervals per Martian
year when the shadow is passing through its latitude and about half a dozen
transits of Phobos can be observed at that geographical location over a couple
of weeks during each such interval. The situation is similar for Deimos, except
only zero or one transits occur during such an interval.

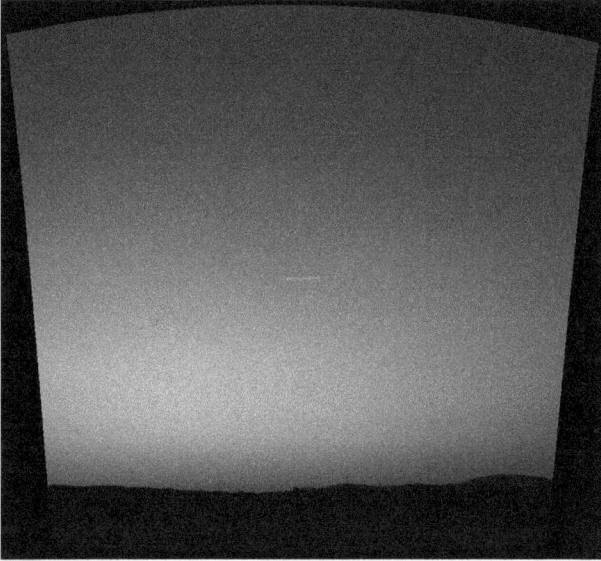

Figure 255: *First meteor photographed from Mars, March 7, 2004, by MER Spirit*

It is easy to see that the shadow always falls on the "winter hemisphere", except when it crosses the equator during the vernal equinox and the autumnal equinox. Thus transits of Phobos and Deimos happen during Martian autumn and winter in the northern hemisphere and the southern hemisphere. Close to the equator they tend to happen around the autumnal equinox and the vernal equinox; farther from the equator they tend to happen closer to the winter solstice. In either case, the two intervals when transits can take place occur more or less symmetrically before and after the winter solstice (however, the large eccentricity of Mars's orbit prevents true symmetry).

Observers on Mars can also view lunar eclipses of Phobos and Deimos. Phobos spends about an hour in Mars's shadow; for Deimos it is about two hours. Surprisingly, despite its orbit being nearly in the plane of Mars's equator and despite its very close distance to Mars, there are some occasions when Phobos escapes being eclipsed.

Phobos and Deimos both have synchronous rotation, which means that they have a "far side" that observers on the surface of Mars can't see. The phenomenon of libration occurs for Phobos as it does for Earth's Moon, despite the low inclination and eccentricity of Phobos's orbit.[390,391] Due to the effect of librations and the parallax due to the close distance of Phobos, by observing at high and low latitudes and observing as Phobos is rising and setting, the

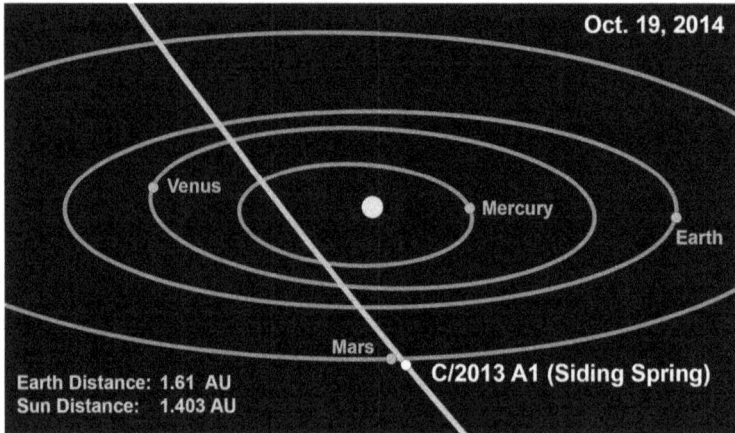

Figure 256: *Projected path of Comet Siding Spring passing Mars on 19 October 2014.*

overall total coverage of Phobos's surface that is visible at one time or another from one location or another on Mars's surface is considerably higher than 50%.

The large Stickney crater is visible along one edge of the face of Phobos. It is easily visible with the naked eye from the surface of Mars.

Comets and meteors

Since Mars has an atmosphere that is relatively transparent at optical wavelengths (just like Earth, albeit much thinner), meteors will occasionally be seen. Meteor showers on Earth occur when the Earth intersects the orbit of a comet, and likewise, Mars also has meteor showers, although these are different from the ones on Earth.

The first meteor photographed on Mars (on March 7, 2004 by the *Spirit* rover) is now believed to have been part of a meteor shower whose parent body was comet 114P/Wiseman-Skiff. Because the radiant was in the constellation Cepheus, this meteor shower could be dubbed the Martian Cepheids.Wikipedia:Citation needed

As on Earth, when a meteor is large enough to actually impact with the surface (without burning up completely in the atmosphere), it becomes a meteorite. The first known meteorite discovered on Mars (and the third known meteorite found someplace other than Earth) was Heat Shield Rock. The first and the second ones were found on the moon by the Apollo missions.[392]

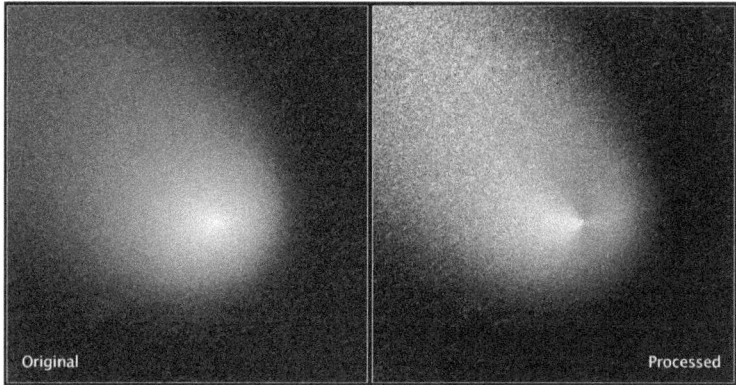

Figure 257: *Comet Siding Spring as seen by Hubble on 11 March 2014.*

On October 19, 2014, Comet Siding Spring passed extremely close to Mars, so close that the coma may have enveloped the planet.   <templatestyles src="Multiple_image/styles.css" />

Comet Siding Spring Mars flyby on October 19, 2014 (artist's concepts)

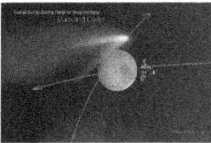

POV: Universe

POV: Comet

POV: Mars

File:Comet-C2013A1-SidingSpring-NearMars-Hubble-20141019.jpg

Close encounter of Comet Siding Spring with the planet Mars

(composite image; Hubble ST; October 19, 2014).

Auroras

Auroras occur on Mars, but they do not occur at the poles as on Earth, because Mars has no planetwide magnetic field. Rather, they occur near magnetic anomalies in Mars's crust, which are remnants from earlier days when Mars did have a magnetic field. Martian auroras are a distinct kind not seen elsewhere in the solar system. They would probably also be invisible to the human eye, being largely ultraviolet phenomena.[393]

Celestial poles and ecliptic

The orientation of Mars's axis is such that its north celestial pole is in Cygnus at R.A. 21^h 10^m 42^s Decl. +52° 53.0′ (or more precisely, 317.67669 +52.88378), near the 6th-magnitude star BD +52 2880 (also known as HR 8106, HD 201834, or SAO 33185), which in turn is at R.A. 21^h 10^m 15.6^s Decl. +53° 33′ 48″.

The top two stars in the Northern Cross, Sadr and Deneb, point to the north celestial pole of Mars. The pole is about halfway between Deneb and Alpha Cephei, less than 10° from the former, a bit more than the apparent distance between Sadr and Deneb. Because of its proximity to the pole, Deneb never

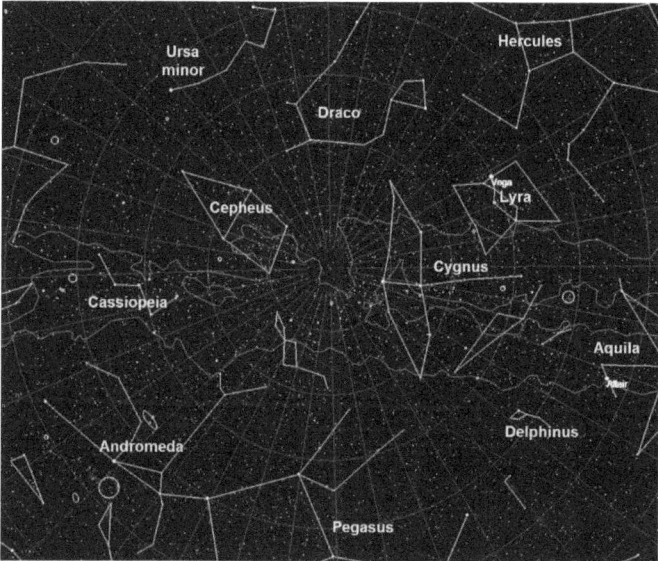

Figure 258: *Celestial north pole on Mars*

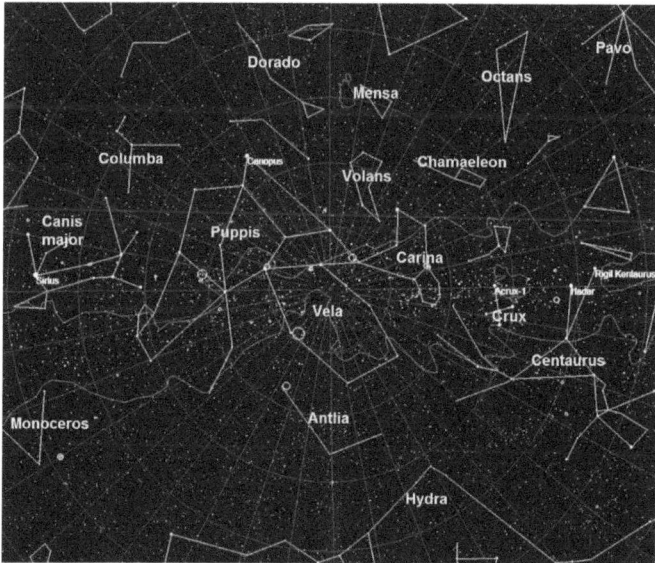

Figure 259: *Celestial south pole on Mars*

sets in nearly all of Mars's northern hemisphere. Except in areas close to the equator, Deneb permanently circles the North pole. The orientation of Deneb and Sadr would make a useful clock hand for telling sidereal time.

Mars's north celestial pole is also only a few degrees away from the galactic plane. Thus the Milky Way, especially rich in the area of Cygnus, is always visible from the northern hemisphere.

The South celestial pole is correspondingly found at $9^h 10^m 42^s$ and $-52° 53.0'$, which is a couple of degrees from the 2.5-magnitude star Kappa Velorum (which is at $9^h 22^m 06.85^s -55° 00.6'$), which could therefore be considered the southern polar star. The star Canopus, second-brightest in the sky, is a circumpolar star for most southern latitudes.

The zodiac constellations of Mars's ecliptic are almost the same as those of Earth — after all, the two ecliptic planes only have a mutual inclination of 1.85° — but on Mars, the Sun spends 6 days in the constellation Cetus, leaving and re-entering Pisces as it does so, making a total of 14 zodiacal constellations. The equinoxes and solstices are different as well: for the northern hemisphere, vernal equinox is in Ophiuchus (compared to Pisces on Earth), summer solstice is at the border of Aquarius and Pisces, autumnal equinox is in Taurus, and winter solstice is in Virgo.

As on Earth, precession will cause the solstices and equinoxes to cycle through the zodiac constellations over thousands and tens of thousands of years.

Long-term variations

As on Earth, the effect of precession causes the north and south celestial poles to move in a very large circle, but on Mars the cycle is 175,000 Earth years rather than 26,000 years as on Earth.

As on Earth, there is a second form of precession: the point of perihelion in Mars's orbit changes slowly, causing the anomalistic year to differ from the sidereal year. However, on Mars, this cycle is 83,600 years rather than 112,000 years as on Earth.

On both Earth and Mars, these two precessions are in opposite directions, and therefore add, to make the precession cycle between the tropical and anomalistic years 21,000 years on Earth and 56,600 years on Mars.

As on Earth, the period of rotation of Mars (the length of its day) is slowing down. However, this effect is three orders of magnitude smaller than on Earth because the gravitational effect of Phobos is negligible and the effect is mainly due to the Sun.[394] On Earth, the gravitational influence of the Moon has a much greater effect. Eventually, in the far future, the length of a day on Earth will equal and then exceed the length of a day on Mars.

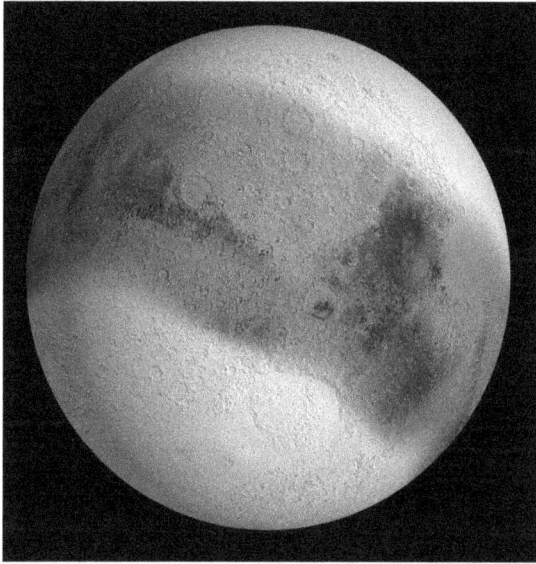

Figure 260: *An illustration of what Mars may have looked like during an ice age about 400,000 years ago caused by a large axial tilt*

As on Earth, Mars experiences Milankovitch cycles that cause its axial tilt (obliquity) and orbital eccentricity to vary over long periods of time, which has long-term effects on its climate. The variation of Mars's axial tilt is much larger than for Earth because it lacks the stabilizing influence of a large moon like Earth's moon. Mars has a 124,000-year obliquity cycle compared to 41,000 years for Earth.

External links

- Analemma on Mars[395]
- Martian time[396]
- NASA - Mars24 Sunclock - Time on Mars[397]

Historical observations

History of Mars observation

The recorded history of observation of the planet Mars dates back to the era of the ancient Egyptian astronomers in the 2nd millennium BCE. Chinese records about the motions of Mars appeared before the founding of the Zhou Dynasty (1045 BCE). Detailed observations of the position of Mars were made by Babylonian astronomers who developed arithmetic techniques to predict the future position of the planet. The ancient Greek philosophers and Hellenistic astronomers developed a geocentric model to explain the planet's motions. Measurements of Mars' angular diameter can be found in ancient Greek and Indian texts. In the 16th century, Nicolaus Copernicus proposed a heliocentric model for the Solar System in which the planets follow circular orbits about the Sun. This was revised by Johannes Kepler, yielding an elliptic orbit for Mars that more accurately fitted the observational data.

The first telescopic observation of Mars was by Galileo Galilei in 1610. Within a century, astronomers discovered distinct albedo features on the planet, including the dark patch Syrtis Major Planum and polar ice caps. They were able to determine the planet's rotation period and axial tilt. These observations were primarily made during the time intervals when the planet was located in opposition to the Sun, at which points Mars made its closest approaches to the Earth.

Better telescopes developed early in the 19th century allowed permanent Martian albedo features to be mapped in detail. The first crude map of Mars was published in 1840, followed by more refined maps from 1877 onward. When astronomers mistakenly thought they had detected the spectroscopic signature of water in the Martian atmosphere, the idea of life on Mars became popularized among the public. Percival Lowell believed he could see a network of artificial canals on Mars. These linear features later proved to be an optical illusion, and the atmosphere was found to be too thin to support an Earth-like environment.

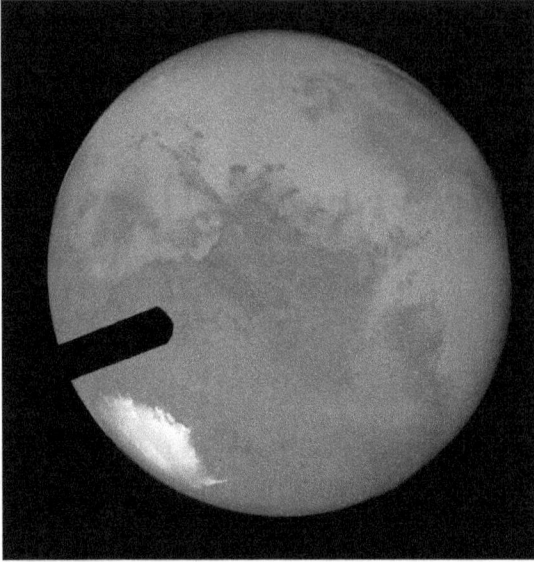

Figure 261: *Hubble's sharpest view of Mars: Although the ACS fastie finger intrudes it achieved a spatial scale of 5 miles, or 8 kilometres per pixel at full resolution.*

Yellow clouds on Mars have been observed since the 1870s, which Eugène M. Antoniadi suggested were windblown sand or dust. During the 1920s, the range of Martian surface temperature was measured; it ranged from –85 to 7 °C (–121 to 45 °F). The planetary atmosphere was found to be arid with only trace amounts of oxygen and water. In 1947, Gerard Kuiper showed that the thin Martian atmosphere contained extensive carbon dioxide; roughly double the quantity found in Earth's atmosphere. The first standard nomenclature for Mars albedo features was adopted in 1960 by the International Astronomical Union. Since the 1960s, multiple robotic spacecraft have been sent to explore Mars from orbit and the surface. The planet has remained under observation by ground and space-based instruments across a broad range of the electromagnetic spectrum. The discovery of meteorites on Earth that originated on Mars has allowed laboratory examination of the chemical conditions on the planet.

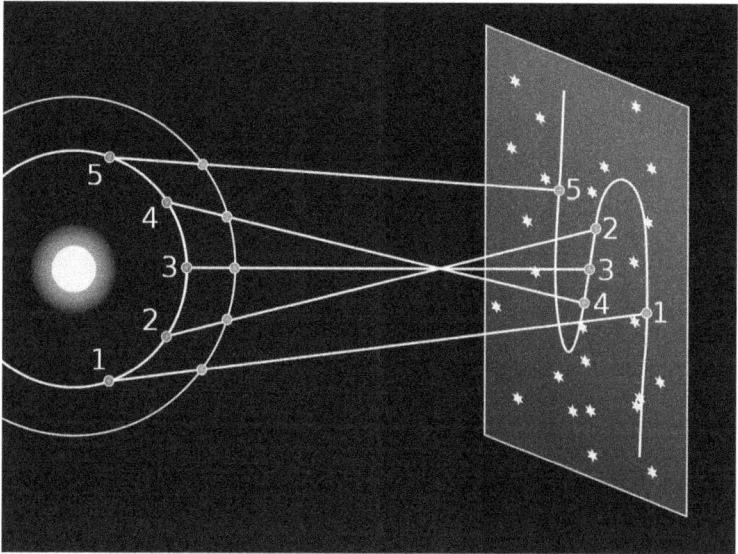

Figure 262: *As Earth passes Mars, the latter planet will temporarily appear to reverse its motion across the sky.*

Earliest records

The existence of Mars as a wandering object in the night sky was recorded by ancient Egyptian astronomers. By the 2nd millennium BCE they were familiar with the apparent retrograde motion of the planet, in which it appears to move in the opposite direction across the sky from its normal progression. Mars was portrayed on the ceiling of the tomb of Seti I, on the Ramesseum ceiling, and in the Senenmut star map. The last is the oldest known star map, being dated to 1534 BCE based on the position of the planets.

By the period of the Neo-Babylonian Empire, Babylonian astronomers were making systematic observations of the positions and behavior of the planets. For Mars, they knew, for example, that the planet made 37 synodic periods, or 42 circuits of the zodiac, every 79 years. The Babylonians invented arithmetic methods for making minor corrections to the predicted positions of the planets. This technique was primarily derived from timing measurements—such as when Mars rose above the horizon, rather than from the less accurately known position of the planet on the celestial sphere.

Chinese records of the appearances and motions of Mars appear before the founding of the Zhou Dynasty (1045 BCE), and by the Qin Dynasty (221

Schema huius præmiſſæ diui

Figure 263: *Geocentric model of the Universe.*

BCE) astronomers maintained close records of planetary conjunctions, including those of Mars. Occultations of Mars by Venus were noted in 368, 375, and 405 CE. The period and motion of the planet's orbit was known in detail during the Tang Dynasty (618 CE).

The early astronomy of ancient Greece was influenced by knowledge transmitted from the Mesopotamian culture. Thus the Babylonians associated Mars with Nergal, their god of war and pestilence, and the Greeks connected the planet with their god of war, Ares. During this period, the motions of the planets were of little interest to the Greeks; Hesiod's *Works and Days* (*c.* 650 BCE) makes no mention of the planets.

Orbital models

The Greeks used the word *plan̄eton* to refer to the seven celestial bodies that moved with respect to the background stars and they held a geocentric view that these bodies moved about the Earth. In his work, *The Republic* (X. 616E–617B), the Greek philosopher Plato provided the oldest known statement defining the order of the planets in Greek astronomical tradition. His list, in order of the nearest to the most distant from the Earth, was as follows: the Moon, Sun, Venus, Mercury, Mars, Jupiter, Saturn, and the fixed stars. In his dialogue *Timaeus*, Plato proposed that the progression of these objects across the skies depended on their distance, so that the most distant object moved the slowest.

Aristotle, a student of Plato, observed an occultation of Mars by the Moon in 365 BCE. From this he concluded that Mars must lie further from the Earth than the Moon. He noted that other such occultations of stars and planets had been observed by the Egyptians and Babylonians.[398] Aristotle used this observational evidence to support the Greek sequencing of the planets. His work *De Caelo* presented a model of the universe in which the Sun, Moon, and planets circle about the Earth at fixed distances. A more sophisticated version of the geocentric model was developed by the Greek astronomer Hipparchus when he proposed that Mars moved along a circular track called the epicycle that, in turn, orbited about the Earth along a larger circle called the deferent.

In Roman Egypt during the 2nd century CE, Claudius Ptolemaeus (Ptolemy) attempted to address the problem of the orbital motion of Mars. Observations of Mars had shown that the planet appeared to move 40% faster on one side of its orbit than the other, in conflict with the Aristotelian model of uniform motion. Ptolemy modified the model of planetary motion by adding a point offset from the center of the planet's circular orbit about which the planet moves at a uniform rate of rotation. He proposed that the order of the planets, by increasing distance, was: the Moon, Mercury, Venus, Sun, Mars, Jupiter, Saturn, and the fixed stars. Ptolemy's model and his collective work on astronomy was presented in the multi-volume collection *Almagest*, which became the authoritative treatise on Western astronomy for the next fourteen centuries.

In the 5th century CE, the Indian astronomical text *Surya Siddhanta* estimated the angular size of Mars as 2 arc-minutes (1/30 of a degree) and its distance to Earth as 10,433,000 km (1,296,600 yojana, where one yojana is equivalent to eight km in the *Surya Siddhanta*). From this the diameter of Mars is deduced to be 6,070 km (754.4 yojana), which has an error within 11% of the currently accepted value of 6,788 km. However, this estimate was based upon an inaccurate guess of the planet's angular size. The result may have been influenced by the work of Ptolemy, who listed a value of 1.57 arc-minutes. Both estimates are significantly larger than the value later obtained by telescope.

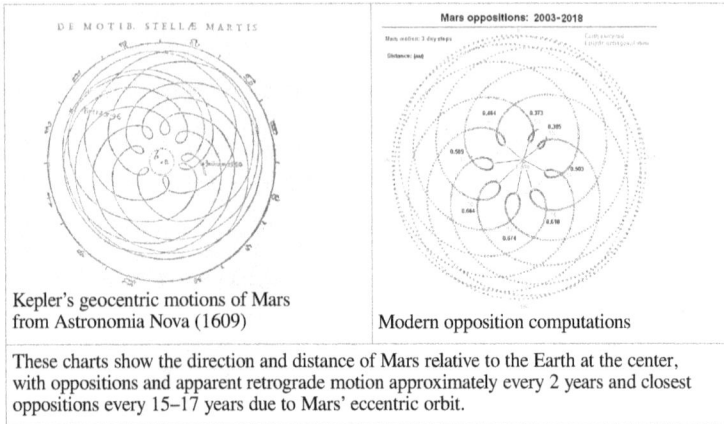

Kepler's geocentric motions of Mars
from Astronomia Nova (1609)

Modern opposition computations

These charts show the direction and distance of Mars relative to the Earth at the center,
with oppositions and apparent retrograde motion approximately every 2 years and closest
oppositions every 15–17 years due to Mars' eccentric orbit.

In 1543, Nicolaus Copernicus published a heliocentric model in his work *De revolutionibus orbium coelestium*. This approach placed the Earth in an orbit around the Sun between the circular orbits of Venus and Mars. His model successfully explained why the planets Mars, Jupiter and Saturn were on the opposite side of the sky from the Sun whenever they were in the middle of their retrograde motions. Copernicus was able to sort the planets into their correct heliocentric order based solely on the period of their orbits about the Sun. His theory gradually gained acceptance among European astronomers, particularly after the publication of the *Prutenic Tables* by the German astronomer Erasmus Reinhold in 1551, which were computed using the Copernican model.

On October 13, 1590, the German astronomer Michael Maestlin observed an occultation of Mars by Venus. One of his students, Johannes Kepler, quickly became an adherent to the Copernican system. After the completion of his education, Kepler became an assistant to the Danish nobleman and astronomer, Tycho Brahe. With access granted to Tycho's detailed observations of Mars, Kepler was set to work mathematically assembling a replacement to the Prutenic Tables. After repeatedly failing to fit the motion of Mars into a circular orbit as required under Copernicanism, he succeeded in matching Tycho's observations by assuming the orbit was an ellipse and the Sun was located at one of the foci. His model became the basis for Kepler's laws of planetary motion, which were published in his multi-volume work *Epitome Astronomiae Copernicanae* (Epitome of Copernican Astronomy) between 1615 and 1621.

Figure 264: *The low albedo feature Syrtis Major is visible at the disk center. NASA/HST image.*

Early telescope observations

At its closest approach, the angular size of Mars is 25 arcseconds (a unit of degree); this is much too small for the naked eye to resolve. Hence, prior to the invention of the telescope, nothing was known about the planet besides its position on the sky. The Italian scientist Galileo Galilei was the first person known to use a telescope to make astronomical observations. His records indicate that he began observing Mars through a telescope in September 1610. This instrument was too primitive to display any surface detail on the planet, so he set the goal of seeing if Mars exhibited phases of partial darkness similar to Venus or the Moon. Although uncertain of his success, by December he did note that Mars had shrunk in angular size. Polish astronomer Johannes Hevelius succeeded in observing a phase of Mars in 1645.

In 1644, the Italian Jesuit Daniello Bartoli reported seeing two darker patches on Mars. During the oppositions of 1651, 1653 and 1655, when the planet made its closest approaches to the Earth, the Italian astronomer Giovanni Battista Riccioli and his student Francesco Maria Grimaldi noted patches of differing reflectivity on Mars. The first person to draw a map of Mars that displayed terrain features was the Dutch astronomer Christiaan Huygens. On November 28, 1659 he made an illustration of Mars that showed the distinct dark region now known as Syrtis Major Planum, and possibly one of the polar ice caps. The same year, he succeeded in measuring the rotation period of the planet, giving it as approximately 24 hours. He made a rough estimate of the diameter of Mars, guessing that it is about 60% of the size of the Earth, which compares well with the modern value of 53%. Perhaps the first definitive mention of Mars's southern polar ice cap was by the Italian astronomer Giovanni Domenico Cassini, in 1666. That same year, he used observations of the surface markings on Mars to determine a rotation period of $24^h 40^m$. This differs from the currently-accepted value by less than three minutes. In 1672, Huygens noticed a fuzzy white cap at the north pole.

After Cassini became the first director of the Paris Observatory in 1671, he tackled the problem of the physical scale of the Solar System. The relative size of the planetary orbits was known from Kepler's third law, so what was needed was the actual size of one of the planet's orbits. For this purpose, the position of Mars was measured against the background stars from different points on the Earth, thereby measuring the diurnal parallax of the planet. During this year, the planet was moving past the point along its orbit where it was nearest to the Sun (a perihelic opposition), which made this a particularly close approach to the Earth. Cassini and Jean Picard determined the position of Mars from Paris, while the French astronomer Jean Richer made measurements from Cayenne, South America. Although these observations were hampered by the quality of the instruments, the parallax computed by Cassini came within 10% of the correct value. The English astronomer John Flamsteed made comparable measurement attempts and had similar results.

In 1704, Italian astronomer Jacques Philippe Maraldi "made a systematic study of the southern cap and observed that it underwent" variations as the planet rotated. This indicated that the cap was not centered on the pole. He observed that the size of the cap varied over time. The German-born British astronomer Sir William Herschel began making observations of the planet Mars in 1777, particularly of the planet's polar caps. In 1781, he noted that the south cap appeared "extremely large", which he ascribed to that pole being in darkness for the past twelve months. By 1784, the southern cap appeared much smaller, thereby suggesting that the caps vary with the planet's seasons and thus were made of ice. In 1781, he estimated the rotation period of Mars as $24^h 39^m$ 21.67^s and measured the axial tilt of the planet's poles to the orbital plane as $28.5°$. He noted that Mars had a "considerable but moderate atmosphere, so that its inhabitants probably enjoy a situation in many respects similar to ours". Between 1796 and 1809, the French astronomer Honoré Flaugergues noticed obscurations of Mars, suggesting "ochre-colored veils" covered the surface. This may be the earliest report of yellow clouds or storms on Mars.

Geographical period

At the start of the 19th century, improvements in the size and quality of telescope optics proved a significant advance in observation capability. Most notable among these enhancements was the two-component achromatic lens of the German optician Joseph von Fraunhofer that essentially eliminated coma—an optical effect that can distort the outer edge of the image. By 1812, Fraunhofer had succeeded in creating an achromatic objective lens 190 mm (7.5 in) in diameter. The size of this primary lens is the main factor in determining the light gathering ability and resolution of a refracting telescope. During the opposition of Mars in 1830, the German astronomers Johann Heinrich Mädler and Wilhelm Beer used a 95 mm (3.7 in) Fraunhofer refracting telescope to launch an extensive study of the planet. They chose a feature located 8° south of the equator as their point of reference. (This was later named the Sinus Meridiani, and it would become the zero meridian of Mars.) During their observations, they established that most of Mars' surface features were permanent, and more precisely determined the planet's rotation period. In 1840, Mädler combined ten years of observations to draw the first map of Mars. Rather than giving names to the various markings, Beer and Mädler simply designated them with letters; thus Meridian Bay (Sinus Meridiani) was feature "a".

Working at the Vatican Observatory during the opposition of Mars in 1858, Italian astronomer Angelo Secchi noticed a large blue triangular feature, which he named the "Blue Scorpion". This same seasonal cloud-like formation was seen by English astronomer J. Norman Lockyer in 1862, and it has been viewed by other observers. During the 1862 opposition, Dutch astronomer Frederik Kaiser produced drawings of Mars. By comparing his illustrations to those of Huygens and the English natural philosopher Robert Hooke, he was able to further refine the rotation period of Mars. His value of $24^h\ 37^m\ 22.6^s$ is accurate to within a tenth of a second.

<templatestyles src="Multiple_image/styles.css" />

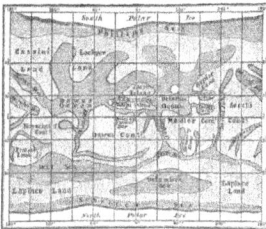

A later version of Proctor's map of Mars, published in 1905

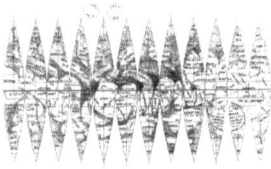

1892 atlas of Mars by the Belgian astronomer Louis Niesten

Father Secchi produced some of the first color illustrations of Mars in 1863. He used the names of famous explorers for the distinct features. In 1869, he observed two dark linear features on the surface that he referred to as *canali*, which is Italian for 'channels' or 'grooves'. In 1867, English astronomer Richard A. Proctor created a more detailed map of Mars based on the 1864 drawings of English astronomer William R. Dawes. Proctor named the various lighter or darker features after astronomers, past and present, who had contributed to the observations of Mars. During the same decade, comparable maps and nomenclature were produced by the French astronomer Camille Flammarion and the English astronomer Nathan Green.

At the University of Leipzig in 1862–64, German astronomer Johann K. F. Zöllner developed a custom photometer to measure the reflectivity of the Moon, planets and bright stars. For Mars, he derived an albedo of 0.27. Between 1877 and 1893, German astronomers Gustav Müller and Paul Kempf observed Mars using Zöllner's photometer. They found a small phase coefficient—the variation in reflectivity with angle—indicating that the surface of Mars is smooth and without large irregularities. In 1867, French astronomer Pierre Janssen and British astronomer William Huggins used spectroscopes to examine the atmosphere of Mars. Both compared the optical spectrum of Mars to that of the Moon. As the spectrum of the latter did not display absorption lines of water, they believed they had detected the presence of water vapor in the atmosphere of Mars. This result was confirmed by German astronomer Herman C. Vogel in 1872 and English astronomer Edward W. Maunder in 1875, but would later come into question.

A particularly favorable perihelic opposition occurred in 1877. The English astronomer David Gill used this opportunity to measure the diurnal parallax of Mars from Ascension Island, which led to a parallax estimate of 8.78 ± 0.01 arcseconds. Using this result, he was able to more accurately determine the distance of the Earth from the Sun, based upon the relative size of the orbits of Mars and the Earth. He noted that the edge of the disk of Mars appeared fuzzy because of its atmosphere, which limited the precision he could obtain for the planet's position.

In August 1877, the American astronomer Asaph Hall discovered the two moons of Mars using a 660 mm (26 in) telescope at the U.S. Naval Observatory. The names of the two satellites, Phobos and Deimos, were chosen by Hall based upon a suggestion by Henry Madan, a science instructor at Eton College in England.

Martian canals

<templatestyles src="Multiple_image/styles.css" />

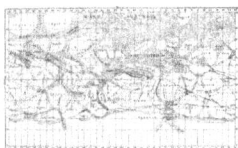

Map of Mars by Giovanni Schiaparelli, compiled between 1877 and 1886, showing *canali* features as fine lines

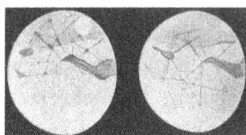

Mars sketched as observed by Lowell sometime before 1914. (South top)

During the 1877 opposition, Italian astronomer Giovanni Schiaparelli used a 22 cm (8.7 in) telescope to help produce the first detailed map of Mars. These maps notably contained features he called *canali*, which were later shown to be an optical illusion. These *canali* were supposedly long straight lines on the surface of Mars to which he gave names of famous rivers on Earth. His term *canali* was popularly mistranslated in English as *canals*. In 1886, the English astronomer William F. Denning observed that these linear features were irregular in nature and showed concentrations and interruptions. By 1895, English astronomer Edward Maunder became convinced that the linear features were merely the summation of many smaller details.

In his 1892 work *La planète Mars et ses conditions d'habitabilité*, Camille Flammarion wrote about how these channels resembled man-made canals, which an intelligent race could use to redistribute water across a dying Martian world. He advocated for the existence of such inhabitants, and suggested they may be more advanced than humans.

Influenced by the observations of Schiaparelli, Percival Lowell founded an observatory with 30-and-45 cm (12-and-18 in) telescopes. The observatory was used for the exploration of Mars during the last good opportunity in 1894 and

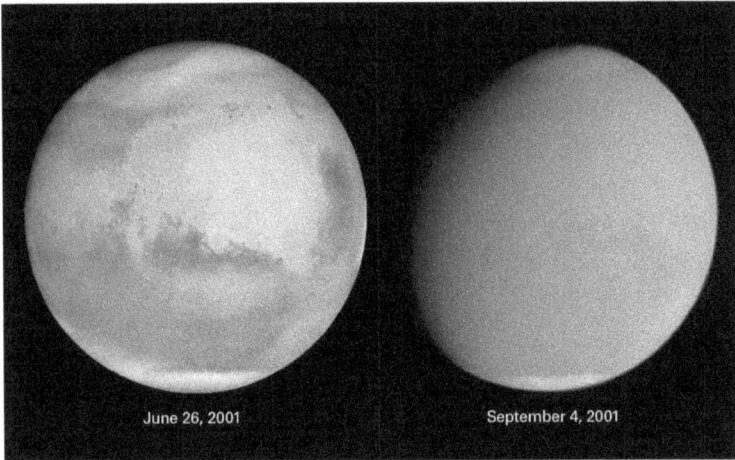

Figure 265: *In the left image, thin Martian clouds are visible near the polar regions. At right, the surface of Mars is obscured by a dust storm. NASA/HST images*

the following less favorable oppositions. He published books on Mars and life on the planet, which had a great influence on the public. The *canali* were found by other astronomers, such as Henri Joseph Perrotin and Louis Thollon using a 38 cm (15 in) refractor at the Nice Observatory in France, one of the largest telescopes of that time.

Beginning in 1901, American astronomer A. E. Douglass attempted to photograph the canal features of Mars. These efforts appeared to succeed when American astronomer Carl O. Lampland published photographs of the supposed canals in 1905. Although these results were widely accepted, they became contested by Greek astronomer Eugène M. Antoniadi, English naturalist Alfred Russel Wallace and others as merely imagined features. As bigger telescopes were used, fewer long, straight *canali* were observed. During an observation in 1909 by Flammarion with a 84 cm (33 in) telescope, irregular patterns were observed, but no *canali* were seen.

Refining planetary parameters

Surface obscuration caused by yellow clouds had been noted in the 1870s when they were observed by Schiaparelli. Evidence for such clouds was observed during the oppositions of 1892 and 1907. In 1909, Antoniadi noted that the presence of yellow clouds was associated with the obscuration of albedo features. He discovered that Mars appeared more yellow during oppositions when the planet was closest to the Sun and was receiving more energy. He suggested windblown sand or dust as the cause of the clouds.

In 1894, American astronomer William W. Campbell found that the spectrum of Mars was identical to the spectrum of the Moon, throwing doubt on the burgeoning theory that the atmosphere of Mars is similar to that of the Earth. Previous detections of water in the atmosphere of Mars were explained by unfavorable conditions, and Campbell determined that the water signature came entirely from the Earth's atmosphere. Although he agreed that the ice caps did indicate there was water in the atmosphere, he did not believe the caps were sufficiently large to allow the water vapor to be detected. At the time, Campbell's results were considered controversial and were criticized by members of the astronomical community, but they were confirmed by American astronomer Walter S. Adams in 1925.

Baltic German astronomer Hermann Struve used the observed changes in the orbits of the Martian moons to determine the gravitational influence of the planet's oblate shape. In 1895, he used this data to estimate that the equatorial diameter was 1/190 larger than the polar diameter. In 1911, he refined the value to 1/192. This result was confirmed by American meteorologist Edgar W. Woolard in 1944.

Using a vacuum thermocouple attached to the 2.54 m (100 in) Hooker Telescope at Mount Wilson Observatory, in 1924 the American astronomers Seth Barnes Nicholson and Edison Pettit were able to measure the thermal energy being radiated by the surface of Mars. They determined that the temperature ranged from –68 °C (–90 °F) at the pole up to 7 °C (45 °F) at the midpoint of the disk (corresponding to the equator). Beginning in the same year, radiated energy measurements of Mars were made by American physicist William Coblentz and American astronomer Carl Otto Lampland. The results showed that the night time temperature on Mars dropped to –85 °C (–121 °F), indicating an "enormous diurnal fluctuation" in temperatures. The temperature of Martian clouds was measured as –30 °C (–22 °F). In 1926, by measuring spectral lines that were redshifted by the orbital motions of Mars and Earth, American astronomer Walter Sydney Adams was able to directly measure the amount of oxygen and water vapor in the atmosphere of Mars. He determined that "extreme desert conditions" were prevalent on Mars. In 1934, Adams and American astronomer Theodore Dunham, Jr. found that the amount of oxygen in the atmosphere of Mars was less than one percent of the amount over a comparable area on Earth.

In 1927, Dutch graduate student Cyprianus Annius van den Bosch made a determination of the mass of Mars based upon the motions of the Martian moons, with an accuracy of 0.2%. This result was confirmed by the Dutch astronomer Willem de Sitter and published posthumously in 1938. Using observations of the near Earth asteroid Eros from 1926 to 1945, German-American astronomer Eugene K. Rabe was able to make an independent estimate the mass of Mars, as well as the other planets in the inner Solar System, from the planet's gravitational perturbations of the asteroid. His estimated margin of error was 0.05%, but subsequent checks suggested his result was poorly determined compared to other methods.

During the 1920s, French astronomer Bernard Lyot used a polarimeter to study the surface properties of the Moon and planets. In 1929, he noted that the polarized light emitted from the Martian surface is very similar to that radiated from the Moon, although he speculated that his observations could be explained by frost and possibly vegetation. Based on the amount of sunlight scattered by the Martian atmosphere, he set an upper limit of 1/15 the thickness of the Earth's atmosphere. This restricted the surface pressure to no greater than 2.4 kPa (24 mbar). Using infrared spectrometry, in 1947 the Dutch-American astronomer Gerard Kuiper detected carbon dioxide in the Martian atmosphere. He was able to estimate that the amount of carbon dioxide over a given area of the surface is double that on the Earth. However, because he overestimated the surface pressure on Mars, Kuiper concluded erroneously that the ice caps could not be composed of frozen carbon dioxide. In 1948, American meteorologist Seymour L. Hess determined that the formation of the thin Martian clouds would only require 4 mm (0.16 in) of water precipitation and a vapor pressure of 0.1 kPa (1.0 mbar).

The first standard nomenclature for Martian albedo features was introduced by the International Astronomical Union (IAU) when in 1960 they adopted 128 names from the 1929 map of Antoniadi named *La Planète Mars*. The Working Group for Planetary System Nomenclature (WGPSN) was established by the IAU in 1973 to standardize the naming scheme for Mars and other bodies.

Remote sensing

The International Planetary Patrol Program was formed in 1969 as a consortium to continually monitor planetary changes. This worldwide group focused on observing dust storms on Mars. Their images allow Martian seasonal patterns to be studied globally, and they showed that most Martian dust storms occur when the planet is closest to the Sun.

Since the 1960s, robotic spacecraft have been sent to explore Mars from orbit and the surface in extensive detail. In addition, remote sensing of Mars from Earth by ground-based and orbiting telescopes has continued across much of

Figure 266: *Photograph of the Martian meteorite ALH84001*

the electromagnetic spectrum. These include infrared observations to determine the composition of the surface, ultraviolet and submillimeter observation of the atmospheric composition, and radio measurements of wind velocities.

The Hubble Space Telescope (HST) has been used to perform systematic studies of Mars and has taken the highest resolution images of Mars ever captured from Earth. This telescope can produce useful images of the planet when it is at an angular distance of at least 50° from the Sun. The HST can take images of a hemisphere, which yields views of entire weather systems. Earth-based telescopes equipped with charge-coupled devices can produce useful images of Mars, allowing for regular monitoring of the planet's weather during oppositions.

X-ray emission from Mars was first observed by astronomers in 2001 using the Chandra X-ray Observatory, and in 2003 it was shown to have two components. The first component is caused by X-rays from the Sun scattering off the upper Martian atmosphere; the second comes from interactions between ions that result in an exchange of charges. The emission from the latter source has been observed out to eight times the radius of Mars by the XMM-Newton orbiting observatory.

In 1983, the analysis of the shergottite, nakhlite, and chassignite (SNC) group of meteorites showed that they may have originated on Mars. The Allan Hills

84001 meteorite, discovered in Antarctica in 1984, is believed to have originated on Mars but it has an entirely different composition than the SNC group. In 1996, it was announced that this meteorite might contain evidence for microscopic fossils of Martian bacteria. However, this finding remains controversial. Chemical analysis of the Martian meteorites found on Earth suggests that the ambient near-surface temperature of Mars has most likely been below the freezing point of water (0 C°) for much of the last four billion years.

External links

> Wikimedia Commons has media related to *Old observations of Mars*.

- "Pop culture Mars"[399]. *Mars Exploration Program*. NASA. May 5, 2008. Retrieved 2012-06-16.
- Snyder, Dave (May 2001). "An observational history of Mars"[400]. Retrieved 2012-06-16.

<indicator name="featured-star"> ⭐ </indicator>

Martian canal

For a time in the late 19th and early 20th centuries, it was erroneously believed that there were **canals on Mars**. These were a network of long straight lines in the equatorial regions from 60° north to 60° south latitude on the planet Mars observed by astronomers using early low-resolution telescopes without photography. They were first described by the Italian astronomer Giovanni Schiaparelli during the opposition of 1877, and confirmed by later observers. Schiaparelli called these *canali*, which was translated into English as "canals". The Irish astronomer Charles E. Burton made some of the earliest drawings of straight-line features on Mars, although his drawings did not match Schiaparelli's. By the early 20th century, improved astronomical observations revealed the "canals" to be an optical illusion, and modern high resolution mapping of the Martian surface by spacecraft shows no such features.

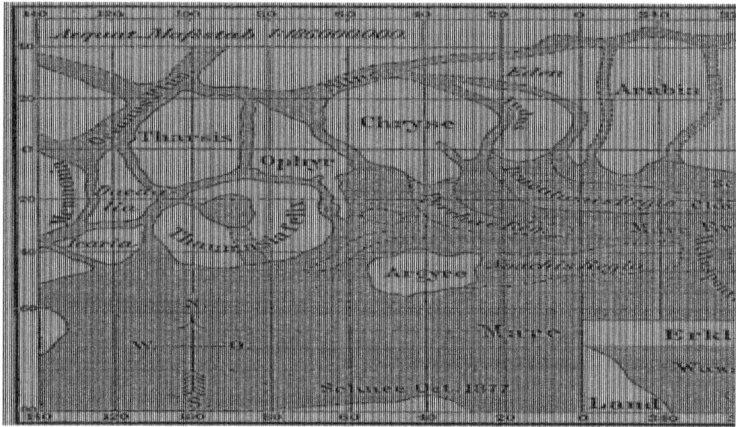

Figure 267: *1877 map of Mars by Giovanni Schiaparelli.*

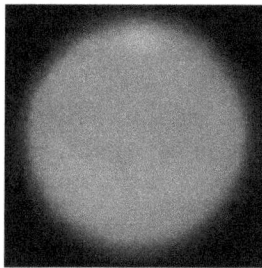

Figure 268: *Mars as seen through 6 inch (15 cm) aperture reflecting telescope, as Schiaparelli may have seen it.*

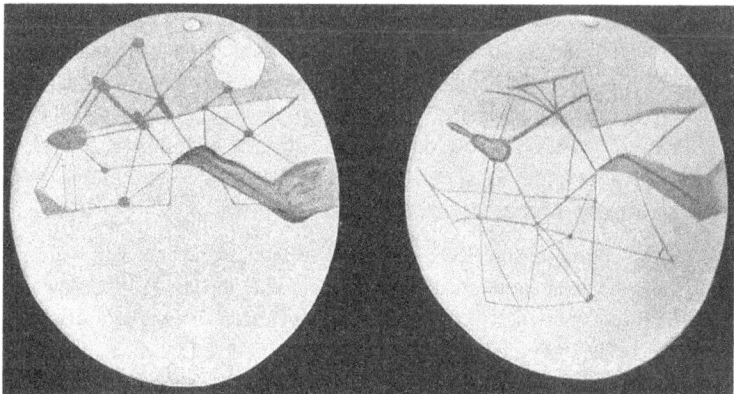

Figure 269: *Martian canals depicted by Percival Lowell*

History

The Italian word *canale* (plural *canali*) can mean "canal", "channel", "duct" or "gully".[401] The first person to use the word *canale* in connection with Mars was Angelo Secchi in 1858, although he did not see any straight lines and applied the term to large features —for example, he used the name "Canale Atlantico" for what later came to be called Syrtis Major Planum.

It is not necessarily odd that the idea of Martian canals was so readily accepted by many. At this time in the late 19th century, astronomical observations were made without photography. Astronomers had to stare for hours through their telescopes, waiting for a moment of still air when the image was clear, and then draw a picture of what they had seen. They saw some lighter or darker albedo features (for instance Syrtis Major) and believed that they were seeing oceans and continents. They also believed that Mars had a relatively substantial atmosphere. They knew that the rotation period of Mars (the length of its day) was almost the same as Earth's, and they knew that Mars' axial tilt was also almost the same as Earth's, which meant it had seasons in the astronomical and meteorological sense. They could also see Mars' polar ice caps shrinking and growing with these changing seasons. It was only when they interpreted changes in surface features as being due to the seasonal growth of plants that life was hypothesized by them (in fact, Martian dust storms are responsible for some of this). By the late 1920s, however, it was known that Mars is very dry and has a very low atmospheric pressure.

In 1889, astronomer Charles A. Young reported that Schiaparelli's canal discovery of 1877 had been confirmed in 1881, though new canals had appeared where there had not been any before, prompting "very important and perplexing" questions as to their origin.[402]

During the favourable opposition of 1892, W. H. Pickering observed numerous small circular black spots occurring at every intersection or starting-point of the "canals". Many of these had been seen by Schiaparelli as larger dark patches, and were termed seas or lakes; but Pickering's observatory was at Arequipa, Peru, about 2400 meters above the sea, and with such atmospheric conditions as were, in his opinion, equal to a doubling of telescopic aperture. They were soon detected by other observers, especially by Lowell.

During the oppositions of 1892 and 1894, seasonal color changes were reported. As the polar snows melted the adjacent seas appeared to overflow and spread out as far as the tropics, and were often seen to assume a distinctly green colour. At this time (1894) it began to be doubted whether there were any seas at all on Mars. Under the best conditions, these supposed 'seas' were seen to lose all trace of uniformity, their appearance being that of a mountainous country, broken by ridges, rifts, and canyons, seen from a great elevation.

These doubts soon became certainties, and it is now universally agreed that Mars possesses no permanent bodies of surface water.

Interpretation as engineering works

During the 1894 opposition the idea that Schiaparelli's *canali* were really irrigation canals made by intelligent beings, was first hinted at, and then adopted as the only intelligible explanation, by American astronomer Percival Lowell and a few others. The visible seasonal melting of Mars polar icecaps fueled speculation that an advanced race built canals to transport the water to drier equatorial regions. Newspaper and magazine articles about Martian canals captured the public imagination. Lowell published his views in three books: *Mars* (1895), *Mars and Its Canals* (1906), and *Mars As the Abode of Life* (1908). He remained a strong proponent for the rest of his life of the idea that the canals were built for irrigation by an intelligent civilization, going much further than Schiaparelli, who for his part considered much of the detail on Lowell's drawings to be imaginary. Some observers drew maps in which dozens if not hundreds of canals were shown with an elaborate nomenclature for all of them. Some observers saw a phenomenon they called "gemination", or doubling - two parallel canals. The late 19th century was a time of construction of giant infrastructure projects of all kinds, and particularly canal building. For instance, the Suez Canal was completed in 1869, and the abortive French attempt to build the Panama Canal began in 1880. It is understandable that 19th century people who accepted the idea of a Mars inhabited by a civilization might interpret the canal features as giant engineering works.

Doubts

Other observers disputed the notion of canals. The observer E. E. Barnard did not see them. In 1903, Joseph Edward Evans and Edward Maunder conducted visual experiments using schoolboy volunteers that demonstrated how the canals could arise as an optical illusion.[403] This is because when a poor-quality telescope views many point-like features (e.g. sunspots or craters) they appear to join up to form lines. In 1907 the British naturalist Alfred Russel Wallace published the book *Is Mars Habitable?* that severely criticized Lowell's claims. Wallace's analysis showed that the surface of Mars was almost certainly much colder than Lowell had estimated, and that the atmospheric pressure was too low for liquid water to exist on the surface; and he pointed out that several recent efforts to find evidence of water vapor in the Martian atmosphere with spectroscopic analysis had failed. He concluded that complex life was impossible, let alone the planet-girding irrigation system claimed by Lowell. The influential observer Eugène Antoniadi used the 83-cm (32.6 inch) aperture telescope at Meudon Observatory at the 1909 opposition of Mars and

Figure 270: *Mars surface by Mariner 4 in 1965*

saw no canals, the outstanding photos of Mars taken at the new Baillaud dome
at the Pic du Midi observatory also brought formal discredit to the Martian
canals theory in 1909,[404] and the notion of canals began to fall out of favor.
Around this time spectroscopic analysis also began to show that no water was
present in the Martian atmosphere. However, as of 1916 Waldemar Kaempf-
fert (editor of *Scientific American* and later *Popular Science Monthly*) was still
vigorously defending the Martian canals theory against skeptics.

Spacecraft evidence

The arrival of the United States' Mariner 4 spacecraft in 1965, which took
pictures revealing impact craters and a generally barren landscape, was the
final nail in the coffin of the idea that Mars could be inhabited by higher forms
of life, or that any canal features existed. A surface atmospheric pressure of
4.1 to 7.0 millibars (410 to 700 pascals), 0.4% to 0.7% of Earth atmospheric
pressure, and daytime temperatures of −100 degrees Celsius were estimated.
No magnetic field or Martian radiation belts were detected.

William Kenneth Hartmann, a Mars imaging scientist from the 1960s to the
2000s, explains the "canals" as streaks of dust caused by wind on the leeward
side of mountains and craters.[405]

In popular culture

Although the concept of the canals had been available since Schiaparelli's 1877 description of them, early fictional descriptions of Mars omitted these features. They receive no mention, for instance, in H. G. Wells' *The War of the Worlds* (1897), which describes a slowly drying Mars, covetous of Earth's resources, but one which still has dwindling oceans such as are depicted on Schiaparelli's maps. Later works of fiction, influenced by the works of Lowell, described an ever-more arid Mars, and the canals became a more prominent feature, though how they were explained varied widely from author to author.

- Camille Flammarion's *Uranie* (1889, published as *Urania* in English in 1890) include descriptions of life on Mars; "They have straightened and enlarged the watercourses and made them like canals, and have constructed a network of immense canals all over the continents. The continents themselves are not bristling all over with Alpine or Himalayan upheavals like those of the terrestrial globe, but are immense plains, crossed in all directions by canals, which connect all the seas with one another, and by streams made to resemble canals."
- Garrett P. Serviss' *Edison's Conquest of Mars* (1898) repeatedly mentions Schiaparellian canals (which play a key part in the denouement of the story), but does not describe them in detail, apparently considering them simply irrigation canals comparable to those on Earth — ignoring the fact that, in that case, they could hardly be visible from Earth. Serviss' Mars also has lakes and oceans.
- George Griffith's *A Honeymoon in Space* (1900) describes the canals as the remnants of gulfs and straits "widened and deepened and lengthened by... Martian labour".
- Carl Jung's inaugural dissertation for his medical degree, *On the Psychology and Pathology of So-Called Occult Phenomena* (1902), describes the recounts of a 15-year-old patient, a medium who encountered supernatural beings during seance: "she told us all the peculiarities of the star-dwellers:... the whole of Mars is covered with canals, the canals are all flat ditches, the water in them is very shallow. The excavating of the canals caused the Martians no particular trouble, as the soil there is lighter than on earth."
- Edgar Rice Burroughs' influential *A Princess of Mars* (1912) describes an almost entirely desert Mars, with only one small body of liquid water on the surface (though swamps and forests appear in the sequels). The canals, or waterways as Burroughs calls them, are still irrigation works, but these are surrounded by wide cultivated tracts of farmland which make their visibility somewhat credible.

- Alexander Bogdanov's *Engineer Menni* (1913) details the social, scientific, and political history of the construction of the Martian canals and the socio-economic ramifications the construction had on Martian society.
- Otis Adelbert Kline's *Outlaws of Mars* (1933) has multiple parallel canals, surrounded by walls and terraces, and describes the construction of the canals by Martian machines.
- In Stanley G. Weinbaum's *A Martian Odyssey* (1934) the lead character Jarvis crosses several canals: One is "a dry ditch about four hundred feet wide, and straight as a railroad on its own company map." Some canals have "mud cities" and vegetation beside them. One appears to be covered with what looks like a nice green lawn, but turns out to be hundreds of small creatures that move out of the way when approached. In the sequel *Valley of Dreams* (1934) it is discovered that the various races on Mars cooperatively maintain the canal system, driving water northward from the southern polar icecap.
- In C. S. Lewis' *Out of the Silent Planet* (1938), the "canals" (*handramit* in Martian) are actually vast rifts in the surface of an almost airless, desert Mars, in which the only breathable atmosphere and water have collected where life is possible, with the rest of Mars being entirely dead. As depicted by Lewis, these were of artificial origin - a vast engineering project undertaken long ago by the Martians to save what was left of their planet, after Mars was attacked and devastated by the evil Guardian Angel of Earth (who, in Lewis' system of theological Science Fiction, is the same as Satan).
- Robert A. Heinlein gave two depictions of the Martian canals:
  - In *The Green Hills of Earth* (1947), the blind poet Rhysling, composes "The Grand Canal", describing the beauty of Mars' main canal as he saw it when first arriving on Mars. Having later become blind, Rhysling does not realize that human colonists have proceeded in short order to heavily pollute the canals with industrial wastes, tear down half of the delicate beautiful structures at the canal side and convert the other half to industrial uses - with the remnant of the indigenous Martians helpless to stop them.
  - In *Red Planet* (1949), colonists use the frozen canals for travel and a seasonal migration (by iceboat during winter when the canals are frozen and by boat when the ice melts during the Martian summer). Teenagers Jim Marlowe and Frank Sutton set out to skate the thousands of miles to their homes on the frozen Martian canals when escaping the Lowell Academy boarding school.
- In Ray Bradbury's *The Martian Chronicles* (1950), the canals are artificial waterways stretching between stone banks, filled with blue water, or sometimes poetically described as full of "green liquors" or "lavender

wine". Bradbury revisited the martian canals in 1967 in his short story "The Lost City of Mars".

- In the BBC radio production *Journey into Space: The Red Planet* (1954–1955), the canals are valleys filled with a plant life resembling giant rhubarbs.
- In *Robinson Crusoe on Mars* (1964) Kit Draper and Friday flee from the enemy aliens through the underground canals on their way to the polar ice cap.
- In Colin Greenland's *Take Back Plenty* (1990), humans arriving on Mars discover a networks of canals in very bad condition due to the long period since the original builders became extinct. Human colonists energetically renovate the canals and put them to renewed use, discover at the Grand Canal the colossal buried city of the original builders, excavate it and build a thriving human city all around it. The human city is named "Schiaparelli".
- The Mars of the steampunk role-playing game *Space: 1889* (1988) is crisscrossed by artificial canals which support cities inhabited by the ancient civilization of the Canal Martians.
- The 1991 computer game Ultima: Martian Dreams features a plot based around Victorian expeditions to Mars. The Martian canals play a very prominent role as the main characters have to find a way to refill them using ice from the polar caps.
- Kim Stanley Robinson's science fiction chronicling of the terraforming of Mars in the Mars trilogy (1993-1999) and 2312 (2012) features the creation of canals on Mars ("burned" into the land with magnified sunlight) with the Lowell maps as inspiration. "Thus a nineteenth-century fantasy forms the basis for the actual landscape."
- In S. M. Stirling's 2008 *In the Courts of the Crimson Kings* alternate history novel Mars is terraformed and seeded with earth life including early humans, at some point in prehistory. The humans of Mars do indeed build a planet wide canal network due to their world's exceptional dryness, however it's left ambiguous whether or not these were what Lowell actually saw in the 19th century.
- Ken Kalfus's 2013 novel, *Equilateral*, is based entirely on the supposed existence of "man"-made Martian canals and on the construction of a vast triangle in the Arabian desert in order to communicate with the Martian beings.
- Scott Walker's "Lullaby" from the 2014 album Soused (with Sunn O)))) contains the lyrics, "Tonight my assistant will hear the canals of Mars." The composition first appeared on Ute Lemper's 2000 album, Punishing Kiss.

- "Seeds of the Dusk" is a short story by Raymond Raymond Z. Gallun about a far-future twilight race of humans and animals on earth that are threatened by a newcomer plant that can defend itself adaptively, even killing an attacker. As it propagates and links in long chains with others of its species, the linked chains of plants begin pumping water through their specially formed inner chambers. At the end it is made apparent that the spores of the plant had drifted to Earth from Mars, and it was beginning to form long canals on Earth as it had done on Mars.

List of canals

The canals were named, by Schiaparelli and others, after real and legendary rivers of various places on Earth or the mythological underworld.

References

- Wallace, A. R. (1907) *Is Mars habitable? A critical examination of Professor Percival Lowell's book "Mars and its canals", with an alternative explanation, by Alfred Russel Wallace, F.R.S., etc.* London, Macmillan and co.
- Antoniadi, E. M. (1910) "Sur la nature des »canaux« de Mars", AN **183** (1910) 221/222[406] (in French)

External links

- http://www.theguardians.com/space/orbitalmech/gm_emoi1.htm
- Martian canals throughout the history[407]

Appendix

References

[1]This image was taken by the *Rosetta* spacecraft's Optical, Spectroscopic, and Infrared Remote Imaging System (OSIRIS), at a distance of ≈240,000 kilometres (150,000 mi) during its February 2007 encounter. The view is centered on the Aeolis quadrangle, with Gale crater, the landing site of the *Curiosity* rover, prominently visible just left of center. The darker, more heavily cratered terrain in the south, Terra Cimmeria, is composed of older terrain than the much smoother and brighter Elysium Planitia to the north. Geologically recent processes, such as the possible existence of a global ocean in Mars's past, could have helped lower-elevated areas, such as Elysium Planitia, retain a more youthful look.

[2]

[3]//tools.wmflabs.org/geohack/geohack.php?pagename=Mars&params=0_N_180_W_globe: Mars

[4]//tools.wmflabs.org/geohack/geohack.php?pagename=Mars&params=3.34_N_219.38_E_ globe:Mars

[5]//tools.wmflabs.org/geohack/geohack.php?pagename=Mars&params=30_N_260_W_globe: Mars

[6]//tools.wmflabs.org/geohack/geohack.php?pagename=Mars&params=0_N_310_W_globe: Mars

[7]Mustard, Jack (July 9, 2009) MEPAG Report to the Planetary Science Subcommittee http://www.lpi.usra.edu/pss/july2009/presentations/05MEPAG.pdf. lpi.usra.edu. p. 3

[8]//en.wikipedia.org/w/index.php?title=Mars&action=edit

[9]https://curlie.org/Science/Astronomy/Solar_System/Planets/Mars/

[10]http://mars.jpl.nasa.gov/

[11]http://www.google.com/mars/

[12]https://www.google.com/maps/space/mars/

[13]http://www.geody.com/?world=mars

[14]http://photojournal.jpl.nasa.gov/targetFamily/Mars

[15]http://mars.jpl.nasa.gov/multimedia/

[16]http://www.msss.com/science-images/

[17]http://hirise.lpl.arizona.edu/

[18]http://sos.noaa.gov/Datasets/dataset.php?id=224

[19]https://www.usgs.gov/media/videos/rotating-globe-mars-geology

[20]https://www.youtube.com/watch?v=Jr1Xu2i-Uc0

[21]http://themis.mars.asu.edu/node/5470

[22]https://www.flickr.com/photos/136797589@N04/32657197761/in/photostream/

[23]https://www.flickr.com/photos/136797589@N04/albums/72157680411399285

[24]http://planetarynames.wr.usgs.gov/Page/MARS/target

[25]http://planetarynames.wr.usgs.gov/Page/mars1to5mTHEMIS

[26]http://pubs.usgs.gov/sim/3292/

[27]http://planetologia.elte.hu/terkep/mars-viking-en.pdf

[28]http://planetologia.elte.hu/terkep/mars-mola-en.pdf

[29]P. Zasada (2013) Generalised Geological Map of Mars, 1:140.000.000, Source Link http://astronomynow.com/2015/02/01/4billion-year-old-meteorite-reveals-mars-darker-side/.

[30]Carr, M.H (2007). Mars: Surface and Interior in *Encyclopedia of the Solar System,* 2nd ed., McFadden, L.-A. et al. Eds. Elsevier: San Diego, CA, p.319

[31]Boyce, J.M. (2008) *The Smithsonian Book of Mars;* Konecky&Konecky: Old Saybrook, CT, p. 13.

[32]Carr, M.H.; Saunders, R.S.; Strom R.G. (1984). *Geology of the Terrestrial Planets;* NASA Scientific and Technical Information Branch: Washington DC, 1984, p. 223. http://www.lpi.usra.edu/publications/books/geologyTerraPlanets/

[33]Kargel, J.S. (2004) Mars: A Warmer Wetter Planet; Springer-Praxis: London, p. 52.

[34] Sheehan, W. (1996). *The Planet Mars: A History of Observation & Discovery;* University of Arizona Press: Tucson, p. 25. http://www.uapress.arizona.edu/onlinebks/mars/contents.htm.

[35] Hale, W.S.; Head, J.W. (1981). Lunar Planet. Sci. XII, pp. 386-388. (abstract 1135). http://www.lpi.usra.edu/meetings/lpsc1981/pdf/1135.pdf

[36] Boyce, J.M. *The Smithsonian Book of Mars;* Konecky&Konecky: Old Saybrook, CT, 2008, p. 203.

[37] http://hirise.lpl.eduPSP_008508_1870

[38] Bleacher, J. and S. Sakimoto. *Pedestal Craters, A Tool For Interpreting Geological Histories and Estimating Erosion Rates.* LPSC

[39] Carr, M. 2006. The Surface of Mars. Cambridge University Press.

[40] Grotzinger, J. and R. Milliken (eds.) 2012. Sedimentary Geology of Mars. SEPM

[41] Patrick Zasada (2013/14): Gradation of extraterrestrial fluvial sediments – related to the gravity. - *Z. geol. Wiss.* **41/42** (3): 167-183. Abstract http://www.zgw-online.de/en/media/167-133.pdf

[42] S. W. Squyres and A. H. Knoll, *Sedimentary Geology at Meridiani Planum, Mars,* Elsevier, Amsterdam, (2005); reprinted from *Earth and Planetary Science Letters, Vol. 240,* No. 1 (2005).

[43] Zasada, P., 2013: *Entstehung des Marsgesteins "Jake Matijevic".* – Sternzeit http://www.sternzeit-online.de/, issue 2/2013 http://www.sternzeit-online.de/?p=archiv&q=2%2F2013: 98 ff. (in German language).

[44] M. Wiseman, J. C. Andrews-Hanna, R. E. Arvidson3, J. F. Mustard, K. J. Zabrusky DISTRIBUTION OF HYDRATED SULFATES ACROSS ARABIA TERRA USING CRISM DATA: IMPLICATIONS FOR MARTIAN HYDROLOGY. 42nd Lunar and Planetary Science Conference (2011) 2133.pdf

[45] DiscoveryChannel.ca - Mars avalanche caught on camera http://dsc.discovery.com/news/2008/11/07/mars-avalanche.html

[46] //en.wikipedia.org/w/index.php?title=Template:MarsRocks&action=edit

[47] http://pubs.usgs.gov/sim/3292/

[48] http://pubs.usgs.gov/sim/3292/pdf/sim3292_map.pdf

[49] https://www.youtube.com/watch?v=quZMhSohIEU

[50] http://www.lpi.usra.edu/resources/mars_maps/1083/index.html

[51] http://mars3d.com/

[52] http://www.impacttectonics.org/SyriaPlanum.html

[53] http://marstrek.jpl.nasa.gov/

[54] http://mars.jpl.nasa.gov/msl/multimedia/images/?ImageID=6012

[55] http://www.nasa.gov/mission_pages/MRO/multimedia/pia15116.html

[56] https://www.youtube.com/watch?v=0t0LWFHB8Qo

[57] Goetz et al. (2007), Seventh Mars Conference http://www.lpi.usra.edu/meetings/7thmars2007/pdf/3104.pdf

[58] http://photojournal.jpl.nasa.gov/figures/PIA16156_fig1.jpg

[59] http://mars.jpl.nasa.gov/msl/images/pia16223-stereoHattah-Mastcam-br2.jpg

[60] https://www.youtube.com/watch?v=Jr1Xu2i-Uc0

[61] Carr, 2006, p. 173.

[62] Carr, 2006, pp 144–147.

[63] NASA Mars Exploration Program Overview. http://www.nasa.gov/mission_pages/mars/overview/index.html.

[64] Sheehan, 1996, p. 35.

[65] hartmann, 2003, p. 20.

[66] Sheehan, 1996, p. 150.

[67] ., p. 120

[68] Meyer, C. (2012) The Martian Meteorite Compendium; National Aronautics and Space Administration. http://curator.jsc.nasa.gov/antmet/mmc/.

[69] Agree, C., et al. 2013. Unique Meteorite from Early Amazonian Mars: Water-Rich Basaltic Breccia Northwest Africa 7034. Science: 339, 780–785.

[70] Matsubara, Yo, Alan D. Howard, and Sarah A. Drummond. "Hydrology of early Mars: Lake basins." Journal of Geophysical Research: Planets 116.E4 (2011).

[71] Parker, T., D. Curie. 2001. *Geomorphology* 37. 303–328.

[72] de Pablo, M., M. Druet. 2002. XXXIII LPSC. Abstract #1032.

[73] de Pablo, M. 2003. VI Mars Conference, Abstract #3037.

[74] Baker, D., J. Head. 2014. 44th LPSC, abstract #1252

[75] Lamb, Michael P., et al. "Can springs cut canyons into rock?." *Journal of Geophysical Research: Planets* (1991–2012) 111.E7 (2006).

[76] Rodriguez, J., et al. 2016. Tsunami waves extensively resurfaced the shorelines of an early Martian ocean. Scientific Reports: 6, 25106.

[77] Cornell University. "Ancient tsunami evidence on Mars reveals life potential." ScienceDaily. ScienceDaily, May 19, 2016. <https://www.sciencedaily.com/releases/2016/05/160519101756.htm>.

[78] Scanlon, K., et al. 2018. The Dorsa Argentea Formation and the Noachian-Hesperian climate transition. Icarus: 299, 339–363.

[79] Head, J, S. Pratt. 2001. Extensive Hesperian-aged south polar ice sheet on Mars: Evidence for massive melting and retreat, and lateral flow and pending of meltwater. J. Geophys. Res.-Planet, 106 (E6), 12275-12299.

[80] Supplementary Materials for: Radar evidence of subglacial liquid water on Mars http://science.sciencemag.org/content/sci/suppl/2018/07/24/science.aar7268.DC1/aar7268_Orosei_SM.pdf. (PDF). R. Orosei, S. E. Lauro, E. Pettinelli, A. Cicchetti, M. Coradini, B. Cosciotti, F. Di Paolo, E. Flamini, E. Mattei, M. Pajola, F. Soldovieri, M. Cartacci, F. Cassenti, A. Frigeri, S. Giuppi, R. Martufi, A. Masdea, G. Mitri, C. Nenna, R. Noschese, M. Restano, R. Seu. *Science*. July 25, 2018

[81] Mars (probably) has a lake of liquid water https://www.sciencenews.org/article/mars-may-have-lake-liquid-water-search-life. Lisa Grossman, *Science News*. July 25, 2018.

[82] Exposed subsurface ice sheets in the Martian mid-latitudes http://science.sciencemag.org/content/359/6372/199. Colin M. Dundas, et al. *Science*, January 12, 2018. Vol. 359, Issue 6372, pp. 199–201.

[83]

[84] Ice cliffs spotted on Mars http://www.sciencemag.org/news/2018/01/ice-cliffs-spotted-mars. *Science News*. Paul Voosen. January 11, 2018.

[85] Dundas, C., S. Bryrne, A. McEwen. 2015. Modeling the development of martian sublimation thermokarst landforms. Icarus: 262, 154–169.

[86] Bramson, A, et al. 2015. Widespread excess ice in Arcadia Planitia, Mars. Geophysical Research Letters: 42, 6566–6574

[87] Stuurman, C., et al. 2016. SHARAD detection and characterization of subsurface water ice deposits in Utopia Planitia, Mars. Geophysical Research Letters: 43, 9484_9491.

[88] Recurring Martian Streaks: Flowing Sand, Not Water? https://www.jpl.nasa.gov/news/news.php?release=2017-299. JPL NASA News. November 20, 2017.

[89] Astrobiology Strategy 2015 http://nai.nasa.gov/media/medialibrary/2016/04/NASA_Astrobiology_Strategy_2015_FINAL_041216.pdf (PDF) NASA.

[90] A Salty, Martian Meteorite Offers Clues to Habitability http://www.astrobio.net/news-exclusive/salty-martian-meteorite-offers-clues-habitability/ By Elizabeth Howell – Astrobiology Magazine (NASA) August 28, 2014

[91] Surviving the conditions on Mars http://www.dlr.de/dlr/en/desktopdefault.aspx/tabid-10081/151_read-3409/ DLR, April 26, 2012

[92] Jean-Pierre de Vera Lichens as survivors in space and on Mars http://www.sciencedirect.com/science/article/pii/S1754504812000098 *Fungal Ecology* Volume 5, Issue 4, August 2012, Pages 472–479

[93] R. de la Torre Noetzel, F.J. Sanchez Inigo, E. Rabbow, G. Horneck, J. P. de Vera, L.G. Sancho Survival of lichens to simulated Mars conditions http://norlx51.nordita.org/~brandenb/astrobiology/EANA2012/single_abstracts/Delatorre.pdf

[94] F.J. Sáncheza, E. Mateo-Martíb, J. Raggioc, J. Meeßend, J. Martínez-Fríasb, L.Ga. Sanchoc, S. Ottd, R. de la Torrea 2012 The resistance of the lichen Circinaria gyrosa (nom. provis.) towards simulated Mars conditions—a model test for the survival capacity of an eukaryotic extremophile http://www.sciencedirect.com/science/article/pii/S0032063312002425 *Planetary and Space Science* Volume 72, Issue 1, Pages 102–110

[95] "Depending on the local solar constant, grain emissivity and thermal conductivity of ice, ice surrounding the dust grain melt for up to few hours a day during the warmest days of summer. For example, for solar constant 350 W/m2, emissivity 0.80, grain size 2 um, and thermal conductivity 0.4 W/mK melting lasts for ∼300 minutes and result in melting of 6 mm of ice." ICE MELTING BY RADIANTLY HEATED DUST GRAINS ON THE MARTIAN NORTHERN POLE http://www.hou.usra.edu/meetings/metsoc2014/pdf/5314.pdf A. Losiak, L. Czechowski and M.A. Velbel, 77th Annual Meteoritical Society Meeting (2014)

[96] Watery niche may foster life on Mars https://www.newscientist.com/article/mg20427373. 700 "According to Möhlmann, the heat from sunlight penetrating into ice or snow should get absorbed by any embedded dust grains, warming the dust and the surrounding ice. This heat mostly gets trapped because ice absorbs infrared radiation."

[97] HABITABILITY OF TRANGRESSING MARS DUNES http://www.lpi.usra.edu/meetings/lpsc2013/pdf/1434.pdf. M Fisk, R Popa, N. Bridges, N. Renno, M. Mischna, J. Moores, R. Wiens, 44th Lunar and Planetary Science Conference (2013)

[98] "Here we show that calderas on five major volcanoes on Mars have undergone repeated activation and resurfacing during the last 20 per cent of martian history, with phases of activity as young as two million years, suggesting that the volcanoes are potentially still active today. Glacial deposits at the base of the Olympus Mons escarpment show evidence for repeated phases of activity as recently as about four million years ago. Morphological evidence is found that snow and ice deposition on the Olympus construct at elevations of more than 7,000 metres led to episodes of glacial activity at this height. Even now, water ice protected by an insulating layer of dust may be present at high altitudes on Olympus Mons." Recent and episodic volcanic and glacial activity on Mars revealed by the High Resolution Stereo Camera http://www.nature.com/nature/journal/v432/n7020/abs/nature03231.html G. Neukum1, R. Jaumann, H. Hoffmann, E. Hauber, J. W. Head, A. T. Basilevsky, B. A. Ivanov, S. C. Werner, S. van Gasselt, J. B. Murray, T. McCord & The HRSC Co-Investigator Team, Nature 432, 971–979 (December 23, 2004) | doi:10.1038/nature03231; Received September 3, 2004; Accepted November 30, 2004

[99] Phoenix Mars Lander Finds Surprises About Planet's Watery Past http://uanews.org/story/phoenix-mars-lander-finds-surprises-about-planet%E2%80%99s-watery-past University of Arizona news, By Daniel Stolte, University Communications, and NASA's Jet Propulsion Laboratory | September 9, 2010

[100] Hunting for young lava flows http://redplanet.asu.edu/?p=501 Red Planet report, Posted on June 1, 2011 by rburnham

[101] The Search For Volcanic Eruptions On Mars Reaches The Next Level http://www.astrobio.net/news-exclusive/search-volcanic-eruptions-mars-reaches-next-level/#sthash.fB6b0fFp.dpuf, Elizabeth Howell – February 12, 2015, Astrobiology Magazine (NASA)

[102] "Ice Towers and Caves of Mount Erebus" https://web.archive.org/web/20100423110657/http://erebus.nmt.edu/imagepages/icetowers/index.html, photographs from the Mount Erebus Observatory

[103] "Giant hollow towers of ice formed by steaming volcanic vents on Ross Island, Antarctica are providing clues about where to hunt for life on Mars." Martian Hot Spots http://www.astrobio.net/topic/solar-system/mars/martian-hot-spots/ Astrobiology Magazine (NASA) – August 7, 2003, Dr Nick Hoffman

[104] The Ice Towers of Mt. Erebus as analogues of biological refuges on Mars, N. Hoffman and P. R. Kyle, Sixth International Conference on Mars (2003)

[105] Grin, E. A., N. A. Cabrol, and C. P. McKay. "The hypothesis of caves on Mars revisited through MGS data; Their potential as targets for the surveyor program." http://www.lpi.usra.edu/meetings/marsmiss99/pdf/2535.pdf Workshop on Mars 2001: Integrated Science in Preparation for Sample Return and Human Exploration. Vol. 1. 1999.

[106], Lasue, Jeremie, et al. "Quantitative assessments of the martian hydrosphere." http://perso.utinam.cnrs.fr/~mousis/papier92.pdf Space Science Reviews 174.1–4 (2013): 155–212.

[107] Michalski, Joseph R., et al. "Groundwater activity on Mars and implications for a deep biosphere." http://www.nature.com/ngeo/journal/v6/n2/abs/ngeo1706.html Nature Geoscience 6.2 (2013): 133–138.

[108], Lasue, Jeremie, et al. "Quantitative assessments of the martian hydrosphere." http://perso. utinam.cnrs.fr/~mousis/papier92.pdf Space Science Reviews 174.1–4 (2013): 155–212.

[109]

[110]Perchlorates on Mars enhance the bacteriocidal effects of UV light http://www.nature.com/ articles/s41598-017-04910-3. Jennifer Wadsworth, and Charles S. Cockell. *Nature - Scientific Reports.* volume 7, Article number: 4662. 6 July 2017.

[111]

[112]http://photojournal.jpl.nasa.gov/figures/PIA16156_fig1.jpg

[113]http://mars.jpl.nasa.gov/msl/images/pia16223-stereoHattah-Mastcam-br2.jpg

[114]https://science.nasa.gov/science-news/science-at-nasa/2012/27sep_streambed/

[115]http://marsoweb.nas.nasa.gov/HiRISE/hirise_images/

[116]https://www.youtube.com/watch?v=HQKnDdB36zY

[117]https://www.youtube.com/watch?v=Jr1Xu2i-Uc0

[118]https://www.youtube.com/watch?v=WH8kHncLZwM

[119]http://www.windows.ucar.edu/tour/link=/mars/places/mars_polar_regions.html&edu=high

[120]http://spaceref.com/mars/new-view-of-mars-south-pole.html

[121]http://m.esa.int/Our_Activities/Space_Science/Mars_Express/Swirling_spirals_at_the_north_pole_of_Mars

[122]http://www.spaceref.com/news/viewpr.html?pid=26493

[123]http://www.spaceref.com/news/viewpr.html?pid=33388

[124]Phillips, R., et al. 2011. Massive CO_2 ice deposits sequestered in the south polar layered deposits of Mars. Science: 332, 638-841

[125]Bierson, C., et al. 2016. Stratigraphy and evolution of the buried CO_2 depositin the Martian south polar cap. Geophysical Research Letters: 43, 4172-4179

[126]Head, J, S. Pratt. 2001. Extensive Hesperian-aged south polar ice sheet on Mars: Evidence for massive melting and retreat, and lateral flow and pending of meltwater. J. Geophys. Res.-Planet, 106 (E6), 12275-12299.

[127]List of U.S. states and territories by area

[128]Scanlon, K., et al. 2018. Icarus: 299, 339-363.

[129]Thomas,P., M. Malin, P. James, B. Cantor, R. Williams, P. Gierasch South polar residual cap of Mars: features, stratigraphy, and changes Icarus, 174 (2 SPEC. ISS.). 2005. pp. 535–559. http://doi.org/10.1016/j.icarus.2004.07.028

[130]Thomas, P., P. James, W. Calvin, R. Haberle, M. Malin. 2009. Residual south polar cap of Mars: stratigraphy, history, and implications of recent changes Icarus: 203, 352–375 http://doi. org/10.1016/j.icarus.2009.05.014

[131]Thomas, P., W.Calvin, P. Gierasch, R. Haberle, P. James, S. Sholes. 2013. Time scales of erosion and deposition recorded in the residual south polar cap of mars Icarus: 225: 923–932 http://doi.org/10.1016/j.icarus.2012.08.038

[132]Thomas, P., W. Calvin, B. Cantor, R. Haberle, P. James, S. Lee. 2016. Mass balance of Mars' residual south polar cap from CTX images and other data Icarus: 268, 118–130 http://doi.org/ 10.1016/j.icarus.2015.12.038

[133]http://www.news.cornell.edu/releases/March00/Mars.NASA.deb.html

[134]Hartmann, W. 2003. A Traveler's Guide to Mars. Workman Publishing. NY NY.

[135]http://hirise.lpl.arizona.edu/PSP_005095_0935

[136]Buhler, Peter, Andrew Ingersoll, Bethany Ehlmann, Caleb Fassett, James Head. 2017. How the martian residual south polar cap develops quasi-circular and heart-shaped pits, troughs, and moats. Icarus: 286, 69-9.

[137]http://hirise.lpl.arizona.edu/PSP_003443_0980

[138]Hansen, C, A. McEwen and HiRISE Team. December 2007. AGU Press Conference Spring at the South Pole of Mars.

[139]http://hirise.lpl.arizona.edu/PSP_004959_0865

[140]//tools.wmflabs.org/geohack/geohack.php?pagename=Geography_of_Mars&params=0_N_180_W_globe:Mars

[141]Carr, M.H., 2006, The Surface of Mars, Cambridge, 307 p.

[142]Greeley, R. and J. Guest. 1987. Geological map of the eastern equatorial region of Mars, scale 1:15,000,000. U. S. Geol. Ser. Misc. Invest. Map I-802-B, Reston, Virginia

[143] Plaut, J. et al. 2008. Radar Evidence for Ice in lobate debris aprons in the Mid-Northern Latitudes of Mars. Lunar and Planetary Science XXXIX. 2290.pdf

[144] Watters, T. et al. 2007. Hemispheres Apart: The Crustal Dichotomy on Mars. Annual Review Earth Planet Science: 35. 621–652

[145] Irwin III, R. et al. 2004. Sedimentary resurfacing and fretted terrain development along the crustal dichotomy boundary, Aeolis Mensae, Mars.: 109. E09011

[146] Tanaka, K. et al. 2003. Resurfacing history of the northern plains of Mars based on geologic mapping of Mars Global surveyor data. Journal of Geophysical Research: 108. 8043

[147] Scott, D. and M. Carr. 1978. Geological map of Mars. U.S. Geol. Surv. Misc. Invest. Map I-803, Reston, Virginia

[148] Watters, T et al. 2007. Hemispheres Apart: The Crustal Dichotomy on Mars. Annu. Rev. Earth Planet. Sci: 35. 621–652.

[149] Jeffrey C. Andrews-Hanna, Maria T. Zuber & W. Bruce Banerdt *The Borealis basin and the origin of the martian crustal dichotomy* Nature 453, 1212–1215 (26 June 2008)

[150] Ley, Willy and von Braun, Wernher *The Exploration of Mars* New York:1956 The Viking Press Pages 70–71 Schiaparelli's original map of Mars

[151] http://www.uapress.arizona.edu/onlinebks/mars/contents.htm

[152] http://mars.google.com

[153] http://planetologia.elte.hu/ipcd/ipcd.html?cim=mars

[154] http://www.lpi.usra.edu/resources/mars_maps/MEC-1/index.html

[155] http://www.staff.science.uu.nl/~gent0113/celestia/martianglobes.htm

[156] http://mars3dmap.com

[157] http://marstrek.jpl.nasa.gov/

[158] https://archive.org/details/TheEtymologyOfMarsCraterNames_201803

[159] Head, J.W. (2007). The Geology of Mars: New Insights and Outstanding Questions in *The Geology of Mars: Evidence from Earth-Based Analogs*, Chapman, M., Ed; Cambridge University Press: Cambridge UK, p. 10.

[160] Carr, 2006, p. 44.

[161] Wilson, L. (2007). Planetary Volcanism in Encyclopedia of the Solar System, McFadden, L.-A. et al., Eds., Academic Press: San Diego, CA, p. 829.

[162] Wilson, M. (1995) *Igneous Petrogenesis;* Chapman Hall: London, 416 pp.

[163] Carr, M.H. (2007) Mars: Surface and Interior in Encyclopedia of the Solar System, McFadden, L.-A. et al., Eds., Academic Press: San Diego, CA, p. 321.

[164] Carr, M.H (2007). Mars: Surface and Interior in *Encyclopedia of the Solar System*, 2nd ed., McFadden, L.-A. et al. Eds. Elsevier: San Diego, CA, p.319

[165] Barlow, N.G. (2008). *Mars: An Introduction to Its Interior, Surface, and Atmosphere;* Cambridge University Press: Cambridge, UK, p. 129.

[166] Head, J.W. (2007). The Geology of Mars: New Insights and Outstanding Questions in *The Geology of Mars: Evidence from Earth-Based Analogs*, Chapman, M., Ed; Cambridge University Press: Cambridge UK, p. 11.

[167] //tools.wmflabs.org/geohack/geohack.php?pagename=Volcanology_of_Mars&params=21. 965_N_197.807_E_globe:Mars

[168] (via archive.org)

[169] Fagents, F.A.; Thordarson, T. (2007). Rootless Volcanic Cones in Iceland and on Mars, in *The Geology of Mars: Evidence from Earth-Based Analogs*, Chapman, M., Ed; Cambridge University Press: Cambridge UK, pp. 151–177.

[170] Chapman, M.G.; Smellie, J.L. (2007). Mars Interior Layered Deposits and Terrestrial Sub-Ice Volcanoes Compared: Observations and Interpretations of Similar Geomorphic Characteristics, in *The Geology of Mars: Evidence from Earth-Based Analogs*, Chapman, M., Ed; Cambridge University Press: Cambridge UK, pp. 178–207.

[171] http://news.stanford.edu/pr/93/93/206Arc3014.html

[172] http://www.chm.bris.ac.uk/webprojects1997/JoeA/welcome.htm

[173] http://www.lukew.com/marsgeo/volcanic.html

[174] Interplanetary Whodunit – Methane on Mars http://www.astrobio.net/news/article1651.html, David Tenenbaum, *Astrobiology Magazine*, NASA, July 20, 2005. (Note: part one of a four-part series.)

[175] https://www.youtube.com/watch?v=fIKxdRFx2Wo

[176] New views of the Martian ionosphere http://sci.esa.int/mars-express/51056-new-views-of-the-martian-ionosphere/. ESA (15 November 2012)

[177] A seasonal ozone layer over the Martian south pole http://sci.esa.int/mars-express/52881-a-seasonal-ozone-layer-over-the-martian-south-pole/. ESA (29 September 2013)

[178] Mars Thermospheric Temperature Sensitivity to Solar EUV Forcing from the MAVEN EUV Monitor http://adsabs.harvard.edu/abs/2017DPS....4951004T. Thiemann, Ed; Eparvier, Francis; Andersson, Laila; Pilinski, Marcin; Chamberlin, Phillip; Fowler, Christopher; MAVEN Extreme Ultraviolet Monitor Team, MAVEN Langmuir Probe and Waves Team. *American Astronomical Society*, DPS meeting #49, id.510.04. Bibliographic Code: 2017DPS....4951004T

[179] ExoMars Trace Gas Orbiter provides atmospheric data during Aerobraking into its final orbit http://adsabs.harvard.edu/abs/2017DPS....4941801S. Svedhem, Hakan; Vago, Jorge L.; Bruinsma, Sean; Müller-Wodarg, Ingo; ExoMars 2016 Team. *American Astronomical Society*, DPS meeting #49, id.418.01. Bibliographic Code: 2017DPS....4941801S. Published October 2017.

[180] Janssen, Huggins, Secchi, Vogel, and Maunder

[181] What happened to early Mars' atmosphere? New study eliminates one theory https://www.sciencedaily.com/releases/2015/09/150903121019.htm. *Science Daily* (September 3, 2015)

[182] Clouds http://marsrovers.jpl.nasa.gov/gallery/press/opportunity/20041213a.html – December 13, 2004 NASA Press release. URL accessed March 17, 2006.

[183] Mars methane rises and falls with the seasons http://science.sciencemag.org/content/359/6371/16.summary. Eric Hand, *Science*, 5 January 2018: Vol. 359, Issue 6371, pp. 16-17

[184] Team Finds New Hope for Life in Martian Crust http://astrobiology.com/2015/06/team-finds-new-hope-for-life-in-martian-crust.html. *Astrobiology.com*. Western University. June 16, 2014.

[185] On Mars, atmospheric methane—a sign of life on Earth—changes mysteriously with the seasons http://www.sciencemag.org/news/2018/01/mars-atmospheric-methane-sign-life-earth-changes-mysteriously-seasons. Eric Hand, *Science Magazine*. 3 January 2018.

[186] India's Mars Orbiter Mission Has a Methane Problem https://www.seeker.com/india-mars-orbiter-mission-methane-detector-flaw-red-planet-2133861312.html. Irene Klotz, *Seeker*, 7 December 2016.

[187] Global Albedo Map of Mars https://www.isro.gov.in/pslv-c25-mars-orbiter-mission/global-albedo-map-of-mars. ISRO. 2017-07-14

[188] Inside ExoMars – Quarterly Newsletter http://exploration.esa.int/science-e/www/object/index.cfm?fobjectid=50371 (May 2012)

[189] McAdam, A. C.; Franz, H.; Archer, P. D.; Freissinet, C.; Sutter, B.; Glavin, D. P.; Eigenbrode, J. L.; Bower, H.; Stern, J.; Mahaffy, P. R.; Morris, R. V.; Ming, D. W.; Rampe, E.; Brunner, A. E.; Steele, A.; Navarro-González, R.; Bish, D. L.; Blake, D.; Wray, J.; Grotzinger, J.; MSL Science Team (2013). "Insights into the Sulfur Mineralogy of Martian Soil at Rocknest, Gale Crater, Enabled by Evolved Gas Analyses" http://adsabs.harvard.edu/abs/2013LPI....44.1751M. 44th Lunar and Planetary Science Conference, held March 18–22, 2013 in The Woodlands, Texas. LPI Contribution No. 1719, p. 1751

[190] NASA GISS: Science Briefs: Reaction of Ozone and Climate to Increasing Stratospheric Water Vapor https://www.giss.nasa.gov/research/briefs/shindell_05/

[191] Flying Observatory Detects Atomic Oxygen in Martian Atmosphere | NASA https://www.nasa.gov/feature/ames/sofia/flying-observatory-detects-atomic-oxygen-in-martian-atmosphere

[192] Planetary Science http://astronomynotes.com/solarsys/s10.htm. Astronomynotes.com. Retrieved 19 August 2013.

[193] http://www.space.com/scienceastronomy/060828_mars_clouds.html

[194] http://hypertextbook.com/facts/2000/LaurenMikulski.shtml

[195] http://www-k12.atmos.washington.edu/k12/resources/mars_data-information/pressure_overview.html

[196] https://web.archive.org/web/20091207081507/http://www.scilogs.eu/en/blog/go-for-launch/2009-12-04/methane_mars_atmosphere_frascati

[197] http://www.scilogs.eu/en/blog/go-for-launch/2009-12-04/methane_mars_atmosphere_frascati

[198] http://sci.esa.int/mars-express/52881-a-seasonal-ozone-layer-over-the-martian-south-pole/

[199] Exploring Mars in the 1700s https://web.archive.org/web/20010220024335/http://www.exploringmars.com/history/1700.html

[200] Exploring Mars in the 1800s http://www.exploringmars.com/history/1800.html

[201] Hartmann, W. 2003. A Traveler's Guide to Mars. Workman Publishing. NY NY.

[202] James E. Tillman Mars – Temperature Overview http://www-k12.atmos.washington.edu/k12/resources/mars_data-information/temperature_overview.html

[203] Extreme Planet Takes its Toll http://marsrover.nasa.gov/spotlight/20070612.html *Jet Propulsion Laboratory Featured Story, June 12, 2007.*

[204] William Sheehan, *The Planet Mars: A History of Observation and Discovery,* Chapter 13 (available on the web http://www.uapress.arizona.edu/onlinebks/mars/chap13.htm)

[205] https://www.sciencedaily.com/releases/2015/09/150903121019.htm

[206] https://www.sciencedaily.com/releases/2015/11/151124170249.htm

[207] http://www.nasa.gov/mission_pages/MRO/multimedia/pia15116.html

[208] https://www.youtube.com/watch?v=0t0LWFHB8Qo

[209] Antares project "Mars Small-Scale Weather" (MSW) http//websrv2.tekes.fi

[210] Mechanism For Mars Dust Storms 1973, Pollack, Leovy, Zurek. http://journals.ametsoc.org/doi/pdf/10.1175/1520-0469%281973%29030%3C0749%3AMFMDS%3E2.0.CO%3B2

[211] Rapidly intensifying, possibly planet-wide dust storm affecting Mars https://watchers.news/2018/06/13/dust-storm-mars/, 13 June 2018.

[212] https://www.sciencenews.org/article/mars-dust-storms-water?mode=topic&context=36

[213] N. Heavens et al. Hydrogen escape from Mars enhanced by deep convection in dust storms. Nature Astronomy. Published online January 22, 2018. doi: 10.1038/s41550-017-0353-4.

[214] https://www.jpl.nasa.gov/news/news.php?release=2018-012&rn=news.xml&rst=7041

[215] G. Landis, et al., "Dust and Sand Deposition on the MER Solar Arrays as Viewed by the Microscopic Imager," 37th Lunar and Planetary Science Conference, Houston TX, March 13–17, 2006. pdf file http://www.lpi.usra.edu/meetings/lpsc2006/pdf/1932.pdf (also summarized in NASA Glenn Research and Technology 2006 http://www.grc.nasa.gov/WWW/RT/2006/RP/RPV-landis2.html report)

[216] Mumma, M. J.; Novak, R. E.; DiSanti, M. A.; Bonev, B. P., "A Sensitive Search for Methane on Mars" http://adsabs.harvard.edu/cgi-bin/nph-bib_query?bibcode=2003DPS....35. 1418M&db_key=AST&data_type=HTML&format= (abstract only). American Astronomical Society, DPS meeting #35, #14.18.

[217] Madeleine, J. et al. 2007. Mars: A proposed climatic scenario for northern mid-latitude glaciation. Lunar Planet. Sci. 38. Abstract 1778.

[218] Madeleine, J. et al. 2009. Amazonian northern mid-latitude glaciation on Mars: A proposed climate scenario. Icarus: 203. 300–405.

[219] Mischna, M. et al. 2003. On the orbital forcing of martian water and CO2 cycles: A general circulation model study with simplified volatile schemes. J. Geophys. Res. 108. (E6). 5062.

[220] Hauber, E., D. Reiss, M. Ulrich, F. Preusker, F. Trauthan, M. Zanetti, H. Hiesinger, R. Jaumann, L. Johansson, A. Johnsson, S. Van Gaselt, M. Olvmo. 2011. Landscape evolution in Martian mid-latitude regions: insights from analogous periglacial landforms in Svalbard. In: Balme, M., A. Bargery, C. Gallagher, S. Guta (eds). Martian Geomorphology. Geological Society, London. Special Publications: 356. 111–131

[221] Madeleine, J., F. Forget, J. Head, B. Levrard, F. Montmessin. 2007. Exploring the northern mid-latitude glaciation with a general circulation model. In: Seventh International Conference on Mars. Abstract 3096.

[222] Malin, M. et al. 2010. An overview of the 1985–2006 Mars Orbiter Camera science investigation. MARS INFORMATICS. http://marsjournal.org

[223] Mars Emerging from Ice Age, Data Suggest http://www.space.com/scienceastronomy/mars_ice-age_031208.html

[224] //doi.org/10.1038/35084184

[225] //www.ncbi.nlm.nih.gov/pubmed/11449285

[226] http://www.physorg.com/news4106.html

[227] https://www.newscientist.com/article.ns?id=dn1660

[228] http://www.msss.com/mars_images/moc/2005/07/13/index.html

[229] http://www.realclimate.org/index.php?p=192

[230] http://mars.jpl.nasa.gov/mgs/msss/camera/images/CO2_Science_rel/index.html

[231] http://news.nationalgeographic.com/news/2007/02/070228-mars-warming.html

[232] https://web.archive.org/web/20120823212559/http://cab.inta-csic.es/rems/marsweather.html

[233] http://cab.inta-csic.es/rems

[234] http//hrscview.fu-berlin.de

[235] Jean Meeus, *Astronomical Formulæ for Calculators*. (Richmond, VA: Willmann-Bell, 1988) 99. Elements by F. E. Ross

[236] In ephemeris days of 86,400 seconds. The sidereal and anomalistic years are 686.980 days and 686.996 days long, respectively. (About a 20 minute difference). The sidereal year is the time taken to revolve around the Sun relative to a fixed reference frame. More precisely, the sidereal year is one way to express the rate of change of the mean longitude at one instant, with respect to a fixed equinox. The calculation shows how long it would take for the longitude to change 360 degrees at the given rate. The anomalistic year is the time span between successive perihelion or aphelion passages. This may be calculated in the same manner as the sidereal year, but the mean anomaly is used.

[237] Jean Meeus, *Astronomical Algorithms* (Richmond, VA: Willmann-Bell, 1998) 238. The formula by Ramanujan is accurate enough.

[238] The averages between 1850 and 2150. The extreme values in that range are 1.66635 and 1.38097 AU

[239] Mars distance and eccentricity, using Solex. By its creator, Aldo Vitagliano

[240] The synodic period may be calculated as $1/(1/p-1/q)$, where p and q are the smaller and larger sidereal periods.

[241] The synodic period of Mars is 92.9 days longer than its sidereal period of 687.0 days. It has then moved forward 92.9/687.0 times 360, or 48.7 degrees. After seven oppositions it has moved forward 341 degrees, and after eight it has advanced 390 degrees; in the first case its longitude is different from one revolution by 19°, and by 30° in the second. So the situations will then be similar. Similar calculations show that the longitude changes only 2° after 37 oppositions.

[242] Mars page, Vitagliano

[243] William Sheehan, *The Planet Mars: A History of Observation and Discovery* (Tucson, AZ: The University of Arizona Press, 1996) Chapter 1

[244] Meeus (1998) pp 269-270

[245] see, for example, Simon et al. (1994) p 681

[246] 2012 version of the *Explanatory Supplement*

[247] As noted in a 2008 JPL Memorandum regarding DE 421, "The error in the Earth and Mars orbits in DE 405 is now known to be about 2 km, which was good accuracy in 1997 but much worse than the current sub-kilometer accuracy." p. 1

[248] "asteroid." Encyclopædia Britannica. Encyclopædia Britannica Online. Encyclopædia Britannica Inc., 2014. Web. 19 Aug. 2014. http://www.britannica.com/EBchecked/topic/39730/asteroid

[249] "The uncertainty in the Mars orbit for a one-year prediction is about 300 m, as required for the Mars Science Laboratory mission, but grows rapidly for times before and after the spacecraft observation time span due to the influence of asteroids with orbits near that of Mars. The predicted orbit and uncertainty depend greatly on the asteroid model used. "

[250] Average distance over times. Constant term in VSOP87. It corresponds to the average taken of many short, equal time intervals.

[251] //en.wikipedia.org/w/index.php?title=Template:Life_in_the_Universe&action=edit

[252] Rover could discover life on Mars – here's what it would take to prove it https://phys.org/news/2018-01-rover-life-mars-proveit.html. Claire Cousins, *PhysOrg*. 5 January 2018.

[253]

[254] Earth analogues for past and future life on Mars: isolation of perchlorate resistant halophiles from Big Soda Lake https//www.cambridge.org. Toshitaka Matsubara, Kosuke Fujishima, Chad W., Satoshi Nakamura. International Journal of Astrobiology. 28 November 2016. Volume 16, Issue 3, July 2017, pp. 218-228. DOI: https://doi.org/10.1017/S1473550416000458

[255] Bacterial growth tolerance to concentrations of chlorate and perchlorate salts relevant to Mars https//www.cambridge.org. Amer F. Al Soudi, Omar Farhat, Fei Chen, Benton C. Clark.

International Journal of Astrobiology. 22 November 2016. DOI: https://doi.org/10.1017/
S1473550416000434

[256] "Silicates Eroded under Simulated Martian Conditions Effectively Kill Bacteria—A Challenge for Life on Mars". Ebbe N. Bak1, Michael G. Larsen1, Ralf Moeller, Silas B. Nissen, Lasse R. Jensen, Per Nørnberg, Svend J. K. Jensen and Kai Finster1. *Front. Microbiol.*, 12 September 2017. https://doi.org/10.3389/fmicb.2017.01709

[257] Why Life on Mars May Be Impossible http://time.com/4845251/mars-life-toxins-microbes/. Jeffrey Kluger. *Time* - Science; Jul 6, 2017.

[258] Mars Soil May Be Toxic to Microbes https://www.space.com/37402-mars-life-soil-toxic-perchlorates-radiation.html. Mike Wall. Space.com. July 6, 2017

[259] Mars soil is likely toxic to cells — does this mean humans won't be able to grow vegetables there? http://www.abc.net.au/news/2017-07-07/mars-toxic-soil-could-make-growing-vegies-harder/8687626. David Coady. *The World Today*. 7 Jul 2017

[260] Ionic Strength Is a Barrier to the Habitability of Mars http://online.liebertpub.com/doi/abs/10.1089/ast.2015.1432. Fox-Powell Mark G., Hallsworth John E., Cousins Claire R., and Cockell Charles S. *Astrobiology*. June 2016, 16(6): 427-442. https://doi.org/10.1089/ast.2015.1432

[261] Nitrogen on Mars: Insights from Curiosity https://ntrs.nasa.gov/archive/nasa/casi.ntrs.nasa.gov/20170002371.pdf (PDF). J. C. Stern, B. Sutter, W. A. Jackson, Rafael Navarro-González, Christopher P. McKay, Douglas W. Ming, P. Douglas Archer, D. P. Glavin1, A. G. Fairen, and Paul R. Mahaffy. Lunar and Planetary Science XLVIII (2017).

[262] Readily available phosphate from minerals in early aqueous environments on Mars http://www.nature.com/ngeo/journal/v6/n10/full/ngeo1923.html?foxtrotcallback=true. C. T. Adcock, E. M. Hausrath & P. M. Forster. *Nature Geoscience*. No. 6, 824–827 (2013) doi:10.1038/ngeo1923

[263] Supplementary Materials for: Radar evidence of subglacial liquid water on Mars http://science.sciencemag.org/content/sci/suppl/2018/07/24/science.aar7268.DC1/aar7268_Orosei_SM.pdf. (PDF). R. Orosei, S. E. Lauro, E. Pettinelli, A. Cicchetti, M. Coradini, B. Cosciotti, F. Di Paolo, E. Flamini, E. Mattei, M. Pajola, F. Soldovieri, M. Cartacci, F. Cassenti, A. Frigeri, S. Giuppi, R. Martufi, A. Masdea, G. Mitri, C. Nenna, R. Noschese, M. Restano, R. Seu. *Science*. 25 July 2018

[264] Compilation of scientific research references on the Nakhla meteorite:

[265] (Audio interview, MP3 6 min.)

[266] Is Searching for Martian Life a Priority for the Mars Community? http://online.liebertpub.com/doi/full/10.1089/ast.2017.1772 Fairén Alberto G., Parro Victor, Schulze-Makuch Dirk, and Whyte Lyle. *Astrobiology*. February 2018, 18(2): 101-107.

[267] Kounaves, S. P. et al., Evidence of martian perchlorate, chlorate, and nitrate in Mars meteorite EETA79001: implications for oxidants and organics, Icarus, 2014, 229, 206-213, ,

[268] http://www.space.com/scienceastronomy/060420_mars_water.html

[269] http://marsprogram.jpl.nasa.gov/overview/

[270] http://news.bbc.co.uk/1/hi/sci/tech/3560867.stm

[271] http://news.bbc.co.uk/1/hi/sci/tech/3896335.stm

[272] http://www.space.com/scienceastronomy/mars_microorganisms_040803.html

[273] http://www.space.com/news/science_top10_041216.html

[274] http://www2.jpl.nasa.gov/snc/nasa1.html

[275] http://www.sciam.com/article.cfm?chanID=sa006&articleID=00073A97-5745-1359-94FF83414B7F0000

[276] http://www.monochrom.at/dark-dune-spots/

[277] 3D Printing With Ice on Mars. http://www.marsicehouse.com/3d-printing-with-ice/ Mars Ice House. 2015. Accessed: 25 August 2018.

[278] Everything SpaceX revealed about its updated plan to reach Mars by 2022 https://techcrunch.com/2017/09/28/everything-spacex-revealed-about-its-updated-plan-to-reach-mars-by-2022/. Darrell Etherington, *TechCrunch*. 29 September 2018

[279] //en.wikipedia.org/w/index.php?title=Template:Space_colonization&action=edit

[280] Extract of page 600 https://books.google.com/books?id=BnPE37Ms5awC&pg=PA600

[281] http://aeon.co/magazine/technology/the-elon-musk-interview-on-mars/

[282] Paper presented at *Mars 2001: Integrated Science in Preparation for Sample Return and Human Exploration*, Lunar and Planetary Institute, Oct. 2–4 1999, Houston, TX.

[283] Presented at *Concepts and Approaches for Mars Exploration*, July 18–20, 2000 Houston, Texas.

[284] Lovelock, James and Allaby, Michael, "*The Greening of Mars*" 1984

[285] Queens University Belfast scientist helps NASA Mars project https://www.bbc.co.uk/news/uk-northern-ireland-27526981 "No-one has yet proved that there is deep groundwater on Mars, but it is plausible as there is certainly surface ice and atmospheric water vapour, so we wouldn't want to contaminate it and make it unusable by the introduction of micro-organisms."

[286] COSPAR PLANETARY PROTECTION POLICY https://science.nasa.gov/media/medialibrary/2012/05/04/COSPAR_Planetary_Protection_Policy_v3-24-11.pdf (20 October 2002; As Amended to 24 March 2011)

[287] When Biospheres Collide – a history of NASA's Planetary Protection Programs http://www.nasa.gov/connect/ebooks/when_biospheres_collide_detail.html#.U_uVh_mwJcQ, Michael Meltzer, May 31, 2012, see Chapter 7, Return to Mars – final section: "Should we do away with human missions to sensitive targets"

[288] Johnson, James E. "Planetary Protection Knowledge Gaps for Human Extraterrestrial Missions: Goals and Scope." (2015) http://www.hou.usra.edu/meetings/ppw2015/pdf/1010.pdf

[289] Safe on Mars page 37 http://www.nap.edu/openbook.php?record_id=10360&page=37 "Martian biological contamination may occur if astronauts breathe contaminated dust or if they contact material that is introduced into their habitat. If an astronaut becomes contaminated or infected, it is conceivable that he or she could transmit Martian biological entities or even disease to fellow astronauts, or introduce such entities into the biosphere upon returning to Earth. A contaminated vehicle or item of equipment returned to Earth could also be a source of contamination."

[290] Minkel, JR. "Sex and Pregnancy on Mars: A Risky Proposition." *Space.com*. Space.com, 11 Feb. 2011. Web. 09 Dec. 2016.

[291] Schuster, Haley, and Steven L. Peck. "Mars Ain't the Kind of Place to Raise Your Kid: Ethical Implications of Pregnancy on Missions to Colonize Other Planets." *Life Sciences, Society and Policy* 12.1 (2016): 1–8. Web. 9 Dec. 2016.

[292] Szocik, Konrad, Kateryna Lysenko-Ryba, Sylwia Banaś, and Sylwia Mazur. "Political and Legal Challenges in a Mars Colony." *Space Policy* (2016): n. pag. Web. 24 Oct. 2016.

[293] *Commercial Space Exploration: [ethics, Policy and Governance]*. , 2015. Print.

[294] https://www.space.com/35394-president-obama-spaceflight-exploration-legacy.html

[295] https://www.nasa.gov/press-release/new-space-policy-directive-calls-for-human-expansion-across-solar-system

[296] http://www.sciencemag.org/news/2018/03/updated-us-spending-deal-contains-largest-research-spending-increase-decade

[297] http//docs.google.com

[298] http://www.praxis-publishing.co.uk/9780387981901.htm

[299] https://www.springer.com/astronomy/space+exploration/book/978-0-387-98190-1

[300] https://www.amazon.com/Martian-Outpost-Challenges-Establishing-Exploration/dp/038798190X

[301] http://news.idg.no/cw/art.cfm?id=CB21C80A-17A4-0F78-3174039834A1E181

[302] http://www.marssociety.org/

[303] http://mmp.planetary.org/index.html

[304] http://www.4FrontiersCorp.com/

[305] http://www.marshome.org/

[306] http://ngm.nationalgeographic.com/big-idea/07/mars

[307] Galileo, Kepler, & Two Anagrams: Two Wrong Solutions Turn Into Two Correct Solutions http://judgestarling.tumblr.com/post/62652246148/galileo-kepler-two-anagrams-two-wrong

[308] MathPages – Galileo's Anagrams and the Moons of Mars http://www.mathpages.com/home/kmath151/kmath151.htm

[309] V. G. Perminov – **The Difficult Road to Mars** (1999) – NASA http://klabs.org/richcontent/Reports/mars/difficult_road_to_mars.pdf

[310] Roscoe Lamont, 'The Moons of Mars' in *Popular Astronomy Vol. 33* (1925), pp496 - 8 http://adsabs.harvard.edu/full/1925PA.....33..496L

[311] William Sheehan, *The Planet Mars: A History of Observation and Discovery* http://www. *uapress.arizona.edu/onlinebks/mars/chap14.htm*

[312] Voltaire explained that since Mars is further from the Sun than Earth is, it could not make do with less than two moons. (Patrick Moore, 2000, *The Wandering Astronomer*)

[313] Gazetteer of Planetary Nomenclature http://planetarynames.wr.usgs.gov/Page/PHOBOS/target USGS Astrogeology Research Program, Phobos

[314] Morley, T. A.; *A Catalogue of Ground-Based Astrometric Observations of the Martian Satellites, 1877-1982* http://adsabs.harvard.edu//full/seri/A+AS./0077//0000220.000.html, Astronomy and Astrophysics Supplement Series, Vol. 77, No. 2 (February 1989), pp. 209–226 (Table II, p. 220: first observation of Phobos on 1877-08-18.38498)

[315] Naval Observatory 26-inch Refractor http://amazing-space.stsci.edu/resources/explorations/groundup/lesson/scopes/naval/index.php

[316] The 26-inch "Great Equatorial" Refractor http://www.usno.navy.mil/USNO/about-us/the-26-inch-refractor

[317] *Jefferson City Post-Tribune* 4 May 1959

[318] Astron. J., 128, 2542–2546 (2004) https://arxiv.org/abs/astro-ph/0409522

[319] *Moon Shadows*: "Somewhere near the martian equator, Phobos eclipses the sun nearly every day." http://www.astrobio.net/news/index.php?name=News&file=article&sid=775

[320] New Map Provides More Evidence Mars Once Like Earth: "... the new map shows evidence of features, transform faults, that are a "tell-tale" of plate tectonics on Earth." http://www.nasa.gov/centers/goddard/news/topstory/2005/mgs_plates.html

[321] Sahife 6: "Deimos orbits far enough away from Mars that it is being slowly pushed farther and farther away from the planet." http://www.edb.utexas.edu/missiontomars/pdf/MarsSSBody.pdf

[322] Burns, J. A. "Contradictory Clues as to the Origin of the Martian Moons," in *Mars*, H. H. Kieffer *et al.*, eds., U. Arizona Press, Tucson, 1992

[323] Landis, G. A. "Origin of Martian Moons from Binary Asteroid Dissociation," American Association for the Advancement of Science Annual Meeting; Boston, MA, 2001; abstract https://ntrs.nasa.gov/search.jsp?R=946501&id=8&qs=No%3D70&N%3D4294808501.

[324] Craddock, R. A.; (1994); *The Origin of Phobos and Deimos*, Abstracts of the 25th Annual Lunar and Planetary Science Conference, held in Houston, TX, 14–18 March 1994, p. 293

[325] 1.Pascal Rosenblatt, Sébastien Charnoz, Kevin M. Dunseath, Mariko Terao-Dunseath, Antony Trinh, Ryuki Hyodo, Hidenori Genda, Stéven Toupin. Accretion of Phobos and Deimos in an extended accretion debris disc stirred by transient moons. Nature Geoscience, 2016; DOI: 10.1038/ngeo2742

[326] CNRS. "A giant impact: Solving the mystery of how Mars' moons formed." ScienceDaily. ScienceDaily, 4 July 2016. <www.sciencedaily.com/releases/2016/07/160704144236.htm>.

[327] OSIRIS-REx II to Mars: Mars Sample Return from Phobos and Deimos http://adsabs.harvard.edu/abs/2012LPICo1679.4017E. Elifritz, T. L. (2012)

[328] http://www.lpi.usra.edu/meetings/phobosdeimos2007/phobosdeimos2007.authorindex.shtml

[329] http://www.lpi.usra.edu/meetings/phobosdeimos2007/pdf/7044.pdf

[330] http://planetarynames.wr.usgs.gov/jsp/SystemSearch2.jsp?System=Mars

[331] David S. F. Portree, *Humans to Mars: Fifty Years of Mission Planning, 1950–2000*, NASA Monographs in Aerospace History Series, Number 21, February 2001. Available as NASA SP-2001-4521 https://history.nasa.gov/monograph21/humans_to_Mars.htm.

[332]

[333] Mars 2 Lander – NASA http://nssdc.gsfc.nasa.gov/nmc/spacecraftDisplay.do?id=1971-045D. Nssdc.gsfc.nasa.gov. Retrieved on 2012-05-10.

[334] Mars 6 – NASA http://nssdc.gsfc.nasa.gov/nmc/spacecraftDisplay.do?id=1973-052A. Nssdc.gsfc.nasa.gov. Retrieved on 2012-05-10.

[335] Space probe performs Mars fly-by http://news.bbc.co.uk/2/hi/6394141.stm. BBC News (2007-02-25). Retrieved on 2012-08-14.

[336] Raeburn, P. (1998) "Uncovering the Secrets of the Red Planet Mars". National Geographic Society. Washington D.C.

[337] December 4, 1996 – First successful Mars Rover – Sojourner – was launched http://todayinspacehistory.wordpress.com/2007/12/04/december-4-1996-first-successful-mars-

rover-sojourner-was-launched-toward-mars/. Todayinspacehistory.wordpress.com (2007-12-04). Retrieved on 2012-08-14.

[338] "Russia's failed Phobos-Grunt space probe heads to Earth" https://www.bbc.co.uk/news/science-environment-16491457, BBC News (2012-01-14).

[339] "Phobos-Grunt: Failed Russian Mars Probe Falls to Earth" http://abcnews.go.com/Technology/phobos-grunt-failed-russian-mars-probe-falls-earth/story?id=15366151. ABC News, January 15, 2012.

[340] "Phobos-Grunt: Failed probe likely to return late Sunday" https://www.bbc.co.uk/news/science-environment-16491457. BBC News (2012-01-15).

[341] Morris Jones (2011-11-17). "Yinghuo Was Worth It" http://www.spacedaily.com/reports/Yinghuo_Was_Worth_It_999.html. Space Daily. Retrieved 19 November 2011.

[342] NASA will send robot drill to Mars in 2016 https//www.washingtonpost.com, Washington Post, By Brian Vastag, Monday, August 20

[343] Concepts and Approaches for Mars Exploration – LPI – USRA (2012) http://www.lpi.usra.edu/meetings/marsconcepts2012/. Lpi.usra.edu. Retrieved on 2012-05-10.

[344] NASA Announces Mars 2020 Rover Payload to Explore the Red Planet as Never Before http://www.nasa.gov/press/2014/july/nasa-announces-mars-2020-rover-payload-to-explore-the-red-planet-as-never-before/. July 31, 2014.

[345] UAE to explore Mars' atmosphere with probe named 'Hope' http://apnews.excite.com/article/20150506/ml--emirates-mars_mission-a48c414829.html. Adam Schreck, *Excite News* 0 May 2015.

[346] Tharoor, Ishaan 2014/16/07. "U.A.E. plans Arab world's first mission to Mars" https://www.washingtonpost.com/blogs/worldviews/wp/2014/07/16/u-a-e-plans-arab-worlds-first-mission-to-mars/.

[347] Planetary Science Decadal Survey Mission & Technology Studies http://sites.nationalacademies.org/SSB/SSB_059331. Sites.nationalacademies.org. Retrieved on 2012-05-10.

[348] Oh, David Y. *et al.* (2009) Single Launch Architecture for Potential Mars Sample Return Mission Using Electric Propulsion https://archive.org/details/singlelauncharchitecture. JPL/Caltech.

[349] Jones, S.M. *et al.* Mars Sample Return at 6 Kilometers per Second: Practical, Low Cost, Low Risk, and Ready http://www.lpi.usra.edu/meetings/msr2008/pdf/4020.pdf. Ground Truth from Mars: Science Payoff from a Sample Return Mission, held April 21–23, 2008, in Albuquerque, New Mexico. LPI Contribution No. 1401, pp. 39–40.

[350] Decadal Survey Document Listing: White Papers http//solarsystem.nasa.gov (NASA)

[351] Balloons – NASA http://mars.jpl.nasa.gov/programmissions/missions/missiontypes/balloons/. Mars.jpl.nasa.gov. Retrieved on 2012-05-10.

[352] Oliver Morton – "'MarsAir'" (January 2000) – Air & Space magazine https://archive.is/20120717061237/http://www.airspacemag.com/space-exploration/mars.html?c=y&page=1. Airspacemag.com. Retrieved on 2012-08-14.

[353] Lupisella, ML. "Human Mars Mission Contamination Issues." http://www.lpi.usra.edu/publications/reports/CB-1089/lupisella.pdf *NASA*.

[354] K.Klaus, M. L. Raftery and K. E. Post (2014) "An Affordable Mars Mission Design" http://www.hou.usra.edu/meetings/lpsc2014/eposter/2258 (Houston, Texas: Boeing Co.)

[355] M. L. Raftery (May 14, 2014) "Mission to Mars in Six (not so easy) Pieces" https://www.dropbox.com/s/0gagd1dbyptnvwg/Raftery_05-14-14.pdf (Houston, Texas: Boeing Co.)

[356] NASA (December 2, 2014) "NASA's Journey to Mars News Briefing" https://www.youtube.com/watch?v=zBoj-1m-qLU *NASA TV*

[357] The "Mars Curse": Why Have So Many Missions Failed? http://www.universetoday.com/2008/03/22/the-mars-curse-why-have-so-many-missions-failed/. Universetoday.com (2008-03-22). Retrieved on 2012-08-14.

[358] "The Depths of Space: The Story of the Pioneer Planetary Probes (2004)" http://www.nap.edu/books/0309090504/html/41.html from The National Academies Press http://www.nap.edu/. URL accessed April 7, 2006.

[359] "Uncovering the Secrets of Mars" http://www.time.com/time/archive/preview/0,10987, 986681,00.html (first paragraph only). *Time* July 14, 1997 Vol. 150 No. 2. URL accessed April 7, 2006.

[360] Matthews, John & Caitlin. "The Element Encyclopedia of Magical Creatures",Barnes & Noble Publishing, 2005.

[361] Mars Exploration Program: Historical Log http://mars.jpl.nasa.gov/programmissions/missions/log/. Mars.jpl.nasa.gov. Retrieved on 2012-08-14.

[362] http://www.jpl.nasa.gov/news/press_kits/odysseyarrival.pdf

[363] http://library.cqpress.com/cqpac/hsdcp03p-229-9844-633819

[364] Mars InSight Launch Press Kit https://www.jpl.nasa.gov/news/press_kits/insight/download/mars_insight_launch_presskit.pdf. (PDF). NASA's JPL.

[365] China says it plans to land rover on Mars in 2020 http://www.kpax.com/story/30431280/china-says-it-plans-to-land-rover-on-mars-in-2020. Shen Lu. CNN, 3 November 2015

[366] China's 2020 Mars probe unveiled https://gbtimes.com/chinas-2020-mars-probe-unveiled. *GB-Times*, 3 November 2015.

[367] http://www.thothx.com/

[368] T. Satoh – MELOS – JAXA http://mepag.jpl.nasa.gov/meeting/feb-12/day_1/09_MELOS_at_MEPAG_feb2012.pdf source http://mepag.jpl.nasa.gov/meeting/feb-12/index.html

[369] Anderson, D. *et al.* The Biological Oxidant and Life Detection (BOLD) Mission: An outline for a new mission to Mars http://autonomy.caltech.edu/publications/journals/BOLD_SPIE_submitted_final.pdf. (PDF) . Retrieved on 2012-08-14.

[370] Korean Mars Mission Design Using KSLV-III https://www.researchgate.net/publication/252577175_Korean_Mars_Mission_Design_Using_KSLV-III. Young-Joo Song, Sung-Moon Yoo, Eun-Seo Park, Byung-Kyo Kim. January 2006.

[371] Design Study of a Korean Mars Mission http://www.koreascience.or.kr/article/ArticleFullRecord.jsp?cn=HGJHC0_2004_v5n2_54. International Journal of Aeronautical and Space Sciences Volume 5, Issue 2; 2004, pp.54–61; Publisher : The Korean Society for Aeronautical & Space Sciences; DOI : 10.5139/IJASS.2004.5.2.054

[372] Советский грунт с Марса http://www.novosti-kosmonavtiki.ru/content/numbers/213/50.shtml . www.novosti-kosmonavtiki.ru

[373] C. Tarrieu, "Status of the Mars 96 Aerostat Development", Paper IAF-93-Q.3.399, 44th Congress of the International Astronautical Federation, 1993.

[374] P.B. de Selding, "Planned French Balloon May Be Dropped", Space News, 17–23 April 1995, pp. 1, 20

[375] Oliver Morton in *To Mars, En Masse*, pp. 1103–04, Science (Magazine) vol. 283, 19 February 1999, ISSN 0036-8075

[376] MIT Mars Airplane Project https://web.archive.org/web/20110410092809/http://www.marsnews.com/missions/airplane/. Marsnews.com. Retrieved on 2012-08-14.

[377] Exploring Mars: Blowing in the Wind? http://www.jpl.nasa.gov/news/features.cfm?feature=486 Jpl.nasa.gov (2001-08-10). Retrieved on 2012-08-14.

[378] http://mars.nasa.gov/

[379] http://www.scientificamerican.com/report/mars-exploration/

[380] http://www.spacedaily.com/news/mars-future-05f.html

[381] http://www.mentallandscape.com/C_CatalogMars.htm

[382] http://www.phy6.org/stargaze/Smars1.htm

[383] http://www.planetary.org/explore/space-topics/mars/

[384] radiative time constant http://pds-atmospheres.nmsu.edu/education_and_outreach/encyclopedia/radiative_time_constant.htm

[385] The Obliquity of Mars http://www.spacedaily.com/news/mars-water-science-00d.html

[386] The Martian Sky: Stargazing from the Red Planet http://starryskies.com/The_sky/events/mars/opposition08.html

[387] Phil Plait's Bad Astronomy: Misconceptions: What Color is Mars? http://www.badastronomy.com/bad/misc/hoagland/mars_colors.html

[388] Mars Global Surveyor MOC2-368 Release http://www.msss.com/mars_images/moc/2003/05/22/

[389] Astronomical Phenomena From Mars http://www.arm.ac.uk/~aac/mars/Information.html

[390] 1990A&A...233..235B Page 235 http://adsbit.harvard.edu/cgi-bin/nph-iarticle_query?1990A%26A...233..235B

[391] 1991BAICz..42..271P Page 271 http://adsbit.harvard.edu/cgi-bin/nph-iarticle_query?1991BAICz..42..271P

[392] Meteoritical Bulletin Database: *Hadley Rille* http://www.lpi.usra.edu/meteor/index.php?code=11469

[393] Hundreds of auroras detected on Mars http://www.physorg.com/news8987.html

[394] 1988BAICz..39..168B Page 168 http://adsbit.harvard.edu/cgi-bin/nph-iarticle_query?1988BAICz..39..168B

[395] https://web.archive.org/web/20041207003949/http://www.giss.nasa.gov/research/intro/allison_02/

[396] https://web.archive.org/web/20050307083952/http://pweb.jps.net/~tgangale/mars/

[397] http://www.giss.nasa.gov/tools/mars24/

[398] In China, astronomers recorded an occultation of Mars by the Moon in 69 BCE. See Price (2000:148).

[399] http://mars.jpl.nasa.gov/allaboutmars/mystique/

[400] http://www.umich.edu/~lowbrows/reflections/2001/dsnyder.7.html

[401] WordReference.com http://www.wordreference.com/iten/canale

[402] Young, Charles A. *"A Textbook of General Astronomy."* 1889. Ginn and Co. Boston.

[403] Evans, J. E. and Maunder, E. W. (1903) "Experiments as to the Actuality of the 'Canals' observed on Mars", MNRAS, **63** (1903) 488 http://adsabs.harvard.edu//full/seri/MNRAS/0063//0000488.000.html

[404] Dollfus, A. (2010) "The first Pic du Midi photographs of Mars, 1909" http://adsbit.harvard.edu//full/2010JBAA..120..240D/0000241.000.html

[405] Robots On Mars Search And Catalog Red Planet https://www.npr.org/templates/story/story.php?storyId=94802645. Audio recording, supporting statement is approx. 34:00 after start.

[406] http://adsabs.harvard.edu//full/seri/AN.../0183//0000117.000.html

[407] http://www.erbzine.com/mag14/1414.html

Article Sources and Contributors

The sources listed for each article provide more detailed licensing information including the copyright status, the copyright owner, and the license conditions.

Mars *Source:* https://en.wikipedia.org/w/index.php?oldid=855733691 *License:* Creative Commons Attribution-Share Alike 3.0 *Contributors:* A2soup, Aelon51, Aeonx, AlphaBetaGamma01, Angilbas, Apolloe, Ashwinr, BD2412, Barek, Barjimoa, BatteryIncluded, Begoon, BenRG, Bender235, Bentogoa, Bgwhite, BiHVolim, Billhpike, Blinndsay, Bmags16, Brian Everlasting, CV9933, Claudio M Souza, ClueBot NG, Colonel Wilhelm Klink, CuriousMind01, DVdm, DanHobley, Daniel.Cardenas, DavideVeloria88, Dawnseeker2000, Dcirovic, Desildorf, Dhksml1, Dlthewave, Dolphin33438, Double sharp, DrKay, Drbogdan, Dream Focus, Earthandmoon, Edulovers, Fgrosshans, Finnusertop, Firth m, Fish567, FlightTime, Fogelmatrix, Fotaun, Frmorrison, G0mx, GKFX, Gap9551, Garfield Garfield, Gelatinxbox, Gildir, Glevum, Gob Lofa, Gulumeemee, HalloweenNight, Headbomb, HolyT, Huntster, Huritisho, Iacobus, Iggy the Swan, InedibleHulk, Isambard Kingdom, JFG, JeanLucMargot, Jerzy, John of Reading, Jon Kolbert, Jondel, JorisvS, Jytdog, K. Badri Vishal, KAP03, Kaldari, Katolophyromai, Kdammers, Khan singh, Kool Chad, L293D, Lasunncty, Loraof, LordOfPens, Maczkopeti, MadeYourReadThis, Magioladitis, Mandruss, Maranello Prime, Marfinan, Mark Schierbecker, MartinZ, Menchi, Michael Hardy, Nemoanon, NewEnglandYankee, Newone, Nfrango, NightShadow23, Nihiltres, Nikkimaria, Nizil Shah, OeBoe, Omanyd, Originalwana, Oshwah, Ost316, Paine Ellsworth, Pavlor, Pdebee, PeacePeace, PhilipTerryGraham, Phoenix7777, Piledhigheranddeeper, Pjposullivan, Pkbwcgs, PlanetUser, Praemonitus, Rain drop 45, Red Director, Rfl0216, RhinoMind, Rivertorch, Rjwilmsi, RomanSpa, Rowan Forest, Sacchipersempr, SamRathbone, SamuelRiv, Saros136, Sbmeirow, SeoMac, Sepguilherme, Serendipodous, Situphobos, Skepticalgiraffe, Skyfall, Skynxnex, SpikeToronto, SteepLearningCurve, Steve03Mills, Tesi1700, The Anome, The Transhumanist, TheFreeWorld, Thecodingproject, Thevideodrome, Thomas H. White, Tranngocnhatminh, Trappist the monk, Twinsday, Unbuttered Parsnip, UnitedStatesian, Urhixidur, Veedubber86, Vgy7ujm, Voello, W like wiki, Wavelength, Weneedwikipedia, White whirlwind, Wiki-whisky, WolfmanSF, Yoyozobi, Zedshort, Šedý, Шуфель 1

Geology of Mars *Source:* https://en.wikipedia.org/w/index.php?oldid=847208621 *License:* Creative Commons Attribution-Share Alike 3.0 *Contributors:* 2Mars4$2Billion, Alan Liefting, Albany NY, Arjayay, BD2412, Badri Vishal 16-09-2006, BatteryIncluded, BeenAroundAWhile, Benbest, BillC, Brainist, Castncoot, Chris the speller, Citation bot 1, ClueBot NG, CommonsDelinker, Cymru.lass, DanHobley, Danrok, Darth Tacker, Dcirovic, Dodshe, Drbogdan, Eumolpo, Ewen, Floquenbeam, Fotaun, Gadget850, Giovanni.leone.pa, Gob Lofa, Highfields, JIP, Jasper Deng, Jerzy, Jimmarsmars, John of Reading, Jonesey95, JorisvS, Keith D, KylieTastic, LeadSongDog, Leschnei, Magioladitis, Mandarax, Marikanessa, Martin451, Maurobio, MeanMotherJr, Mgiganteus1, Mogism, Niceguyedc, Nihiltres, Nono64, NotWith, Nyttend, Op47, Pinethicket, PinkAmpersand, RHaworth, RJaumann, RSStockdale, Raindows, Rjwilmsi, RockMagnetist, Rogermw, RomanSpa, Ruslik0, Sae1962, Sb2s3, Schaffman, Sean.hoyland, Sgbanham, SheriffIsInTown, Sidelight12, Solar-Wind, Surajt88, Tabletop, Tamfang, Telekenesis, The Banner, Tom.Reding, Trappist the monk, Tycho Magnetic Anomaly-1, Ugog Nizdast, Wareh, WolfmanSF, 44 anonymous edits 45

Martian soil *Source:* https://en.wikipedia.org/w/index.php?oldid=855346255 *License:* Creative Commons Attribution-Share Alike 3.0 *Contributors:* A2-33, Againme, Alkylightsword, Amphioxys, AnakngAraw, BatteryIncluded, Blast Ulna, Bryan Derksen, ChrisGualtieri, Citation bot 1, Citation bot 4, ClueBot NG, Colonies Chris, DadaNeem, Dah31, DanHobley, Drbogdan, Drxenocide, Edittoproveapoint, Eniagrom, FoCuSandLeArN, Fraggle81, Fristunghello, Gluons12, GoingBatty, Huntster, J04n, JFG, Jayron32, Jimmarsmars, K6ka, Kaini, Keith D, L Kensington, Lappspira, Lowellian, Mack2, Magioladitis, Markebson, Materialscientist, Nemo bis, Nihlus, Novangelis, NuclearWarfare, Originalwana, PhilipTerryGraham, Pkbwcgs, Quebec99, Rjwilmsi, SGBailey, Shotgunscribe, Skurremuk, Suniti karunatillake, Tom.Reding, Tysonman777, Vanesschoten9, Z22, 24 anonymous edits 69

Water on Mars *Source:* https://en.wikipedia.org/w/index.php?oldid=856170615 *License:* Creative Commons Attribution-Share Alike 3.0 *Contributors:* A Great Catholic Person, Adûnâi, Alcazar84, Art LaPella, BD2412, Balon Greyjoy, BatteryIncluded, Bereket Addis, Billhpike, Biplab Anand, BoyBeast, Bronte2004, CA2MI, CV9933, Cailalanier, Cannolis, Chris the speller, Chrissymad, Cls14, ClueBot NG, Coffeeandcrumbs, Conthuckway, Cyberbot II, DPdH, DarTar, David R MacKay, David Schopenhauer, Dawnseeker2000, DesertPipeline, Drbogdan, Dthomsen8, Elmidae, Eumolpo, Fazaex, FlightTime, Flyer22 Reborn, FourViolas, Fourthords, GoingBatty, GünniX, Haakonsson, Hairy Dude, Headbomb, Huntster, Hydrargyrum, Iridescent, IronGargoyle, J 1982, JFG, Jimmarsmars, Jodosma, John of Reading, Keiiri, Ketiltrout, Kintetsubuffalo, Lamborghini Urus, Lewis Goudy, Lotu, Magioladitis, Manpoudar, Marcus Cyron, Masem, Materialscientist, Mgiganteus1, Mirokado, Modest Genius, Nixinova, Nthep, PCHS-NJROTC, Paste, PhilipTerryGraham, Philipp.t42, Pkbwcgs, PlyrStar93, PohranicniStraze, Prinsgezinde, Quantanew, Quebec99, Rajakhanchapter1, Randy Kryn, Rjwilmsi, Robertinventor, RockyMasum, Rowan Forest, Salto F, SemiHypercube, Serols, Shellwood, Simcard15, Some Gadget Geek, Stephen, Suffusion of Yellow, The Proffesor, Titodutta, Tlhslobus, Trappist the monk, TwentyTwoPilots, User1696, Wbm1058, WolfmanSF, YuriNikolai, Zadguy, 68 anonymous edits 85

Martian polar ice caps *Source:* https://en.wikipedia.org/w/index.php?oldid=856202884 *License:* Creative Commons Attribution-Share Alike 3.0 *Contributors:* Arado, Arsia Mons, Athaler, BD2412, Bearcat, Biblioworm, Capitalismojo, Chris the speller, Citation bot 1, ClueBot NG, Code1420, Dandelany, Dcirovic, Deneb in Cygnus, Diannaa, Donner60, Dougmcdonell, Drbogdan, Excirial, Flyer22 Reborn, Goustien, Headbomb, Hifivebro, HopsonRoad, J S Ayer, Jimmarsmars, Khazar2, Labtek00, Lenticel, Materialscientist, Maye, Muadd, Mwaidele, N2e, Newyork1501, Quantanew, Rich Farmbrough, Rjwilmsi, Rp2006, SJ Defender, Scarlettail, Serols, Solar-Wind, Telecineguy, Tom.Reding, Trappist the monk, Whoop whoop pull up, WolfmanSF, Yamaguchi先生 , 55 anonymous edits 141

Geography of Mars *Source:* https://en.wikipedia.org/w/index.php?oldid=852522033 *License:* Creative Commons Attribution-Share Alike 3.0 *Contributors:* A412, AstroLynx, Avenue, BD2412, Babymissfortune, Benhocking, Boing! said Zebedee, Bryan Derksen, Caknuck, Carmen56, Chmee2, Chris the speller, ChrisGualtieri, Clean Copy, ClueBot NG, Cmglee, CommonsDelinker, DN-boards1, DanHobley, Dawnseeker2000, Dbigwood, Dogcow, Drbogdan, Edcolins, Fabio Bettani, Florian Blaschke, G310artur, Graphium, Hibernian, Hmainsbot1, Hut 8.5, Ian Pitchford, IanOsgood, Io Herodotus, Iohannes Animosus, Isentropiclift, Ivanbelchev, Japeo, Jdaloner, Jehorn, Jespley, Jimmarsmars, Jimmy Pitt, JorisvS, Joseph Solis in Australia, K6ka, Keraunos, Kris Schnee, Kuralyov, Kwamikagami, Lunokhod, MER-C, Mack2, Mark Schierbecker, Maurreen, Mike Dillon, Mlm42, Morwen, Murgh, Mxn, N2e, NE2, Nealmcb, Nehomesley, Oefe, Oracle125, Paintman, PoccilScript, Poulpy, Ptbotgourou, RA0808, RJHall, Reaverdrop, RedWolf, Remember, Riffsyphon1024, RisingStar, Rsrikanth05, Ruslik0, Sardanaphalus, Sbandrews, Serpinium, Shanes, SimonP, SpK, Stebulus, SteepLearningCurve, SteveMcCluskey, Stone, Tamfang, The Singing Badger, Thirdright, Tillman, Tomruen, Twthmoses, Ulao, Vcxzfdsa, VoABot II, WDGraham, WOSlinker, Wareh, Wikianon, Wknight94, WolfmanSF, Xeno, Xession, Yogi de, Σ, 66 anonymous edits 155

Volcanology of Mars *Source:* https://en.wikipedia.org/w/index.php?oldid=852147634 *License:* Creative Commons Attribution-Share Alike 3.0 *Contributors:* BD2412, BatteryIncluded, Bgwhite, Billhpike, Billinghurst, Castncoot, Chmee2, Chris the speller, ChrisGualtieri, Citation bot 1, Citation bot 4, ClueBot NG, Dcirovic, Dl2000, Drbogdan, EdwardLane, FoCuSandLeArN, Fotaun, GünniX, HazelAB, Huntster, IVAN3MAN, Jim1138, Jimmarsmars, Jon Kolbert, Jonesey95, JorisvS, Kaobear, Keith D, Khazar2, KylieTastic, LilHelpa, Lonet, Martarius, Miroslav Jícha, Niceguyedc, Nitekatt, Nono64, Op47, Originalwana, Ozric14, Quebec99, Rjwilmsi, RomanSpa, Ruslik0, Sb2s3, Schaffman, Sean.hoyland, SheriffIsInTown, Sidelight12, Tabletop, Titodutta, Trappist the monk, Volcanoguy, Vuo, Wareh, WereSpielChequers, WikHead, WolfmanSF, Z22, 42 anonymous edits 167

Atmosphere of Mars *Source:* https://en.wikipedia.org/w/index.php?oldid=853967230 *License:* Creative Commons Attribution-Share Alike 3.0 *Contributors:* A2soup, Alaney2k, Ale jrb, Alexparent, Anders Feder, Arslankiller1234, Auric, BatteryIncluded, Bender235, BlackBeast, Blueboy5252, CLCStudent, CanadianLinuxUser, Cloudswrest, ClueBot NG, Crystallizedcarbon, DH800, DVdm, DavidLeighEllis, Dawnseeker2000, DemocraticLuntz, Dougmcdonell, Download, Drbogdan, Drpickem, ERROR 760, ExperiencedArticleFixer, Fotaun, Gadget850, HanotLo, Headbomb, HowlingAngel, Huntster, I dream of horses, IagoQnsi, IanOfNorwich, Ieditwiki432, Inks.LWC, J 1982, JBeags, Jar3dl, Jcpag2012, Jehochman, Jimmarsmars, JorisvS, KGirlTrucker81, Kaiserkarl13, Katieh5584, Liam McM, Manul, Marcus Cyron, Materialscientist, Melonkelon, Mikhail Ryazanov, Monolithica, N2e, NSH002, Namoroka, Newone, Originalwana, Oshwah, PlyrStar93, Pokemon1000bc, Postdlf, Pratyya Ghosh, Quinton Feldberg, Reggie121, Rjwilmsi, Rowan Forest, Sean4993, Shellwood, Simplexity22, SludgyTortoise, Stamptrader, StuHarris, Super48paul, Thnidu, Tom.Reding, Torchiest, Trappist the monk, Troll25255, Vgy7ujm, WOSlinker, Waifscrest, Watibc, WolfmanSF, Wpollard, Wtmitchell, Yamaguchi先生 , Yoshi24517, Zedshort, 128 anonymous edits 191

Climate of Mars *Source:* https://en.wikipedia.org/w/index.php?oldid=854810335 *License:* Creative Commons Attribution-Share Alike 3.0 *Contributors:* A2soup, Alaney2k, Alfredovega25, Anders Feder, Arado, Archon 2488, Bammesk, BatteryIncluded, Bearcat, Bgwhite, Bigeski, Blueowl42, Boundarylayer, Brandmeister, CAPTAIN RAJU, Cannolis, ClueBot NG, Code1420, D Eaketts, DHeyward, Dan Koehl, Darkwind, Dawnseeker2000, Dcirovic, Dewritech, DinosaursLoveExistence, Discospinster, Donomanh9999, Drbogdan, Entranced98, Excirial, Florian Blaschke, Flowerpotman, Flyer22 Reborn, Frietjes, Frosty, Fugitron, Gap9551, Gerrit, Gilliam, Gob Lofa, Gronk Oz, Hairhorn, Hayman30, Headbomb, Helenabella, Hello71, Holdoffhunger, HunterxEditor, Huntster, IHaveAMastersDegree, Jarble, Jbdonnelly12, Jburk, Jekowl, Jim.henderson, Jimmarsmars, K6ka, Keith D, Kku, KylieTastic, Laura9993, Loraof, Lotus 50, MCTales, MONGO, MadGuy7023, Makyen, Materialscientist, Max.kit, Mcarling, Mfb, Minimoy34, Mojoworker, N2e, NewByzantine, NewEnglandYankee, Newyork1501, Ogsrotag2423, Olli Niemitalo, Onel5969, Orenburg1, Overone2, Peter Gulutzan, PhilipTerryGraham, PlyrStar93, Postdlf, Qmzptyusdfggggghhjkkklllngfd, Quondum, Reatlas, Rhododendrites, Rjwilmsi, RomanSpa, Rowan Forest, Roxy the dog, Rsrikanth05, Rusl, SantiLak, Shellwood, Simplexity22, Skyd4ncer33, Snaily, Stephan Schulz, Thiridaz, Thor Dockweiler, Tom.Reding, Trappist the monk, Tymon.r, Vieque, Widr, Wikiuser13, William M. Connolley, Wtmitchell, Yamaguchi先生 , Zamaster4536, Zziccardi, 189 anonymous edits 209

Orbit of Mars *Source:* https://en.wikipedia.org/w/index.php?oldid=855180579 *License:* Creative Commons Attribution-Share Alike 3.0 *Contributors:* Brandmeister, CASSIOPEIA, Copyeditor42, Finell, Gob Lofa, IronGargoyle, JorisvS, Lasunncty, Loraof, Matt7899, Mitch Ames, Modest Genius, N2e, Nyttend, Rich Farmbrough, Saros136, Tom.Reding, Trappist the monk, Twinsday, Usernamekiran (AWB), 22 anonymous edits 239

Life on Mars *Source:* https://en.wikipedia.org/w/index.php?oldid=855662453 *License:* Creative Commons Attribution-Share Alike 3.0 *Contributors:* 3primetime3, 72, A2soup, AManWithNoPlan, Agljones, AndreyKva, Another Believer, Apokrif, Ashton Leatherwood, BatteryIncluded, Bkell, C.Fred, CASSIOPEIA, CFCF, Candido, Cardcapturs, Ceannlann gorm, ClueBot NG, DN-boards1, DanHobley, Daniel.Cardenas, David.moreno72, Dcirovic, Denniscabrams, DiscoveryTracy, Dr.K., Drbogdan, EdJohnston, El cid, el campeador, Espresso Addict, Esszet, Excirial, FoCuSandLeArN, Geogene, Gilliam, Gob Lofa, GrandAdmiralThrawn, Hamiltondaniel, Hcps-walkerdc, Headbomb, Here2help, Huntster, Ifnord, Isambard Kingdom, J 1982, Jkyle06, John, John of Reading, KGirlTrucker81, Kondo125, Kortoso, Kraxler, LjL, Magioladitis, MainlyTwelve, Maria Kappatou, Martijn Oei, Masem, Materialscientist, MerscratianAce, Modest Genius, Morferehwon, Mxmsj, My Chemistry romantic, NeilN, NewEnglandYankee, Newone, Nimbex, Nixinova, Nwbeeson, Oshwah, Ost316, Pavel Krupička, Peyre, Piramidion, Prnj, Postdlf, Prinsipe Ybarro, Prokaryotes, Promotional Attack, Quebec99, R'n'B, Rjwilmsi, Robertinventor, RockyMasum, Rowan Forest, ScrapIronIV, Shellwood, Simplexvir, Snowweatyh, Srnec, StarHOG, Theinstantmatrix, Three-quarter-ten, Trappist the monk, Troll23451, Tuna Cactus, USSAvenger, VQuakr, Vicsar, Voidxor, Vsmith, WOSlinker, Weegeerunner, Westley Turner, Widr, Wiki winter berry, Xanikk999, Yaakovaryeh, Yellowdesk, Zzuuzz, 140 anonymous edits 243

Colonization of Mars *Source:* https://en.wikipedia.org/w/index.php?oldid=856171544 *License:* Creative Commons Attribution-Share Alike 3.0 *Contributors:* 2012Commander, A Lazy Shisno, Aeonx, Agoncharov5, Alaney2k, Almightycat, Andyjsmith, Arado, Archon 2488, Asgrrr, Asta0004, At least I try, Aza2000, BatteryIncluded, Bhartiyahainhum, Billbpike, Bobjan.harlem, CarmenRodriguez91, Chai-alice, CharmaElizabeth, ClueBot NG, DVdm, Daniel.Cardenas, Dawnseeker2000, Dcirovic, Deneb in Cygnus, Dermeister, Dmol, Donner60, Dougmcdonell, Drbogdan, EEJJLL, EditronMan, Ehrenkater, Entranced98, Erudite Manatee, ExperiencedArticleFixer, Flyer22 Reborn, FoCuSandLeArN, Fotaun, Frauke80, Fuortu, GeneralizationsAre-Bad, Grayfell, Gulumeemee, GünniX, Hack-Man, Haydenbunker, Headbomb, Hertz1888, HolyT, Icarus of old, J 1982, JFG, JaconaFrere, Jak525, John of Reading, Jonah and the Whale, Kees08, Kingst19, Kleuske, Lake Matthew Team, Lake Matthew Team - Cole, LocalLaddie, Lordtobi, Magioladitis, MainlyTwelve, Me, Myself, and I are Here, Mikael Häggström, Millennium bug, Mindmatrix, Mitchwhite5, Morferehwon, Mortee, N2e, Narky Blert, Neil916, Neko-chan, Nerfer, NickPenguin, Orenburg1, Pek∼enwiki, Phantom in ca, Praemonitus, Prinsgezinde, ProprioMe OW, Psychotic17, Quantanew, Redling2016, Rhoark, RileyBugz, Robertinventor, Rod57, Rowan Forest, Ruyter, Schroffp, Shellwood, SimmerALPHA, Sophivorus, Soumya-8974, Speedbump20, Staszek Lem, Stefan.K., Stonnman, SwagGangster, TAnthony, Tduso01, ThebeastYt, Trappist the monk, Urdadisafaggy, Uuuuuuuuuuu-uuuuuuuuuuuuuuuuuuuu, Velella, Vgy7ujm, William M. Connolley, Woodlot, Word dewd544, Zplizzi, Šedý, Михаилъ Јовановиħ, 137 anonymous edits 275

Moons of Mars *Source:* https://en.wikipedia.org/w/index.php?oldid=852191561 *License:* Creative Commons Attribution-Share Alike 3.0 *Contributors:* 1-555-confide, Acroterion, Adam9007, Amaury, Astredita, Atakdoug, BatteryIncluded, Berntisso, Bgwhite, Biker Biker, Bri, BrownHairedGirl, Butler1234, CLCStudent, ClueBot NG, Cordenea4, DBigXray, DN-boards1, DVdm, Dagger8981, Dl2000, Double sharp, Drbogdan, Dudeman5685, Ebehn, Evercat, Excirial, Eyesnore, F6Zman, Faizan, Finnusertop, Finnyhaha, Flyer22 Reborn, Fotaun, Funguts, Gadget850, Gap9551, Geremy.Hebert, Giant-blackbooty, Gilliam, Gob Lofa, Goustien, Harlock81, Headbomb, HowardMorland, Huntster, Icairns, J 1982, JJohnston2, JaconaFrere, Japanese Rail Fan, Jhertel, Jimmarsmars, Joeinwiki, JoelDick, John of Reading, JorisvS, Jsharpminor, KOOMFF, Kaldari, Kuralyov, Kwamikagami, KylieTastic, Lanthanum-138, Legojer3, MC Scout, MONGO, MacTire02, Male Sailor Guardian of Iris, Materialscientist, Matthewrbowker, Mattnmax, McSly, Mcc1789, Mickey G da man, Mike Rosoft, Mmortal03, Modest Genius, Monty845, Murgh, MusikAnimal, Nardog, Nem1yan, Nicearle, Niceguyedc, Nikkimaria, Njardarlogar, Omeganian, Orionist, Oshwah, Oversaturn!, P. S. F. Freitas, Param Mudgal, Pepper, Polylerus, Prinsipe Ybarro, RA0808, RDBury, RSStockdale, Reatlas, Rfassbind, Rscheeer2, Ruslik0, ScottyBerg, Scwlong, Serols, Shayan Majumder, Shellwood, Sideways713, Situphobos, Skizzik, Str1977, Swister-Twister, SyahirSQRT2, The Phase Master, Theinstantmatrix, Thincat, Thingg, Titus III, Tlhslobus, Tom.Reding, Usb10, UtDicitur, Warner32, Watchduck, Weqqsrand, Wen D House, Whouk, Wiae, Widr, WikiPuppies, Wikijens, Wikipelli, Xanzzibar, Xession, Xhaju, Xxprestigeousxx, 166 anonymous edits 299

Exploration of Mars *Source:* https://en.wikipedia.org/w/index.php?oldid=856305076 *License:* Creative Commons Attribution-Share Alike 3.0 *Contributors:* Abelmoschus Esculentus, AlexSpace1, Alexconnorbrown, AnonMoos, Apokryltaros, Arjayay, BatteryIncluded, Brilliantwiki2, Byteflush, Chonachona, ClueBot NG, Cocopops&vodka, Cowmaster1111, CuriousMind01, Czolgolz, Daniel Zsenits, Dawnseeker2000, Dcirovic, Diannaa, Djmanton, Dougmcdonell, Drbogdan, Ebola=Gabriel, Elk Salmon, Erisie, Eyesnore, Fotaun, Frietjes, Gladamas, Grafen, Gulumeemee, Hibernian, Hms1103, ItTakesTime, JFG, JJGC13, JJMC89, Jburk, Jim1138, John of Reading, JzG, Kees08, Kencf0618, Kirchoid, Kinetic37, Kitchen Knife, LilHelpa, Ludtie, LuigiPortaro29, Mfb. N2e, Narky Blert, Necessary Evil, Nergaal, Netsnipe, Ninney, Openhearted99, PaoVac, Peaceray, Phoenix7777, Postdlf, Quadrplax, Quyxz, Randy Kryn, RatRai, RedPanda25, Rjwilmsi, Robertinventor, Robotics7777, Rowan Forest, SanmingFin, Shellwood, Socksandsandals2016, Srednuas Lenoroc, Tharsis 1, TheArmchairSoldier, Thejavis86, Thereisnous, Thewolfchild, TutoCX, TwoTwoHello, Vgy7ujm, WereSpielChequers, WikiEH1998, Wikishovel, Wllmevans, Zocke1r, Михаилъ Јовановиħ, 刘佳 4499, 71 anonymous edits 311

Astronomy on Mars *Source:* https://en.wikipedia.org/w/index.php?oldid=854578569 *License:* Creative Commons Attribution-Share Alike 3.0 *Contributors:* 84user, Aesopos, Alaney2k, Alansohn, ArnoldReinhold, AxelHurricane001, B.d.mills, Barticus88, Bryan Derksen, CALR, Cendour∼enwiki, Chris the speller, CielProfond, ClueBot NG, CommonsDelinker, CosineKitty, Curps, Danrok, Dcirovic, Discospinster, Dkreisst, Doloco, Donner60, Drbogdan, Duz, Eluchil404, FKmailliW, Flxmghvgvk, Fotaun, Franamax, Frietjes, Geofbob, HaeB, Hall Monitor, Himerish, HolyT, Icairns, Illegitimate Barrister, J 1982, J.delanoy, Jamesb1995, Jikybebna, JimVC3, Joelbellman, John, Jonlighthall, Kanashimi, Karl Naylor, Kuralyov, Kura, LionKimbro, Loopy30, Lowellian, MTSbot∼enwiki, MacTire02, Marcok, Martin S Taylor, Mathrick, Mike s, Neighborhoodist, Nicscatapolous, Ngerric, Shadowsill, Sherurcij, Shizhao, Sillyfolkboy, SkeletorUK, Soap, Sobolewski, SoxBot III, Specs112, Superborsuk, The Anome, TheB1FFS, TheOtherSiguy, Tom.Reding, TomS TDotO, Tomruen, Ugog Nizdast, Urhixidur, Veena, Wareh, Washi, Widr, Wiki13, Woohookitty, Yintan, Z10x, 78 anonymous edits 353

History of Mars observation *Source:* https://en.wikipedia.org/w/index.php?oldid=844428025 *License:* Creative Commons Attribution-Share Alike 3.0 *Contributors:* 84user, After Midnight, Another Believer, Bakeland5, Bencherlite, Bender235, Beyblion, Braindustry, CAPTAIN RAJU, Carendinator11, Casliber, Charles Matthews, Chorleypie, Chris the speller, Citation bot 1, ClueBot NG, Cryptic C62, Daniel Robert Sum, Daryl Janzen, DrKay, Drbogdan, Dthomsen8, Duncan7670, Earthandmoon, Excirial, Fotaun, Gah4, Gene Nygaard, George Ponderevo, Gilliam, Hms Inconceivable, Hamiltonclause, Haon 2.0, Headbomb, Hmains, Jack Greenmaven, Jagged 85, John of Reading, Josve05a, Khirurg, Koavf, Kogge, Krenair, Kurttran64, LilHelpa, Lusanaherandraton, Mark Arsten, Matthew Proctor, Modest Genius, Mr Stephen, Myasuda, Nick Number, Ohms law, Optimale, Oshwah, Piledhigheranddeeper, Pinethicket, Postdlf, Praemonitus, RJHall, Radiphus, Reem Al-Kashif, Rfassbind, Rich Farmbrough, Rjwilmsi, Rotcaeroib, Seaphoto, Shakerb, Smd75jr, Staszek Lem, Svick, Swagle, Syncategoremata, Tewapack, Tom.Reding, Tomruen, Tony Fox, Trusilver, UtherSRG, Wareh, Widr, Williamjt, WolfmanSF, Yiosie2356, عمر, 45 anonymous edits 371

Martian canal *Source:* https://en.wikipedia.org/w/index.php?oldid=856210987 *License:* Creative Commons Attribution-Share Alike 3.0 *Contributors:* A.amitkumar, Academic Challenger, Alakshak, Andrewrutherford, Andy M. Wang, Astrobiologist, BatteryIncluded, Beland, BenjaminBaillaud, Blanche of King's Lynn, Boleslav1∼enwiki, Bryan Derksen, CNichols, Caltas, Captainbeefart, Chetvorno, Chris Capoccia, Citation bot 1, ClueBot NG, Crystallizedcarbon, Curps, DHN-bot∼enwiki, DanHobley, David Edgar, Drbogdan, Drobertpowell, Easphi, Egil, Ferna186, Finlay McWalter, Flowerpotman, Flyguy649, Fotaun, FrancisTyers, Frobulate, Furrykef, Geofbob, HaeB, Hall Monitor, Himerish, HolyT, Icairns, Illegitimate Barrister, J 1982, J.delanoy, Jamesb1995, Jikybebna, JimVC3, Joelbellman, John, Jonlighthall, Kanashimi, Karl Naylor, Kuralyov, Kura, LionKimbro, Loopy30, Lowellian, MTSbot∼enwiki, MacTire02, Marcok, Martin S Taylor, Mathrick, Mike s, Neighborhoodist, Nicsatapolous, Ngerric, Rjwilmsi, Roadrunner, Rusty Cashman, Ryulong, SHCarter, Sagqs, Sardanaphalus, Sbandrews, Serendipodous, Sgerbic, Shadowsill, Sherurcij, Shizhao, Sillyfolkboy, SkeletorUK, Soap, Sobolewski, SoxBot III, Specs112, Superborsuk, The Anome, TheB1FFS, TheOtherSiguy, Tom.Reding, TomS TDotO, Tomruen, Ugog Nizdast, Urhixidur, Veena, Wareh, Washi, Widr, Wiki13, Woohookitty, Yintan, Z10x, 78 anonymous edits 386

Image Sources, Licenses and Contributors

The sources listed for each image provide more detailed licensing information including the copyright status, the copyright owner, and the license conditions.

Image *Source:* https://en.wikipedia.org/w/index.php?title=File:Padlock-silver.svg *Contributors:* AzaToth, BotMultichill, BotMultichillT, Gurch, Jarekt, Kallerna, Multichill, Perhelion, Rd232, Riana, Sarang, Siebrand, Steinsplitter, 4 anonymous edits 1
Image *Source:* https://en.wikipedia.org/w/index.php?title=File:OSIRIS_Mars_true_color.jpg *Contributors:* A2soup, Arjuno3, B dash, Huntster, Jdx, Moheen Reeyad, Mtrova, Mx. Granger, Sanandros, TheFreeWorld, WolfmanSF, Z7504, رافي, 2 anonymous edits 1
Image *Source:* https://en.wikipedia.org/w/index.php?title=File:Loudspeaker.svg *License:* Public Domain *Contributors:* User:Dbenbenn, User:Optimager, User:Tsca, User:Dbenbenn, User:Optimager, User:Tsca, User:Dbenbenn, User:Optimager, User:Tsca 1
Image *Source:* https://en.wikipedia.org/w/index.php?title=File:Mars,_Earth_size_comparison.jpg *License:* Public Domain *Contributors:* User:Jcpag2012 4
Image *Source:* https://en.wikipedia.org/w/index.php?title=File:Mars.ogv *License:* Public Domain *Contributors:* NASA/Goddard Space Flight Center 4
Image *Source:* https://en.wikipedia.org/w/index.php?title=File:GMM-3_Mars_Gravity.webm *License:* Public Domain *Contributors:* NASA's Scientific Visualization Studio 4
Figure 1 *Source:* https://en.wikipedia.org/w/index.php?title=File:USGS-MarsMap-sim3292-20140714-crop.png *License:* Public Domain *Contributors:* Chmee2, Drbogdan, Jarnsax, Kopiersperre, Mapmarks 6
Figure 2 *Source:* https://en.wikipedia.org/w/index.php?title=File:Eso1509a_-_Mars_planet.jpg *Contributors:* ESO/M. Kornmesser 7
Figure 3 *Source:* https://en.wikipedia.org/w/index.php?title=File:Spirit_Mars_Silica_April_20_2007.jpg *License:* Public Domain *Contributors:* BotMultichill, DragonFire1024, Foroa, Huntster, Kilom691, 1 anonymous edits 8
Figure 4 *Source:* https://en.wikipedia.org/w/index.php?title=File:Nasa_mars_opportunity_rock_water_150_eng_02mar04.jpg *License:* Public Domain *Contributors:* NASA/JPL/US Geological Survey 9
Figure 5 *Source:* https://en.wikipedia.org/w/index.php?title=File:PIA16791-MarsCuriosityRover-Composition-YellowknifeBayRocks.png *License:* Public Domain *Contributors:* Huntster, Mindmatrix, Ras67 11
Image *Source:* https://en.wikipedia.org/w/index.php?title=File:Martian_north_polar_cap.jpg *License:* Public Domain *Contributors:* NASA/JPL/ Malin Space Science Systems 11
Image *Source:* https://en.wikipedia.org/w/index.php?title=File:South_Polar_Cap_of_Mars_during_Martian_South_summer_2000.jpg *License:* Public Domain *Contributors:* NASA/JPL/MSSS 12
Figure 6 *Source:* https://en.wikipedia.org/w/index.php?title=File:Mars_topography_(MOLA_dataset)_with_poles_HiRes.png *License:* Public Domain *Contributors:* Billinghurst, PhilipTerryGraham, WolfmanSF 13
Figure 7 *Source:* https://en.wikipedia.org/w/index.php?title=File:PIA11176_-_A_Recent_Cluster_of_Impacts.jpg *License:* Public Domain *Contributors:* Huntster, PhilipTerryGraham 14
Figure 8 *Source:* https://en.wikipedia.org/w/index.php?title=File:PIA15038_Spirit_Lander_and_Bonneville_Crater_in_Color.jpg *License:* Public Domain *Contributors:* Huntster, Kenmayer, Marcus Cyron, O'Dea, PhilipTerryGraham 15
Figure 9 *Source:* https://en.wikipedia.org/w/index.php?title=File:PIA18381-Mars-FreshAsteroidImpact2012-Before27March-After28March.jpg *License:* Public Domain *Contributors:* Drbogdan, Huntster 16
Figure 10 *Source:* https://en.wikipedia.org/w/index.php?title=File:Olympus_Mons_alt.jpg *License:* Public Domain *Contributors:* Image by NASA, modifications by Seddon 17
Figure 11 *Source:* https://en.wikipedia.org/w/index.php?title=File:016vallesmarineris_reduced0.25.jpg *License:* Public Domain *Contributors:* INeverCry, PhilipTerryGraham, Terrific Dunker Guy, Vetranio, WolfmanSF 17
Figure 12 *Source:* https://en.wikipedia.org/w/index.php?title=File:Mars_atmosphere_2.jpg *Contributors:* 19
Figure 13 *Source:* https://en.wikipedia.org/w/index.php?title=File:PIA19088-MarsCuriosityRover-MethaneSource-20141216.png *License:* Public Domain *Contributors:* Drbogdan, Huntster, Iranoia 20
Figure 14 *Source:* https://en.wikipedia.org/w/index.php?title=File:PIA18613-MarsMAVEN-Atmosphere-3UV-Views-20141014.jpg *License:* Public Domain *Contributors:* Drbogdan, JorisvS, 1 anonymous edits 20
Figure 15 *Source:* https://en.wikipedia.org/w/index.php?title=File:PIA22487-Mars-BeforeAfterDust-20180719.gif *License:* Public Domain *Contributors:* Drbogdan, Huntster 23
Image *Source:* https://en.wikipedia.org/w/index.php?title=File:PIA16450-MarsDustStorm-20121118.jpg *License:* Public Domain *Contributors:* Badseed, Chmee2, Drbogdan, Huntster, PhilipTerryGraham, Russavia 22
Image *Source:* https://en.wikipedia.org/w/index.php?title=File:PIA16454_Regional_Dust_Storm_Weakening,_Nov._25,_2012.jpg *License:* Public Domain *Contributors:* Ain92, Chmee2, Drbogdan, Huntster, Jarnsax, PhilipTerryGraham 23
Image *Source:* https://en.wikipedia.org/w/index.php?title=File:PIA22329-Mars-DustStorm-20180606.jpg *License:* Public Domain *Contributors:* Drbogdan, Huntster 23
Figure 16 *Source:* https://en.wikipedia.org/w/index.php?title=File:Marsorbitsolarsystem.gif *License:* Creative Commons Attribution-Sharealike 3.0 *Contributors:* User:Lookang 24
Figure 17 *Source:* https://en.wikipedia.org/w/index.php?title=File:Mars_Viking_11d128.png *License:* Public Domain *Contributors:* "Roel van der Hoorn (Van der Hoorn)" 25
Figure 18 *Source:* https://en.wikipedia.org/w/index.php?title=File:PIA19673-Mars-AlgaCrater-ImpactGlassDetected-MRO-20150608.jpg *License:* Public Domain *Contributors:* Drbogdan, Huntster, Tillman 26
Figure 19 *Source:* https://en.wikipedia.org/w/index.php?title=File:Mars-Curiosity-RockStructures-20180102.jpg *License:* Public Domain *Contributors:* ComputerHotline, Drbogdan, Huntster 29
Figure 20 *Source:* https://en.wikipedia.org/w/index.php?title=File:Mars-SubglacialWater-SouthPoleRegion-20180725.jpg *License:* Public Domain *Contributors:* Drbogdan, Huntster 29
Image *Source:* https://en.wikipedia.org/w/index.php?title=File:Phobos_colour_2008.jpg *License:* Public Domain *Contributors:* NASA / JPL-Caltech / University of Arizona 29
Image *Source:* https://en.wikipedia.org/w/index.php?title=File:Deimos-MRO.jpg *License:* Public Domain *Contributors:* NASA/JPL-caltech/ University of Arizona 29
Figure 21 *Source:* https://en.wikipedia.org/w/index.php?title=File:Orbits_of_Phobos_and_Deimos.gif *License:* Creative Commons Attribution-Sharealike 2.5 *Contributors:* Jahobr, Leyo, Maksim, Ruslik0, Wieralee, Zaccarias, 3 anonymous edits 30
Figure 22 *Source:* https://en.wikipedia.org/w/index.php?title=File:HiRISE_image_of_MSL_during_EDL_(refined).png *License:* Public Domain *Contributors:* NASA/JPL/University of Arizona/HiRISE Team 31
Figure 23 *Source:* https://en.wikipedia.org/w/index.php?title=File:Orion_docked_to_Mars_Transfer_Vehicle.jpg *License:* Public Domain *Contributors:* Allusion.Doze, Craigboy, Haplochromis, Huntster, 1 anonymous edits 32
Image *Source:* https://en.wikipedia.org/w/index.php?title=File:PIA21260_-_Earth_and_Its_Moon,_as_Seen_From_Mars.jpg *License:* Public Domain *Contributors:* Gereon K., Huntster, PhilipTerryGraham, Tomruen 33
Image *Source:* https://en.wikipedia.org/w/index.php?title=File:15-ml-06-phobos2-A067R1.jpg *License:* Public Domain *Contributors:* Bryan Derksen, ComputerHotline, Huntster, Kristaga, Li-sung, RHorning 33
Image *Source:* https://en.wikipedia.org/w/index.php?title=File:PIA19801-TrackingSunspotsOnTheSunFromMars-20150708.gif *License:* Public Domain *Contributors:* Drbogdan 34
Figure 24 *Source:* https://en.wikipedia.org/w/index.php?title=File:Apparent_retrograde_motion_of_Mars_in_2003.gif *License:* GNU Free Documentation License *Contributors:* Eugene Alvin Villar (seav) 34
Figure 25 *Source:* https://en.wikipedia.org/w/index.php?title=File:Mars_orbit_around_Earth.gif *License:* *Contributors:* User:Phoenix7777 35
Figure 26 *Source:* https://en.wikipedia.org/w/index.php?title=File:Galileo.arp.300pix.jpg *License:* Public Domain *Contributors:* ABF, Alefisico, Allforrous, Alno, BotMultichill, BotMultichillT, Bukk, Daniele Pugliesi, David J Wilson, Deadstar, Dirk Hünniger, G.dallorto, Gary King, Herbythyme, Huntster, Kam Solusar, Liberal Freemason, Michael Bednarek, Pérez~commonswiki, Pérez~commonswiki, Quadell, Ragesoss, Schaengel89~commonswiki, Semnoz, Shakko, Silverije, Trijnstel, Túrelio, Un1c0s bot~commonswiki, Yonatanh, Ángel el Astrónomo, 28 anonymous edits 37
Image *Source:* https://en.wikipedia.org/w/index.php?title=File:Karte_Mars_Schiaparelli_MKL1888.png *License:* Public Domain *Contributors:* BLueFiSH.as, Badseed, Bryan Derksen, David Kernow~commonswiki, Mapmarks, Marcok, Romuelio, Sophus Bie, Startaq, Stefan Kühn, W!B:, 2 anonymous edits 38

Figure 92 *Source:* https://en.wikipedia.org/w/index.php?title=File:Warm_Season_Flows_on_Slope_in_Newton_Crater_(animated).gif *License:* Public Domain *Contributors:* NASA/JPL-Caltech/Univ. of Arizona 116
Figure 93 *Source:* https://en.wikipedia.org/w/index.php?title=File:Branched_gullies.jpg *License:* Public Domain *Contributors:* Jim Secosky modified nasa image. 117
Figure 94 *Source:* https://en.wikipedia.org/w/index.php?title=File:Deep_Gullies.jpg *License:* Public Domain *Contributors:* Jim Secosky modified nasa image. 117
Figure 95 *Source:* https://en.wikipedia.org/w/index.php?title=File:Scamander_Vallis_from_Mars_Global_Surveyor.jpg *License:* Public Domain *Contributors:* NASA/JPL/Malin Space Science Systems 125
Figure 96 *Source:* https://en.wikipedia.org/w/index.php?title=File:Streamlined_Islands_in_Maja_Valles.jpg *License:* Public Domain *Contributors:* Jim Secosky modified NASA image. 125
Figure 97 *Source:* https://en.wikipedia.org/w/index.php?title=File:Hematite_region_Sinus_Meridiani_sur_Mars.jpg *License:* Public Domain *Contributors:* NASA/JPL 126
Figure 98 *Source:* https://en.wikipedia.org/w/index.php?title=File:Nanedi_channel.JPG *License:* Public Domain *Contributors:* Jim Secosky modified nasa image. 127
Figure 99 *Source:* https://en.wikipedia.org/w/index.php?title=File:Semeykin_Crater_Drainage.JPG *License:* Public Domain *Contributors:* Jim Secosky modified NASA image. 128
Figure 100 *Source:* https://en.wikipedia.org/w/index.php?title=File:Blocks_in_Aram.JPG *License:* Public Domain *Contributors:* Jim Secosky modified NASA image 129
Figure 101 *Source:* https://en.wikipedia.org/w/index.php?title=File:Phoenix_Sol_0_horizon.jpg *License:* Public Domain *Contributors:* NASA/JPL-Caltech/University of Arizona 130
Figure 102 *Source:* https://en.wikipedia.org/w/index.php?title=File:PIA10741_Possible_Ice_Below_Phoenix.jpg *License:* Public Domain *Contributors:* NASA/Jet Propulsion Lab-Caltech/University of Arizona/Max Planck Institute 131
Figure 103 *Source:* https://en.wikipedia.org/w/index.php?title=File:Opportunity_photo_of_Mars_outcrop_rock.jpg *License:* Public Domain *Contributors:* NASA/JPL/Cornell/US Geological Survey 132
Figure 104 *Source:* https://en.wikipedia.org/w/index.php?title=File:Opp_layered_sol17-B017R1_br.jpg *License:* Public Domain *Contributors:* BotMultichill, Bryan Derksen, File Upload Bot (Magnus Manske), OgreBot 2, Rodhullandemu, 1 anonymous edits 133
Figure 105 *Source:* https://en.wikipedia.org/w/index.php?title=File:07-ml-3-soil-mosaic-B019R1_br.jpg *License:* Public Domain *Contributors:* NASA/JPL/Cornell/USGS 133
Figure 106 *Source:* https://en.wikipedia.org/w/index.php?title=File:Nasa_mars_opportunity_rock_water_150_eng_02mar04.jpg *License:* Public Domain *Contributors:* NASA/JPL/US Geological Survey 134
Figure 107 *Source:* https://en.wikipedia.org/w/index.php?title=File:Springs_in_Vernal_Crater.jpg *License:* Public Domain *Contributors:* Jim Secosky modified NASA image. 135
Figure 108 *Source:* https://en.wikipedia.org/w/index.php?title=File:Asimov_Layers_Close-up.JPG *License:* Public Domain *Contributors:* Jim Secosky modified nasa image. 136
Figure 109 *Source:* https://en.wikipedia.org/w/index.php?title=File:PIA16156-Mars_Curiosity_Rover-Water-AncientStreambed.jpg *License:* Public Domain *Contributors:* ComputerHotline, Drbogdan, Huntster, Romkur 137
Figure 110 *Source:* https://en.wikipedia.org/w/index.php?title=File:PIA16189_fig1-Curiosity_Rover-Rock_Outcrops-Mars_and_Earth.jpg *License:* Public Domain *Contributors:* ComputerHotline, Drbogdan, Karmakolle, Romkur 137
Image *Source:* https://en.wikipedia.org/w/index.php?title=File:Commons-logo.svg *License:* logo *Contributors:* Anomie, Callanecc, CambridgeBayWeather, Jo-Jo Eumerus, RHaworth 140
Image *Source:* https://en.wikipedia.org/w/index.php?title=File:Wikinews-logo.svg *License:* Creative Commons Attribution-Sharealike 3.0 *Contributors:* Vectorized by Simon 01:05, 2 August 2006 (UTC) Updated by Time3000 17 April 2007 to use official Wikinews colours and ap 140
Figure 111 *Source:* https://en.wikipedia.org/w/index.php?title=File:ESP_054515_2595icecaplayers.jpg *License:* Public Domain *Contributors:* Jimmarsmars 143
Figure 112 *Source:* https://en.wikipedia.org/w/index.php?title=File:ESP_054515_2595icecaplayerspartial.jpg *License:* Public Domain *Contributors:* Jimmarsmars 143
Figure 113 *Source:* https://en.wikipedia.org/w/index.php?title=File:PIA22546-Mars-AnnualCO2ice-N&SPoles-20180806.gif *License:* Public Domain *Contributors:* Drbogdan 149
Figure 114 *Source:* https://en.wikipedia.org/w/index.php?title=File:Mars_South_Polar_Layers.JPG *License:* Public Domain *Contributors:* Jim Secosky (Jimmarsmars at en.wikipedia?) modified NASA image. 149
Figure 115 *Source:* https://en.wikipedia.org/w/index.php?title=File:Close-up_of_McMurdo_Crater_Layers.JPG *License:* Public Domain *Contributors:* Jim Secosky modified nasa image. 150
Figure 116 *Source:* https://en.wikipedia.org/w/index.php?title=File:North_pole_layers.JPG *License:* Public Domain *Contributors:* Jim Secosky modified NASA image. 150
Figure 117 *Source:* https://en.wikipedia.org/w/index.php?title=File:Chasma_Boreale_Streamined_Feature.JPG *License:* Public Domain *Contributors:* Jim Secosky modified nasa image. 150
Figure 118 *Source:* https://en.wikipedia.org/w/index.php?title=File:Chasma_Boreale.jpg *License:* Public Domain *Contributors:* Jim Secosky found image. 151
Figure 119 *Source:* https://en.wikipedia.org/w/index.php?title=File:North_Pole_Layers.jpg *License:* Public Domain *Contributors:* Jim Secosky modified nasa image. 152
Figure 120 *Source:* https://en.wikipedia.org/w/index.php?title=File:Chasma_Boreale_Channels.jpg *License:* Public Domain *Contributors:* Jim Secosky modified nasa image 152
Figure 121 *Source:* https://en.wikipedia.org/w/index.php?title=File:Mars_Viking_MDIM21_1km_plus_poles.jpg *License:* Public Domain *Contributors:* WolfmanSF 156
Figure 122 *Source:* https://en.wikipedia.org/w/index.php?title=File:Karte_Mars_Schiaparelli_MKL1888.png *License:* Public Domain *Contributors:* BLueFiSH.as, Badseed, Bryan Derksen, David Kernow~commonswiki, Mapmarks, Marcok, Romuello, Sophus Bie, Startaq, Stefan Kühn, W!B:, 2 anonymous edits 156
Figure 123 *Source:* https://en.wikipedia.org/w/index.php?title=File:Mars_topography_(MOLA_dataset)_with_poles_HiRes.jpg *License:* Public Domain *Contributors:* Billinghurst, PhilipTerryGraham, WolfmanSF 157
Figure 124 *Source:* https://en.wikipedia.org/w/index.php?title=File:Mars_elevation.stl *Contributors:* User:Cmglee 158
Figure 125 *Source:* https://en.wikipedia.org/w/index.php?title=File:Mars_Hubble.jpg *License:* Public domain *Contributors:* NASA and The Hubble Heritage Team (STScI/AURA) 158
Figure 126 *Source:* https://en.wikipedia.org/w/index.php?title=File:Mars_NPArea-PIA00161.jpg *License:* Public Domain *Contributors:* NASA/JPL/USGS 159
Figure 127 *Source:* https://en.wikipedia.org/w/index.php?title=File:PIA17932-Mars-NewImpactCrater-MRO-HiRISE-20131119.jpg *License:* Public Domain *Contributors:* Drbogdan, Huntster, Terriffic Dunker Guy 161
Figure 128 *Source:* https://en.wikipedia.org/w/index.php?title=File:Fretted_terrain_of_Ismenius_Lacus_taken_with_MGS.JPG *License:* Public Domain *Contributors:* jim secosky modified NASA photo 162
Figure 129 *Source:* https://en.wikipedia.org/w/index.php?title=File:Steep_cliff_in_Ismenius_Lacus_taken_with_MGS.JPG *License:* Public Domain *Contributors:* jim secosky modified NASA photo 162
Figure 130 *Source:* https://en.wikipedia.org/w/index.php?title=File:Wide_view_of_Debris_Apron.jpg *License:* Public Domain *Contributors:* Jim Secosky modified nasa image. 163
Figure 131 *Source:* https://en.wikipedia.org/w/index.php?title=File:Face_of_Lobate_Debris_Apron.jpg *License:* Public Domain *Contributors:* Jim Secosky modified nasa image. 163
Figure 132 *Source:* https://en.wikipedia.org/w/index.php?title=File:Olympus_Mons_M9_PIA02999.jpg *Contributors:* - 168
Figure 133 *Source:* https://en.wikipedia.org/w/index.php?title=File:Lava_flow_from_Arsia_Mons_in_Daedalia_Planum.jpg *License:* Public Domain *Contributors:* Jim Secosky (Jimmarsmars at en.wikipedia) modified NASA image. Photo credit NASA/JPL/ASU. 168
Figure 134 *Source:* https://en.wikipedia.org/w/index.php?title=File:Using_Earth_to_Understand_How_Water_May_Have_Affected_Volcanoes_on_Mars.webm *License:* Public Domain *Contributors:* NASA's Goddard Space Flight Center 169
Figure 135 *Source:* https://en.wikipedia.org/w/index.php?title=File:Fractional_crystallization.svg *License:* Creative Commons Attribution-Sharealike 3.0,2.5,2.0,1.0 *Contributors:* Woudloper 169
Figure 136 *Source:* https://en.wikipedia.org/w/index.php?title=File:PIA16463-MarsVolatiles-20121102.jpg *License:* Public Domain *Contributors:* Apocheir, Chmee2, Drbogdan, Mikhail Ryazanov, 1 anonymous edits 171
Figure 137 *Source:* https://en.wikipedia.org/w/index.php?title=File:PIA16217-MarsCuriosityRover-1stXRayView-20121017.jpg *License:* Public Domain *Contributors:* ComputerHotline, Drbogdan, O'Dea, PhilipTerryGraham, Ruff tuff cream puff 172
Figure 138 *Source:* https://en.wikipedia.org/w/index.php?title=File:Tharsis_-_Valles_Marineris_MOLA_shaded_colorized_zoom_32.jpg *License:* Public Domain *Contributors:* Huntster, PhilipTerryGraham, WolfmanSF 174
Figure 139 *Source:* https://en.wikipedia.org/w/index.php?title=File:Tharsis_mons_Viking.jpg *License:* Public Domain *Contributors:* Viking I 174

418

License

Index